P9-AGK-168

Community Ecology

Community Ecology
Processes, Models, and Applications

EDITED BY

Herman A. Verhoef
VU University, Amsterdam, Department of Ecological Science, the Netherlands

Peter J. Morin
Rutgers University, Department of Ecology, Evolution & Natural Resources, USA

OXFORD
UNIVERSITY PRESS

Great Clarendon Street, Oxford OX2 6DP

Oxford University Press is a department of the University of Oxford.
It furthers the University's objective of excellence in research, scholarship,
and education by publishing worldwide in

Oxford New York

Auckland Cape Town Dar es Salaam Hong Kong Karachi
Kuala Lumpur Madrid Melbourne Mexico City Nairobi
New Delhi Shanghai Taipei Toronto

With offices in

Argentina Austria Brazil Chile Czech Republic France Greece
Guatemala Hungary Italy Japan Poland Portugal Singapore
South Korea Switzerland Thailand Turkey Ukraine Vietnam

Published in the United States
by Oxford University Press Inc., New York

First published 2010

Reprinted 2010

British Library Cataloguing in Publication Data
Data available

Library of Congress Cataloging in Publication Data
Library of Congress Control Number: 2009934138

Cover illustration by Janine Mariën
Typeset by SPI Publisher Services, Pondicherry, India
Printed in Great Britain
on acid-free paper by
CPI Antony Rowe, Chippenham, Wiltshire

ISBN 978–0–19–922897–3 (Hbk.)
ISBN 978–0–19–922898–0 (Pbk.)

3 5 7 9 10 8 6 4 2

Contents

Preface

In 2001 H.A.V. started a course for second-year undergraduate biology students at the Vrije Universiteit Amsterdam entitled Community Biology. This course has now been running successfully for 8 years. The course was obligatory for all biology students, and it differed from other courses in that it was multidisciplinary and provided the students with opportunities to perform their own research. The multidisciplinarity was emphasized by the different disciplines of the teachers on the course: soil ecology, plant ecology, systems ecology, microbial physiology and theoretical biology. The important task of finding a textbook that could link all disciplines and encourage participating lecturers to deliver a unified course was solved by using *Community Ecology* by P.J.M. That book linked the different subjects of community ecology, and integrated the more theoretical parts on modelling with the empirical studies, including topics such as biodiversity and applied studies. Subsequently, H.A.V. and P.J.M. met at an international meeting on food webs, and discussed the possibility of participating in a similarly themed graduate-level course. And, thus, our current collaboration began. In The Netherlands PhD students from different universities are organized into interdisciplinary thematic groups, called research schools, that provide an intellectual support base for instruction and research. For example, students working in the field of socioeconomic and natural sciences of the environment belong to the Research School SENSE. In 2005, H.A.V., André de Roos, Claudius van de Vijver and Johan Feenstra organized a PhD course on Community Ecology for the SENSE PhD programme. During this 1 week course held in Zeist, leading researchers in the field of Community Ecology from Europe and the USA were asked to deliver lectures on recent and often unpublished developments in their areas of expertise. The lecturers were accompanied by some of their PhD students, creating an international group of community ecologists. The course was not intended to be encyclopaedic, but rather it focused on the areas of expertise of the invited speakers, many of which share the theme of patterns and processes emerging from ecological networks. Participants addressed the state of the art in theory and applications of community ecology, with special attention to topology, dynamics, the importance of spatial and temporal scale and the applications of community ecology to emerging problems in human-dominated ecosystems, including the restoration and reconstruction of viable communities. The course finished with speculations about future research directions. During the course, it became clear that this international group of students appreciated the information presented by the various lecturers, despite the fact that research topics exhibited great diversity. It was during this very stimulating course that the idea for this book took form. H.A.V. and P.J.M., the editors, convinced most lecturers to transform their lectures into book chapters, and asked other colleagues to fill in some gaps. The result captures much of the excitement about community ecology expressed during the course, and expands the coverage of topics beyond what we were able to discuss in an intensive week-long course. We recognize at the outset that certain subdisciplines of community ecology are not covered here, and we do not claim otherwise. We know that the topics addressed here will be of interest to advanced students and practitioners of community ecology. Ultimately, 19 colleagues participated in writing this book. We thank them all for their important contributions. Writing book chapters, strangely enough, is less valued than writing articles for scientific journals in some academic circles. Still, like the multidisciplinary course mentioned above, we find that the interactive writing that happens when people from different subdisciplines work together is a fascinating, synergistic and productive process.

We would like to thank friends and colleagues who were indispensable during the process of writing: H.A.V. thanks Nico van Straalen, who by writing his book *Ecological Genomics* for Oxford University Press acted as an instigator for this book. H.A.V. also thanks his colleagues who made the Community Biology course a success for so many years: Wilfred Röling, Bob Kooi, Matty Berg, Wilfried Ernst, Tanja Scheublin, Diane Heemsbergen, Stefan Kools, Marcel van der Heijden, Susanne de Bruin, Lothar Kuijper, Rully Nugroho and Henk van Verseveld. H.A.V. acknowledges colleagues who were directly involved in the organization of the PhD course: André de Roos, Claudius van de Vijver, Johan Feenstra and Ad van Dommelen. H.A. V. is very grateful to his critical friend, John Ashcroft (Durham), for supportive focusing.

P.J.M. thanks the many students who participated in the Zeist Community Ecology Course, as well as the students who have taken the community ecology course that he has taught at Rutgers University since 1983. Their collective comments and feedback have helped to refine his perspectives about the nature of community ecology over the years. Thanks also go to participants in a recent seminar on Ecological Networks for critical feedback on some of the writing that appears here, including Mike Sukhdeo, Maria Stanko, Wayne Rossiter, Tavis Anderson, Faye Benjamin, Denise Hewitt, Kris Schantz and Chris Zambel.

H.A.V. and P.J.M. both thank Ian Sherman of Oxford University Press, who was immediately enthusiastic about this book project, and Helen Eaton, who as assistant commissioning editor played a crucial role in the development of the book.

H.A.V. thanks Emilie Verhoef, without whom this book probably would never have been produced. P.J.M. thanks Marsha Morin for her understanding and support during another extended writing project.

Herman A. Verhoef, Amsterdam
Peter J. Morin, New Brunswick

List of Contributors

Jan P. Bakker, Community and Conservation Ecology Group, University of Groningen, PO Box 14, 9750 AA Haren, The Netherlands, Email: j.p.bakker@rug.nl

Janne Bengtsson, Department of Ecology and Crop Production Science, PO Box 7043, Swedish University of Agricultural Sciences, SE-750 07 Uppsala, Sweden, Email: Jan.Bengtsson@ekol.slu.se

Matty P. Berg, VU University, Amsterdam, Department of Ecological Science, De Boelelaan 1085, 1081 HV Amsterdam, The Netherlands, Email: matty.berg@falw.vu.nl

Ulrich Brose, Darmstadt University of Technology, Department of Biology, Schnittspahnstr. 10, 64287 Darmstadt, Germany; Pacific Ecoinformatics and Computational Ecology Lab, 1604 McGee Avenue, Berkeley, CA 94703, USA, Email: brose@bio.tu-darmstadt.de

Jonathan M. Chase, Department of Biology and Tyson Research Center, Box 1229, Washington University in Saint Louis, Saint Louis, MO, USA, Email: jchase@wustl.edu; chase@biology2.wustl.edu

J. Emmett Duffy, School of Marine Science and Virginia Institute of Marine Science, The College of William and Mary, Gloucester Point, VA 23062-1346, USA, Email: jeduffy@vims.edu

Jennifer A. Dunne, Santa Fe Institute, 1399 Hyde Park Road, Santa Fe, NM 87501; Pacific Ecoinformatics and Computational Ecology Lab, 1604 McGee Avenue, Berkeley, CA 94703, USA, Email: jdunne@santafe.edu

Jacintha Ellers, VU University, Amsterdam, Department of Ecological Science, De Boelelaan 1085, 1081 HV Amsterdam, The Netherlands, Email: jacintha.ellers@falw.vu.nl

Tadashi Fukami, Department of Biology, Stanford University, Stanford, CA 94305, USA, Email: tfukami@hawaii.edu

E. Toby Kiers, VU University, Amsterdam, Department of Ecological Science, De Boelelaan 1085, 1081 HV Amsterdam, The Netherlands, Email: toby.kiers@falw.vu.nl

David Kothamasi, Centre for Environmental Management of Degraded Ecosystems, University of Delhi, Delhi 110007, India, Email: dmkothamasi@cemde.du.ac.in

Dries P.J. Kuijper, Mammal Research Institute, Polish Academy of Sciences, ul. Waszkiewicza 1c, 17–230 Białowieża, Poland, Email: dkuijper@zbs.bialowieza.pl

Nicolas Loeuille, Laboratoire d'Ecologie, UMR7625, Université Paris VI, 7 quai St Bernard, F75252 Paris Cedex 05, France, Email: nicolas.loeuille@normalesup.org

Michel Loreau, McGill University, Department of Biology, 1205 avenue du Docteur Penfield, Montréal, Québec, Canada, H3A 1B1, Email: michel.loreau@mcgill.ca

Peter J. Morin, Rutgers University, Department of Ecology, Evolution, & Natural Resources, 14 College Farm Road, New Brunswick, NJ 08901, USA, Email: pjmorin@rci.rutgers.edu

Han Olff, University of Groningen, Community and Conservation Ecology Group, PO Box 14, 9750 AA Haren, The Netherlands, Email: h.olff@rug.nl

Owen L. Petchey, University of Sheffield, Department of Animal and Plant Sciences, Alfred Denny Building, Western Bank, Sheffield S10 2TN, UK, Email: o.petchey@sheffield.ac.uk

Julia Stahl, Landscape Ecology Group, University of Oldenburg, PO Box 2593, D-26111 Oldenburg, Germany,
Email: julia.stahl@uni-oldenburg.de

Marcel G.A. van der Heijden, VU University, Amsterdam, Department of Ecological Science, De Boelelaan 1085, 1081 HV Amsterdam, The Netherlands; Agroscope Reckenholz-Tänikon, Research Station ART, Reckenholzstrasse 191, 8046 Zürich, Switzerland,
Email: marcel.vanderheijden@art.admin.ch

Wim H. van der Putten, Netherlands Institute of Ecology, Centre for Terrestrial Ecology (NIOO-KNAW) PO Box 40, 6666 ZG Heteren; Laboratory of Nematology, Wageningen University, PO Box 8123, 6700 ES Wageningen, The Netherlands,
Email: w.vanderputten@nioo.knaw.nl

Herman A. Verhoef, VU University, Amsterdam, Department of Ecological Science, De Boelelaan 1085, 1081 HV Amsterdam, The Netherlands,
Email: herman.verhoef@falw.vu.nl

Introduction

Herman A. Verhoef and Peter J. Morin

First of all, we must clarify what we consider to be the subject matter of community ecology. In his classic textbook, Krebs (1972) described a community as 'an assemblage of populations of living organisms in a prescribed area or habitat'. He described this as 'the most general definition one can give'. The 'prescribed area' suggests that we are dealing with some sort of spatially bounded systems, and it is the somewhat arbitrary nature of possible boundaries that has caused continuing debate about the nature of communities (e.g. Ricklefs 2008). It is debatable whether community ecology is a proper discipline at all if communities do not exist as natural definable units (Begon *et al.* 1996). At the beginning of the 20th century there was much debate about the nature of communities. The driving question was whether the community was a self-organized system of co-occurring species or simply a haphazard collection of populations with minimal functional integration. At that time, two extreme views prevailed: one view considered a community as a superorganism whose member species were tightly bound together by interactions that contributed to repeatable patterns of species abundance in space and time. This concept led to the assumption that natural communities exist as fundamental units, and this made it possible to classify communities in a manner comparable to the Linnaean taxonomy of species. One of the leading proponents of this approach was the Nebraskan botanist Frederick E. Clements. His view became known as the *organismic concept of communities*. It assumes a common evolutionary history for the integrated species (Clements 1916).

The opposite view has been termed the *individualistic continuum concept* and was advocated by the American botanist H.A. Gleason. His focus was on the traits of individual species that allow each to live within specific habitats or geographical ranges. A community can then be seen as an assemblage of populations of different species whose traits allow them to persist in a prescribed area (Gleason 1926). This is a much more arbitrary unit than that envisioned by Clements. In Gleason's view, spatial boundaries of communities are not sharp and the species assemblages may change considerably over time and space.

We feel that the current broadly accepted view about the nature of communities is much closer to Gleason's opinion. A given species can occur in rather different collections of species, or communities, under different circumstances. Community ecology can, therefore, be described as the study of the community level of organization, rather than of a spatially and/or temporally definable unit (Begon *et al.* 1996).

As suggested by Morin (1999), we include within community ecology the study of two or more species at a location, including predator–prey interactions, interspecific competition, interactions of mutualism and parasitism. We recognize that some ecologists will argue that such pair-wise interactions are simply aspects of population biology, but we also point out that they provide the basic interactions that contribute to patterns involving larger numbers of species. We also recognize that, even in well-studied communities, much of the complexity and daunting diversity involving bacteria, fungi, protists and small invertebrates remains poorly known. Nonetheless, fascinating patterns among well-studied groups of organisms demand our attention and beg for explanation.

In this book, we have grouped the various chapters into several areas: (1) community shape, structure and dynamics, (2) communities over space and time, (3) applications of community ecology and (4) future directions of community research. We start with shape and structure, with a focus on the *topology* of networks of interacting organisms in ecological systems. Chapter 1 deals with subsets of the full network of interactions: competition networks, mutualistic networks and consumption networks (food webs). It involves recent studies and important questions about the structure and the processes or mechanisms that may result in their structure. Chapter 2 follows with dynamics, dealing with the *trophic dynamics of communities.* The focus is on types of dynamics that can be distinguished, the dynamics of simple and complex interactions and the internal and external determinants of food web dynamics. Chapter 3 considers *modelling of the dynamics of complex food webs,* with a focus on whether embedding simple modules in complex food webs affects the dynamics of those modules. The dynamic analysis of complex food webs is organized around the relationships between complexity and stability and diversity and stability. Body-size-dependent scaling of biological rates of populations emerge as a possible solution to the instability predicted for complex food webs. Chapter 4 forms a bridge between dynamics and the next section on space and time. The subject of this chapter is *community assembly dynamics in space,* with a focus on three elements of assembly dynamics: the rate of immigration to local communities, the degree to which the species pool is external to local community dynamics and the amount of variation in immigration history among local communities. The hypothesis is that these elements together determine the degree of historical contingency. Chapter 5, on *increasing spatio-temporal scales: metacommunity ecology,* discusses whether a metacommunity perspective can be helpful in synthesizing community ecology across spatial scales, and whether it can provide insights into the spatial mechanisms causing variation in species coexistence, strengths of species interactions and patterns of local and regional diversity. Chapter 6 deals with *spatio-temporal structure in soil communities and ecosystem processes,* and highlights the complexity exhibited by soil communities at different scales.

Community ecology has many real-world applications deriving from the fact that the abundance of species strongly depends on competitive and predatory interactions. These interactions are major drivers of ecosystem processes, and they are the key to the delivery of ecosystem goods and services. Chapter 7 focuses on *applications of community ecology approaches in terrestrial ecosystems: local problems, remote causes.* In terrestrial ecosystems, there is a clear link between belowground and aboveground subsystems, which is to be compared with benthic–pelagic coupling in aquatic or marine systems. These above/belowground linkages influence many major processes, such as ecosystem restoration following land abandonment or the fate of introduced exotic alien species. Some examples of above/belowground interactions in relation to changing land usage and biological invasions are discussed and related to climate warming. Chapter 8 reviews the consequences of industrial-scale fishing on marine communities and is entitled *sea changes: structure and functioning of emerging marine communities.* Growing evidence from models and empirical time series suggests that fishing pressure can cause relatively rapid regime shifts into distinct semi-stable states that resist change even after fishing has been relaxed. Although theory predicts that a preponderance of weak interactions stabilizes food web dynamics, these weak interactions may be unlikely to buffer marine ecosystems from the impacts of intense exploitation by humans. Thus, the systematic reduction in average food chain length documented in oceanic and coastal ecosystems can initiate regime shifts to alternate semi-stable states, and is already affecting the provision of marine ecosystem services to society. In Chapter 9, which focuses on *applied (meta)community ecology: diversity and ecosystem services at the intersection of local and regional processes,* several expectations from metacommunity theory on the effects of land use intensification are suggested, based on the fact that both local and regional processes are important for diversity and ecosystem functioning. Examples drawn from research on organic farming and biological control illustrate that metacommunity theory combined with good knowledge of the system under management is useful to understand how human-dominated ecosystems can be managed at the larger spatial scales that are relevant to managers.

Community-level management is described in Chapter 10 for *salt marshes* in North-West Europe. It deals with interactions among sedimentation, nutrient availability, plant growth and herbivores. Natural succession in these systems results in the dominance of a single tall grass species to the detriment of the natural herbivores such as geese and hares. Introductions of large herbivores can reverse succession to facilitate geese and hares. Without this management, these systems would develop into low-diversity systems of reduced value.

Finally, we consider *future directions of community research*. These chapters include the study of the evolution of community processes and patterns, ranging from local dynamics to large-scale diversity gradients, and focus on the role of mutualisms in community ecology. Chapter 11 focuses on *evolutionary processes in community ecology* to try to bridge the gap between community and evolutionary ecology by considering the reciprocal effects of individual trait variation and community characteristics. Central principles from evolutionary biology are used to extrapolate the consequences of genetic diversity to community level and illustrate the effects of genetic variation on community properties and vice versa. It is stated that more studies have been performed on the effects of genotypic and phenotypic diversity on community properties than on the effects of community diversity on genetic and phenotypic diversity of single species. This fascinating new field of integrative evolutionary community ecology is still in its infancy, but holds great promise. Chapter 12 deals with the *emergence of complex food web structure in community evolution models* and aims at synthesizing how evolutionary dynamics may help the understanding of food web structures and dynamics. Community evolution models incorporate both the dynamical components of food webs and the complexity necessary to understand empirical food web data. It is shown how this model matches topological properties of empirical data, while giving information on the dynamics of the food web. It also discusses the relevance of the allometric theory of ecology to the debate on the relationship between stability and diversity, as well as the evolution of niche breadth and non-trophic interactions. Another promising line of research involves mutualisms. Chapter 13 deals with the role of *mutualisms and community organization*. Key evolutionary events, such as the origin of the eukaryotic cell, the invasion of land by plants and the radiation of angiosperms, are linked to mutualisms. Mutualism probably brings order to the organization, structure and function of communities by regulating the acquisition of resources and ameliorating stresses. Finally, Chapter 14 speculates about *emerging frontiers of community ecology*. Among various emerging areas, a focus on the important topic of biotic invasions seems crucial, since they alter the world's natural communities as well as their ecological character. Further emerging areas concern questions surrounding the linking of the structure of ecological networks to empirically measured community dynamics.

PART I

SHAPE AND STRUCTURE

CHAPTER 1

The topology of ecological interaction networks: the state of the art

Owen L. Petchey, Peter J. Morin and Han Olff

1.1 Introduction

1.1.1 What do we mean by the 'topology' of ecological networks?

Many diverse systems can be described as a network of linked nodes (Barbarasi 2002); for example, transportation networks of cities linked by roads, networks of interacting populations on a landscape, networks of interacting individuals in a social network, networks of neurones linked by synapses, networks of interacting genes in the genome or networks of biochemical transformations within the cell. In the paradigm used in this chapter, species are the 'nodes' of ecological networks, and interactions among species are the links.

Ecological communities are immensely complex; variation exists at every level of organization (individuals, populations, species) and these entities interact with each other in any number of ways (consumption, competition, mutualism, facilitation, modification). Ecological networks provide a tractable simplification of this complexity – they can be constructed, modelled, experimentally manipulated and analysed with available tools and resources (Proulx *et al.* 2005). This chapter focuses on the networks of interactions among the species in ecological communities to highlight the advances in ecological understanding that research about these networks can provide.

The links in ecological networks can be characterized through three main aspects: (1) their topology, (2) their geometry and (3) the direction and strength of their interactions (Fig. 1.1). The topology of an ecological network is a description of the patterns of interactions ('who interacts with whom'). Topology is thus the study of the arrangement of links from an information/organization perspective. The distances and angles between nodes have no meaning for the topology; they are often chosen so that the network can be conveniently graphically represented (Fig. 1.1a). If the geometry of the network is included, the Euclidean distance between nodes has a meaning (Fig. 1.1b); for example, in the genetic or trait similarity of species or as the physical distance between species within a landscape (e.g. Thompson *et al.* 2001) such as in a metacommunity context (Leibold *et al.* 2004). If the interaction strength is also included (Fig. 1.1c), all interactions are not equivalent – some are stronger than others.

In this chapter, we focus on the origins and consequences of different network topologies. Chapter 2 will address the origin and consequences of variation in interaction strength within interaction networks. Fundamental questions about the topology of ecological networks include how many nodes (species) there are, how many links (interactions) there are in total and how interactions are distributed among species pairs (Fig. 1.1a).

The possible interactions that can be represented in ecological networks theoretically include *all* of the basic pair-wise interactions among species, including competition via various mechanisms, predator–prey and host–parasite interactions, interactions

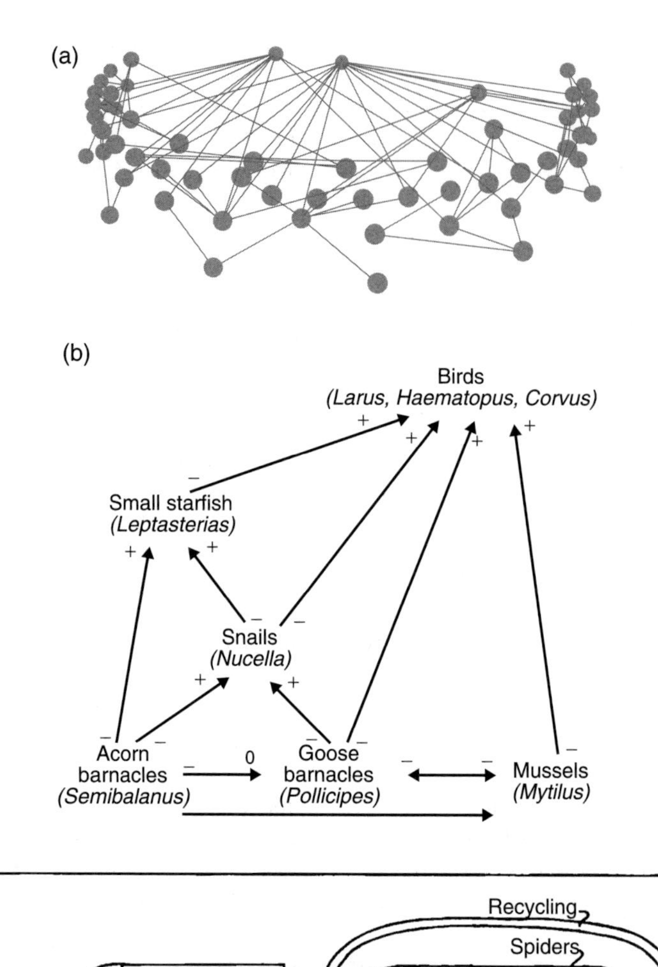

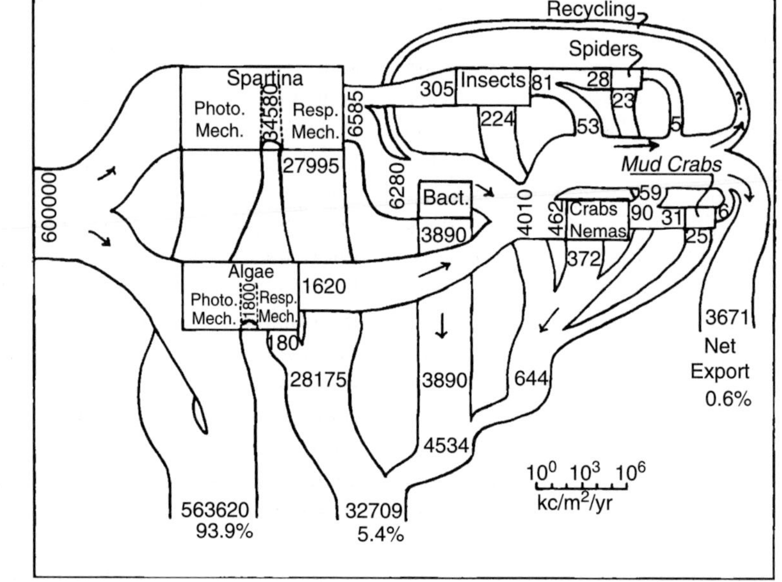

Figure 1.1 Examples of three kinds of ecological networks. (a) A food web for a portion of the grassland community near Silwood Park, UK. Reprinted with permission from www.foodwebs.org using data from Dawah *et al.* (1995). (b) An interaction network showing the signs of interactions among a subset of species in a community from the rocky intertidal zone of the Pacific Coast of northwestern USA. Reprinted from Wootton (1994), with permission of the Ecological Society of America. (c) A network of energy flow from a salt marsh in southeastern USA. Reprinted from Teal (1962), with permission of the Ecological Society of America.

through behavioural interference and positive interactions including mutualisms and commensalisms. In practice, few representations of ecological networks include all of these kinds of interactions, partly for historical reasons and partly because, for reasons of pragmatism and feasibility, the ecologists who describe these networks often focus on only a subset of possible interactions, e.g. when they study the importance of a particular type of interaction in structuring the ecological community of interest.

1.1.2 Different types of ecological networks

What might constitute the links among species in an ecological network? Very broadly, one can think of two types of information represented by a link. The first is when a link represents one specific biological mechanism or process, the second is when a link represents the net effect of a variety of mechanisms or processes. Links that represent biological mechanisms or processes can be recognized by the transfer of something tangible (such as biomass, energy, nutrients, information or combinations of these) between the linked entities, where this usually requires close physical proximity between the organisms involved in time and space.

Food webs are an example of an ecological network with this type of link, in which the consumer–resource interactions represent transfer of energy and material and require a physical interaction between individuals. Food webs are perhaps the most commonly encountered type of ecological network (Elton 1927; Cohen 1978; Pimm 1982; Polis and Winemiller 1996; de Ruiter *et al.* 2005). An example of a food web, the Silwood Park, UK, grassland food web (Dawah *et al.* 1995), is shown in Fig. 1.1a and depicts species as 'balls' and feeding links as 'sticks'. In this particular food web, there are plants, herbivores, primary parasitoids of the herbivores, and hyperparasitoids. Trophic position increases with height and is also coded by colour (from www.foodwebs.org). This is a relatively simple food web (87 species) in which the identity of each resource and consumer is resolved to the species level and sufficient sampling effort ensured that virtually all species and feeding links were recorded (Dawah *et al.* 1995). However, many food webs are more complex and less complete, with less well-resolved taxa, especially in the smaller, relatively inconspicuous and taxonomically difficult groups such as microbes and microinvertebrates.

One should note the very important assumption in the kind of the topological approach shown in Fig. 1.1: every species is treated equally in the network from an interaction perspective. For example, species that are very abundant in biomass and species that are very rare are treated exactly the same. Only if these differences in abundance result in a difference in their number of interactions with other species will those differences show up in the topology of the web. In other words, the central assumption of the approach is that the key point for understanding the food web is *that* species interact, while information on how abundant different species are and how strong these interactions are (e.g. expressed in per capita effects of predators on prey, and vice versa) can be disregarded. This is of course a strongly simplifying and yet poorly tested assumption. Efforts to attach more information to nodes have led to significant advances in understanding (Cohen *et al.* 2003). Nevertheless, a huge body of research concerns the search for patterns in food web topology (Cohen 1978).

Plant–pollinator networks also use links to represent transfers of something tangible among organisms that share close proximity in space and time. Pollen moves from plant to pollinator, and again to the plant, which requires repeated physical contact. These interactions involve different currencies. From the plant perspective, the exchange of genetic information through cross-fertilization is the most important aspect. From the pollinator perspective, the resource rewards of visiting the flower are the driving force. Interest in these so-called two-layer networks is a rather recent development, at least compared with the long study of food webs, and the networks typically describe which plants are visited by which pollinators or seed dispersers (Bascompte *et al.* 2003; Jordano *et al.* 2003; Rezende *et al.* 2007).

Networks of energy and material flow also document this first broad type of link, and they have a venerable place in ecology, dating back at least to

Lindeman (1942). Fig. 1.1c shows the quantification of the patterns of energy flow in a salt marsh in North America among the major functional components (Teal 1962). Owing to measurement constraints and the assumed functional equivalence of different species involved with respect to nutrient and energy transformation, these networks typically have low taxonomic resolution, and have mostly been developed as a way to understand and describe energy fluxes and material in ecosystems, and the associated ecosystem functions and services (such as primary and secondary productivity, carbon sink capacity, etc.). That is, the focus is the dynamics of energy and material, rather than the dynamics of species and their interaction structure. An exception is the work by Ulanowicz (1997), who is one of the rare authors who has extensively considered origins and consequences of different topological patterns of energy and nutrient flows in food webs. In his work, he stresses the importance of positive feedback among species, resulting in autocatalytic loops within subsystems of whole webs, which may even lead to ecosystem-level competition between such alternative loops. Also, the work on soil foods webs by de Ruiter, Moore and others (e.g. de Ruiter *et al.* 1998) has tried to map the flows of nutrients and energy from a much more topological perspective than is done in classic ecosystem science. The outcome of this work will be addressed further in Chapter 2.

The second broad type of information represented by a link is the *effect* of one entity on another. For example, if individuals of two species each have a positive effect on the growth and reproduction of individuals of the other, they would be linked. This positive–positive interaction is termed mutualism and a range of biological mechanisms might underlie this link. Pollination is an example of a mechanism that can result in a positive–positive interaction. Associational resistance, in which a species gains protection from its consumers by spatially associating with its potential competitors, is another example that can lead to a net positive interaction between species (Olff *et al.* 1999). Facilitation, such as often found among grazing herbivores (Huisman and Olff 1998; Arsenault and Owen-Smith 2002), also belongs to this class of interactions. A negative–negative interaction is termed competition, and can result from consumption (a biological process) by two species of a shared resource, as well as from a number of other non-consumptive mechanisms, such as direct behavioural interference among species (Schoener 1983).

When links represent *effects* of one species on another, they may result from direct effects, indirect effects or net effects. Competition via consumption of a shared resource can be represented as a negative–negative link arising from an indirect interaction between two competing species mediated by their joint consumption of a third resource species; that is, there need not be a transfer of anything tangible from one competitor to the other, yet changes in the abundance of one will alter the abundance of the other. Other indirect effects (or interactions) include apparent competition (negative–negative effects caused by the presence of a shared consumer; e.g. Holt 1977) and apparent mutualism (positive–positive effects, e.g. in a trophic cascade; Pace *et al.* 1999).

Ecological networks in which links represent the net effect of one species on another (the outcome of all direct and indirect effects) are often, and perhaps confusingly, termed *interaction webs*. As this also describes the general class of ecological networks in which species interact (irrespective of the nature of their interactions) they should be more correctly called *interaction-sign networks*. They depict the sign (positive, negative, zero) and sometimes the magnitude of the net impact of changes in the abundance of one species on another. These net impacts can be the result of all kinds of direct and indirect interspecific interactions. Interaction webs are less commonly depicted in the literature, perhaps because of the practical difficulty of determining the sign of net interactions, especially within trophic levels, for which elaborate field experiments are often required. However, sometimes the nature of interactions can be inferred from large-scale natural disturbances to ecosystems. A classic example is the effect of rinderpest on the food web structure and ecosystem functioning of the Serengeti (Sinclair 1979). In this work, abiotic components and processes, such as fire, are also taken into account as 'nodes' in the food web, because of the strong effect *biotic* feedbacks that they receive. One good example of an interaction-sign web comes from Wootton's (1994) work on interactions among some of the species living in the rocky intertidal zone of Washington State in the USA (see Fig. 1.1c). The search for patterns in interaction networks has only begun rather recently. For a recent review, see Ohgushi (2005), who focuses on indirect interactions in plant-based terrestrial food webs.

1.1.3 Three general questions

For each type of ecological network, we suggest that there are three general types of questions: those about whether and what structural regularities exist within and among observed networks; those about which mechanisms are responsible for the structure of the networks and of any structural regularities that exist; and lastly, those about the gaps in our knowledge. The aim here is to define a set of general contemporary questions that we will use to organize each of the following sections.

1.2 Competitive networks

1.2.1 Structural regularities

Real-world competitive networks remain poorly studied and documented. A fundamental question about their structure remains quite unresolved: given a set of co-occurring species that appear to consume a similar class of resources (such as different-sized seeds, forming a potential guild, *sensu* Root 1967), how many pairs of these species really compete? That is, how many of the plausible links are actually realized in competitive networks?

Groups of taxonomically similar species are often assumed to be potential competitors, but experiments show that only a small fraction of the potential competitors actually compete. Sometimes this can be generalized to simple integrative traits, building on the early work on limiting similarity (for a review, see Brown 1981). For savanna large herbivores, Prins and Olff (1997), for example, found that large herbivores overlap more in resource use when they are more similar in body size. Thus, more similar-sized species are more likely to interact.

Hairston (1981) studied a group of six species of plethodontid salamanders found in moist forests of the mountains of North Carolina, USA. Based on their morphological similarity and similar lifestyles, all being carnivores feeding on small invertebrates, it seemed plausible that all six species might compete for food (Fig. 1.2). Over a 5 year period Hairston removed each of the two most common species at this site. Only one of the remaining five species responded positively to each removal, and this was the most common species at the site. These results imply that only the two most common species compete. Hairston suggested that competition might be for nest sites, rather than for shared prey, but subsequent research showed that direct territorial (interference) interactions between salamanders constituted the primary mechanism of competition (Nishikawa 1985). In this case, the network of competitors is considerably simpler than the completely connected system that would result if all species shared and competed for common resources (Fig. 1.2).

Another general and important question about the structure of competitive networks is whether the interactions form a hierarchy, or an intransitive loop. In a hierarchy, for example, species A outcompetes species B and C, B outcompetes C, and C cannot outcompete either A or B. This will mostly lead to predominance of a single competitively superior species (A) and therefore will foster relatively low biodiversity. In an intransitive loop, in which A outcompetes B, B outcompetes C, and C outcompetes A, all three species can potentially coexist. Consequently, mechanisms of interaction that allow intransitive competitive networks might explain high biodiversity in some ecosystems.

At first hand, intransitive competitive networks, especially within one trophic level, would be unlikely to emerge, based on the theory of life history and competitive trade-offs. An organism's life history generally reflects a compromise in the use of available resources. Species generally specialize along environmental axes of resources axes and stress factors in such a way that investing in some life history/functional trait (e.g. the ability to crack big seeds in birds) occurs at the expense of some other function (e.g. the efficiency with which small seeds can be collected). In plants, for example, the

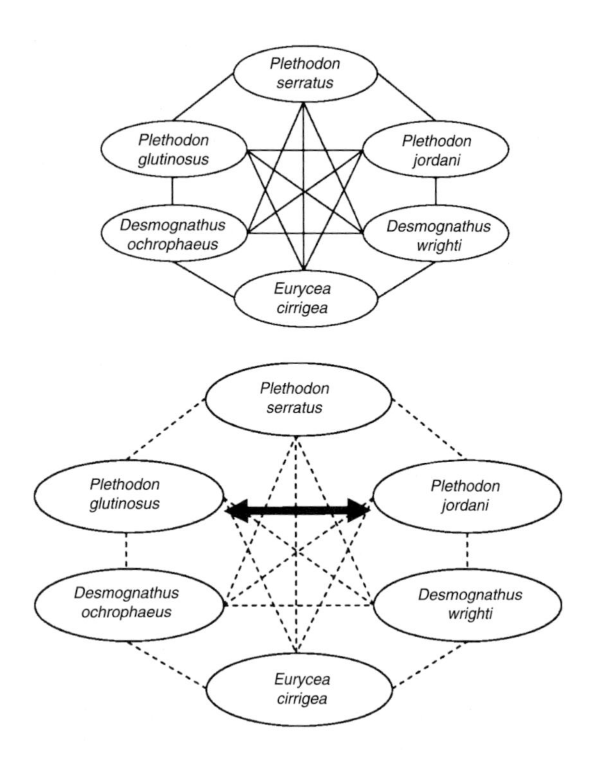

Figure 1.2 Potential (top) and realized (bottom) networks of competing salamanders in the study of Hairston (1981). Dashed lines in the lower graph indicate unrealized competitive interactions; the heavy solid line indicates that only two species, *Plethodon glutinosus and Plethodon jordani*, actually compete.

ability to compete well for nutrients has been suggested to happen at the expense of the ability to compete for light (Tilman 1990). Also, competing needs for growth and reproduction result in different optimal combinations in different species, depending on the conditions to which they are adapted. Such evolution-driven trade-offs make competitive hierarchies much more likely than transitive networks in most cases. Said simply, it is unlikely that the species adapted to the 'left-tail' of some underlying niche axis will outcompete a species from the 'right-tail' or vice versa.

An important class of exceptions exists in organisms in which the outcome of interactions is not driven by resource competition, but by chemical warfare, such as allelopathy. In this case, few trade-offs are expected, as the defence and competition is much more strongly information-based (what is the opponent sensitive to) than energy- or nutrient-based (which requires morphological adaptations, such as allocation shifts).

One example of such an intransitive competitive network involves different strains of bacteria competing in spatially structured environments. Kerr *et al.* (2002) explored how a mixture of mechanisms can allow three strains of *Escherichia coli* to coexist via a set of intransitive interactions that are akin to a game of rock–paper–scissors. There are three key features of this system – one strain produces a toxin, called colicin, whereas the two other strains are either sensitive or resistant to the toxin (Kirkup and Riley 2004). The three strains interact as an intransitive network in the following way. All of the strains compete consumptively for a carbon source, such as glucose, but they differ in competitive ability such that colicin-sensitive strains are the best consumptive competitors, followed by colicin-resistant strains followed by colicin-producing strains. However colicin-producing strains can displace colicin-sensitive strains by poisoning them, but colicin-producing strains can in turn be displaced by competitively superior colicin-resistant strains. These patterns of sequential displacements occur only in an unmixed spatially complex environment, such as the surface of a culture plate. In a well-mixed spatially homogeneous environment, such as a well-stirred liquid culture, only one strain persists. It appears that limited dispersal can promote diversity by allowing an intransitive competitive network to exist (Fig. 1.3). A similar role is played by dispersal limitation in maintaining diversity in neutral communities (i.e. where all species are assumed to be functionally equivalent; e.g. Hubbell 2001).

Qualitatively similar but much more complex networks may explain the high diversity of bacteria that manage to coexist in soils, where dispersal is limited. For example, Torsvik *et al.* (1990) made initial estimates using DNA hybridization techniques that suggested that thousands of bacterial taxa could occur per 30 g of soil. These conservative estimates were re-evaluated by Dykhuizen (1998), who estimated that 30 g of forest soil could contain over half a million species, depending on the assumptions made in the analysis.

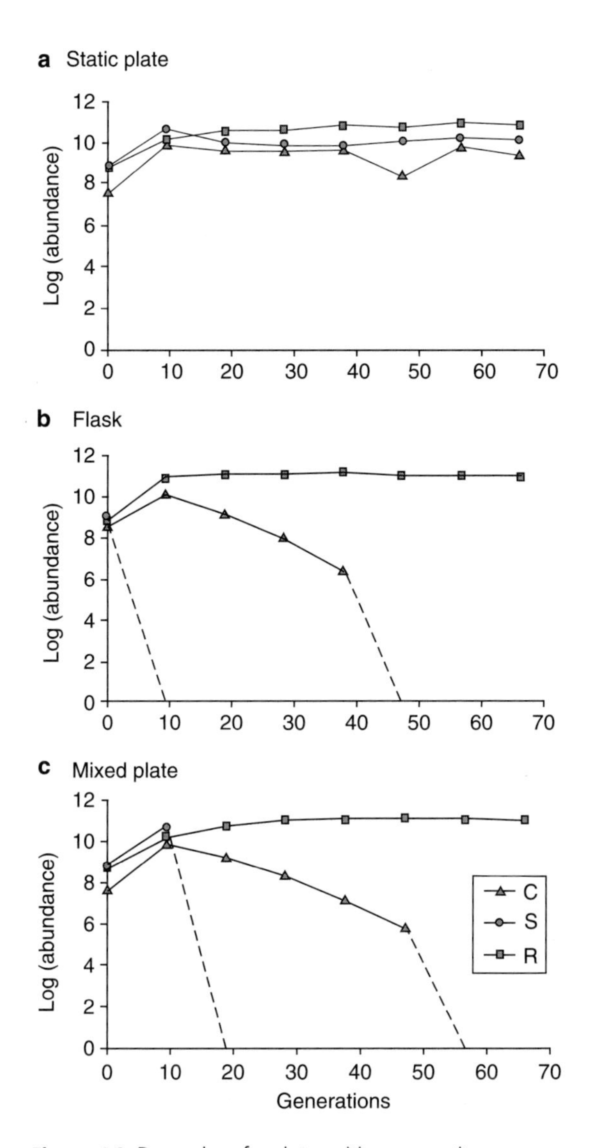

Figure 1.3 Dynamics of an intransitive competitor network. C is a colicin (toxin producing) bacterial strain, which excludes S, a colicin-sensitive strain, which excludes R, a colicin-resistant strain, which in turn competitively displaces C. All three strains persist in a static two-dimensional habitat (a), but R eventually predominates in mixed environments (b and c). Reprinted with permission from Macmillan Publishers Ltd: *Nature*, 171–4, B. Kerr *et al.* © 2002.

Intransitive networks involving multiple toxins and multiple resistance factors to those toxins, along with trade-offs between the ability to produce or resist toxins and to compete consumptively for resources, might provide a mechanism to explain how so many bacterial taxa can coexist in soils (Czárán *et al.* 2002). These authors present a theoretical analysis showing that, by including multiple toxins and multiple toxin resistance genes, a large number of taxa will persist in a two-dimensional model grid of interacting bacteria. They find that a system with a few toxin genes and many resistance genes can support up to 1000 different strains in a 180 × 180 spatial grid. This is obviously somewhat less than the diversity estimates obtained for soils, but it is of the right order of magnitude.

1.2.2 Mechanisms

Intransitive networks of competitive interactions may explain the high diversity of some systems of competitors. It appears that high diversity can be maintained by intransitive competitive networks only in spatially structured environments, such as soils or two-dimensional surfaces (Reichenbach *et al.* 2007). In well-mixed environments without opportunities for much spatial structuring, such as aquatic systems, other mechanisms must be invoked to explain the coexistence of large networks of competitors.

What maintains competitor diversity in well-mixed habitats with only few limiting resources? One possibility has been suggested by Huisman and Weissing (1999). Their models show that large networks of many resource competitors, when competing for several resources (four to six), can persist due to chaotic fluctuations in species abundances (Fig. 1.4). They suggest that switches along alternative transitive networks also seem to play a role. The presence of multiple limiting nutrients in their model prevents the formation of clear competitive hierarchies. This might provide an explanation for the coexistence of large numbers of phytoplankton species in well-mixed environments such as lakes and oceans, a problem that has interested ecologists for many years (Hutchinson 1961). Recently, Huisman and colleagues found experimental evidence for similar chaotic dynamics in a closed experimental multitrophic plankton system that was observed for a large number of generations (Beninca *et al.* 2008). This seems to be the first well-documented example

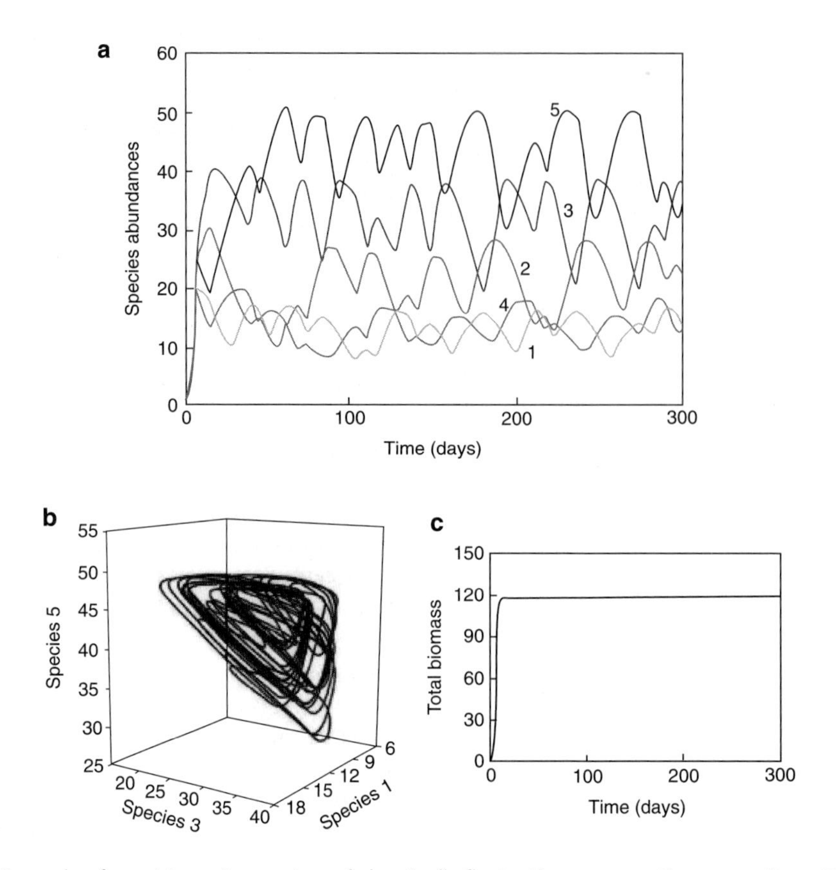

Figure 1.4 (a) Example of coexistence in a system of chaotically fluctuating consumptive competitors. Reprinted with permission from Macmillan Publishers Ltd: *Nature*, 407–10, J. Huisman & F.J. Weissing, © 1999. (b) Populations of three species orbit around a strange attractor, typical of chaotic dynamics. (c) Although the populations fluctuate chaotically, total biomass summed over all species is invariant over time.

of chaotic dynamics in a food web. Whether such chaotic dynamics also frequently occur in natural plankton communities still needs to be seen though, as the experimental set-up prevented many types of interactions between biota and abiotic factors (such as sediments). One can imagine that fluctuations in external forcing factors, such as light and temperature, may overrule the potential for chaotic dynamics in such systems.

Recently, another way out of the 'paradox of the plankton', or the 'principle of competitive exclusion' (no more species are expected to coexist than the number of limiting resources), has been offered by the same research group. Classic explanations for coexistence of plankton species have explored light, nitrogen, phosphorus and silicon as the main limiting resources, leading to an expectation of only few species to coexist, each limited by a different resource. However, different algae species seem to have specialized on different wavelengths of the light spectrum, which therefore is in fact a whole class of resources (Stomp *et al.* 2007a,b). These 'colourful niches' seem to play a yet underestimated role in the coexistence of plankton species.

Of course, other mechanisms that minimize the importance of competition among species can also operate to maintain diversity in spatially structured or well-mixed environments. Perhaps the best known of these mechanisms is keystone predation (Paine 1966), in which consumers feed preferentially on competitively superior species and prevent competitive exclusions of weaker competitors (Leibold 1996).

1.2.3 Unresolved issues

Except for a very few well-studied and relatively simple examples, we know little about the structure of real competitive networks. This is partly due to the difficulty of clearly establishing whether species actually compete, if so to what degree, and of identifying which mechanisms are in play. The necessary experiments are often difficult to perform, except under simplified conditions in the laboratory.

How would inclusion of competitive networks into food webs (networks in which links are consumption) alter our predictions about web stability? We can guess, in part, that inclusion of competitive interactions into a community matrix based only on predator–prey interactions would tend to destabilize the network, to the extent that May's (1972, 1973) analysis holds true.

How common are subnetworks of intransitive competitive interactions, and do they explain the persistence of high species richness? Although they have long been recognized as a potential scenario for maintaining diversity in systems (see Connell 1978), we still have very few well-documented examples of non-transitive competitive networks of any sort. Minimally, such networks would require that competitors interact through more than one kind of competitive mechanism (Schoener 1983). The recurring problem is that, based on traditions, expertise and practical limitations, almost all studies still focus on subsets of entire food webs, such as beetle communities, plant communities, pollinator communities, soil food webs, soil microbial communities, but hardly ever study the entire system with the same level of aggregation. This will require large interdisciplinary efforts and most likely new theoretical approaches to deal with the resulting data.

1.3 Mutualistic networks

1.3.1 Structural regularities

For the most part, mutualisms do not figure prominently in traditional depictions of ecological networks. This reflects, in part, short shrift given to reciprocal positive interactions in community ecology (Boucher 1985), at least until recently (Brooker *et al.* 2008). Some of the earliest works consider plant–pollinator systems (Feinsinger 1976; Petanidou and Ellis 1993; Fonseca and John 1996; Waser *et al.* 1996), and recently there has been a growth in the use of the network perspective to analyse the greater amounts of data collected on mutualistic networks (see following references in this section). For the main part, the two types of mutualistic networks examined are plant–pollinator networks and plant–seed disperser networks. The convention used to depict these networks is to show interactions in a two-layer network between one group of species (plants) and their mutualists (pollinators or dispersers), as shown in Fig. 1.5. The species are not depicted on different trophic levels, as in food webs, and only interactions between the two groups of mutualists are considered (no direct competitive interactions are included).

Mutualistic networks appear to have several types of structural regularity. First, there are nested sets of interactions (Bascompte *et al.* 2003; Lewinsohn *et al.* 2006). That is, more specialized mutualists tend to interact with a proper subset of the species that more generalist mutualists interact with. One consequence is that there is a set of species that form a highly connected 'core' in mutualistic networks. This makes the number of links across the network required to connect any two species rather short (Olesen *et al.* 2006). Indeed, mutualistic networks tend to be more nested than food webs and to have shorter paths between any two species than food webs (Bascompte *et al.* 2003). This 'core' of highly connected interactors appears to have consequences for how mutualistic networks respond to disturbance, and seems to make them more robust to potential perturbations (Bascompte *et al.* 2003).

Another structural feature of mutualistic networks is of asymmetric patterns of interactions (Bascompte *et al.* 2006; Vazquez *et al.* 2007). In general, species with high numbers of connections tend to interact with those that are connected to relatively few species. This asymmetry in connections within mutualism webs is consistent with the features of models that confer greater stability on these networks (Bascompte *et al.* 2006).

Similar to competitive networks, mutualistic networks can also show an intransitive structure, also called hypercycles. Three species may be arranged

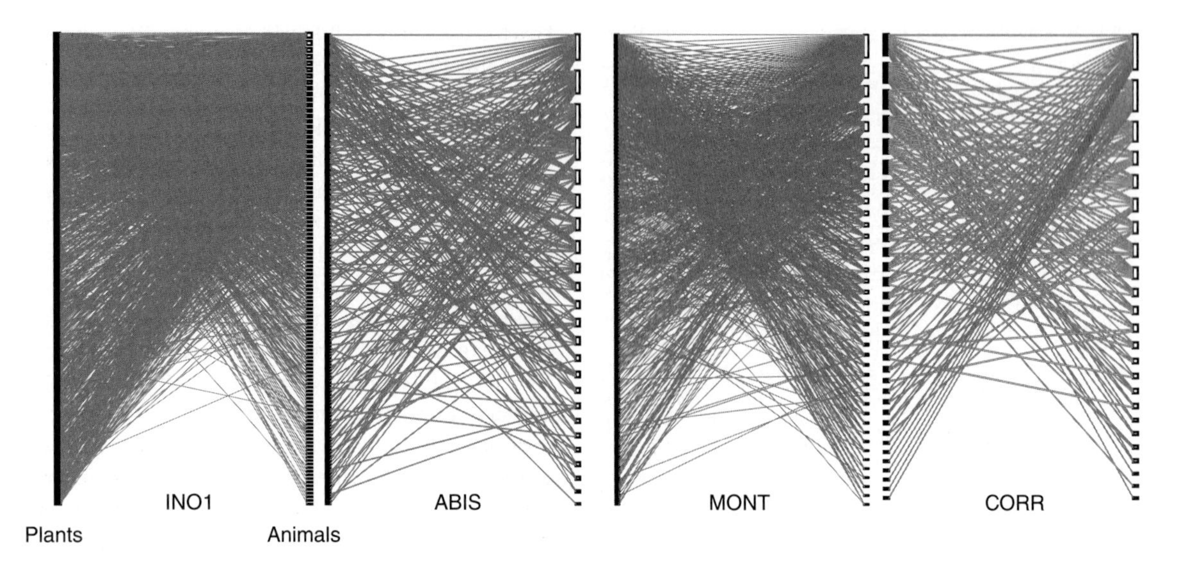

Figure 1.5 Examples of networks of positive interactions (mutualisms) between plants and their pollinators or dispersers in four different kinds of communities. In each graph, nodes on the left correspond to different plant species, and nodes on the right correspond to animals that pollinate the plants or disperse their seeds. Lines connect positively interacting species. Reprinted from Jordano *et al.* (2003), with permission from Wiley-Blackwell.

in an autocatalytic loop, in which species A promotes B, B promotes C, and C promotes A, resulting in indirect mutualism (A helps C through B). This structure can also be seen as positive feedback of species A on its own growth, through species B and C. For example, Ulanowicz (1995) describes and models how the carnivorous waterplant *Utricularia* excretes sugars towards its leaf surface, which promotes the growth of microepiphytobenthos (algae, diatoms and bacteria) on the leaves, which then attract zooplankton grazers (as copepods), which are in turn caught by the *Utricularia* plant for food. Such hypercycles may have played a role in early prebiotic evolution, in which space and limited dispersal again seem to stabilize the interaction, e.g. against the invasion of parasites in the mutualistic network (Boerlijst and Hogeweg 1991).

1.3.2 Mechanisms

Information about the drivers of the structure of mutualistic networks is beginning to emerge. The clearest example of this is a recent study about relationships between the shared evolutionary history of the species in mutualistic networks (Rezende *et al.* 2007). Across a compilation of 36 plant–pollinator and 23 plant–frugivore networks, there was a significant effect of phylogeny on the number of links of a species in 25–39% of the networks. That is, more closely related species tended to both be specialists (or generalists) than more distant ones. The amount of phylogenetic similarity of two species was also found to predict the ecological similarity (measured as the standardized number of interactions in common) in nearly 50% of the networks. These analyses provide good evidence that evolutionary history at least partly explains some of the structural regularities of mutualistic networks.

1.3.3 Unresolved issues

The patterns described above raise additional questions. Why does phylogenetic history apparently influence the structure of mutualism networks? Perhaps evolution, and specifically coevolutionary processes, have a stronger role in structuring networks of mutualists. There is considerable variation unexplained by phylogeny in the webs considered above, so obviously phylogenetic history is not the only factor contributing to structure in these networks (Rezende *et al.* 2007). It is also interesting to ask to what extent the nestedness in mutualism networks is a consequence of the structure of phylogenies, or of more ecological constraints?

So far, the mutualisms depicted in these networks are restricted to interactions involving transport of pollen or fruits/seeds by animals and the reciprocal energetic or nutritional reward obtained by the animals that consume nectar, pollen or fruits. These are only a fraction of the kinds of positive interactions that characterize mutualisms. Other kinds of positive interactions, such as defensive/domicile mutualisms involving the defence of host plants by insects (Janzen 1966), or trading in energy and nutrients between plants and mycorrhizal fungi (Schwartz and Hoeksema 1998), have not yet received this kind of analysis.

Another problem is that effects of mutualisms on the dynamics of the larger ecological networks in which they embedded are essentially unknown. Most models of mutualisms focus on the dynamics of a pair of species involved in a mutualistic interaction (Dean 1983; Wolin 1985). Consequences of mutualisms for the stability of the larger community network in which they occur remain largely unexplored (but see Ringel *et al.* 1996).

1.4 Food webs

1.4.1 Structural regularities

Food webs are perhaps the oldest and most frequently studied type of ecological network. The links in food webs are determined by observing evidence of a resource–consumer interaction between two species. This can be done in a qualitative way (presence/absence of a trophic interaction) or quantitative way (e.g. through feeding rates, predation rates, conversion efficiencies). Much effort has been expended collecting this type of information, especially the qualitative type, for constructing food webs, and examining whether they exhibit particular structure. One of the earliest ideas was that food chain lengths (the number of links from a species at the bottom of the food web to one at the top) were shorter than expected by chance (Pimm and Lawton 1977; Pimm 1982). Another was a relative paucity of omnivores, species that feed on resources at different trophic levels (Pimm and Lawton 1978; Pimm 1980). These and other apparent regularities have been challenged many times (Yodzis 1984; Paine 1988; Polis 1991); some are supported by more recent and higher quality data, others are not (Pimm 1991; Martinez 1994; Warren 1994). Rather than review the literature concerning all of the patterns, we will focus on one of the most important properties of food webs with 'qualitative' (presence/absence) links: connectance. (Note that there are quantitative versions of some properties, including connectance (e.g. Bersier *et al.* 2002).)

In a food web of S species, connectance is the number of realized links (L) divided by the number of possible links (S^2 or $S(S-1)$ if excluding cannibalistic links), i.e. when all species interact. It has strong effects on many other structural features, such as the frequency distribution of the number of links per species (Dunne *et al.* 2002; Stouffer *et al.* 2005), and has long been known to affect the stability of model food webs (Gardner and Ashby 1970; May 1972; DeAngelis 1975; Pimm 1984). One might say that recording and explaining patterns of connectance is a first step towards understanding many other aspects of food web structure. Earlier researchers outlined a number of ideas about how structure depends on the number of species in a food web. In particular, there was much debate about whether connectance remains constant or increases or decreases as the number of species in food webs increases.

There has been a great deal of research about the magnitude of connectance in food webs and how it changes with the overall species richness of the web (Sugihara *et al.* 1989; Winemiller 1989; Havens 1992, 1993; Martinez 1992, 1993; Bengtsson 1994). Some of this early discussion was hampered by problems with the source and quality of the data used to construct food webs. In contrast, over the last two decades or so, a large amount of data was collected with the explicit goal of constructing highly detailed food webs, and we will focus on 15 exceptionally well-characterized food webs.

In these webs, connectance varies from about 0.01 to 0.35, that is, from 1% to 35% of all possible links are realized (Fig. 1.6). Furthermore, connectance appears to decrease with increasing species richness, and this pattern seems to be well supported by recent analyses (Murtaugh and Kollath 1997; Schmid-Araya *et al.* 2002; Montoya and Sole 2003).

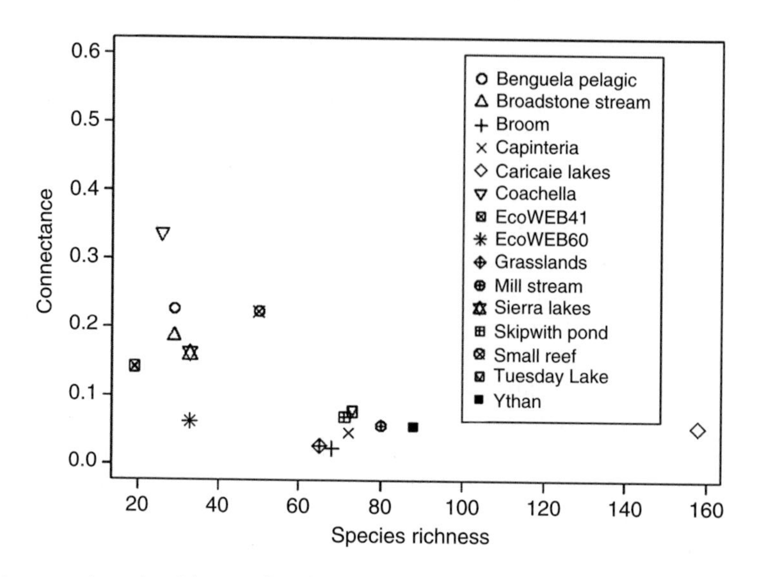

Figure 1.6 Connectance and species richness of 15 food webs from Petchey *et al.* (2008). Note how connectance declines with increasing species richness.

Among the 15 food webs analysed, however, only four are from terrestrial ecosystems (Broom, Coachella, EcoWEB60 and Grasslands); three have rather low connectance; and one, Coachella, has very high connectance. This variation among terrestrial webs is unlikely to be due to the nature of data collection, and has more to do with the types of species in the food webs (e.g. seed eaters versus parasitoids).

As in competitive and mutualistic networks, transitive structures play an interesting role also in food webs, based on consumer–resource interactions. Soil food webs have been shown to contain trophic loops that are suggested to contribute to their stability (Neutel *et al.* 2002). Also known as hypercycles, such interaction structures may even have played an important role in prebiotic evolution (Boerlijst and Hogeweg 1991).

1.4.2 Mechanisms

What mechanisms determine connectance, how does it differ between food webs and how does it vary with species richness? Broadly, one can separate the candidate mechanisms into two classes: those related to foraging ecology and those affecting overall network stability. Foraging-based mechanisms focus on why consumers eat the number of resources that they do, and how this depends on traits of the consumer (such as body size; e.g. Woodward *et al.* 2005) and the diversity of available resources. Stability-based mechanisms focus on the effects of connectance and species richness on dynamic stability, with the idea that unstable combinations of connectance and species richness will be less common in natural systems than stable combinations. Clearly these are not mutually exclusive mechanisms, though from here on we will focus on foraging-based mechanisms.

Qualitative foraging-based explanations of connectance rely on the fact that connectance is derived from the summed number of resources of each of the consumers in a food web. Consequently, if there are many generalist consumers connectance will be higher than if there are many specialists. An example of this type of explanation of connectance is provided by a study of the natural enemies of aphid species in a meadow (Van Veen *et al.* 2008). Aphids are attacked by parasitoids, pathogens and predators, and the hypothesis was made that parasitoid webs would have lower connectance than pathogen ones, which in turn would have lower connectance than predator ones. The idea was that the intimate and prolonged interactions between parasitoid and host would lead to specialization. Whereas parasitoids can often suspend their development in the absence of appropriate prey, the fungal pathogens studied do not, and so might be expected to have a broader host range. Finally, the predators have a less intimate and prolonged interaction with their resources and so may be even more generalized. Observations supported these explanations of connectance. The parasitoid web showed the lowest connectance, the pathogen web was intermediate, and the predator web had highest connectance (when based on quantitative measures of connectance). Examination of parasite–host feeding interactions suggests that food webs constructed of these also are quite different from those focused on predation (Leaper and Huxham 2002).

Quantitative foraging-based explanations of food web structure require a prediction of the number of resource species each consumer species will feed upon. One classic paradigm for predicting species' diets is optimal foraging theory (MacArthur and Pianka 1966; Schoener 1971). Here, species are assumed to forage on the suite of resources species (items) that maximize their rate of energy intake. While optimal foraging theory has its critics (Pierce and Ollason 1987), for example that it poorly deals with species interactions, it does seem to make reasonable predictions about the connectance of a suite of real food webs (Beckerman *et al.* 2006), and it seems likely that models of diet choice or constraint have value in explaining and predicting food web connectance and structure.

The determinants of diet breadth are central to understanding how connectance scales with species richness. If consumers feed on a constant proportion of all available species, then average diet breadth is equal to kS, where k is the average proportion of resources fed upon by consumers. Here, the total number of links L will be kS^2, and connectance, L/S^2, will be constant at k. In contrast, if consumers can feed only on a particular number of resources, say n (the average across consumers), regardless of how many different ones are available, then $L = nS$ and connectance will be n/S. So whether consumers eat a constant proportion of all available species or a fixed number of species defines how connectance scales with species richness. Presumably, this will depend on the mechanism of foraging and resource selection, on the outcome of evolution and on physical constraints of consumption, and these could differ between broad classes of species and consumption (e.g. Warren 1990; Van Veen *et al.* 2008). Analyses of food web data suggest that a range of relationships may exist, but that the number of links scales with S^x, where x is not 1 or 2 but somewhere in between (Murtaugh and Kollath 1997; Schmid-Araya *et al.* 2002; Montoya and Sole 2003). Most likely there is not a single relation between connectance and species richness, but rather a range of possible relations, depending on the taxonomic identity and range of organisms considered.

Stability-based explanations of connectance originate from observations that model systems with large numbers of species and large numbers of connections tend to be less stable than systems with fewer species or connections (Gardner and Ashby 1970; May 1972). However, many good examples of species-rich and well-connected systems have been discovered (e.g. DeAngelis 1975; Neutel *et al.* 2002; Brose *et al.* 2006; Rooney *et al.* 2006). Furthermore, studies suggest that there are many ways for highly connected speciose communities to be stable (McCann *et al.* 1998; Neutel *et al.* 2002; Rooney *et al.* 2006; Otto *et al.* 2007).

Another enduring problem in food web ecology concerns the factors that limit the length of food chains embedded in larger trophic webs. Although simple linear food chains are a convenient abstraction and are absent from most natural systems, one can still trace chains of energy flow within complex food webs. The fundamental question is what limits the length of food chains.

A well-accepted and intuitive explanation for the short length of food chains based on the inefficiency of energy transfer between trophic levels began to be questioned in the late 20th century. Simple models of food chains suggested that the slow recovery from perturbations seemed to characterize longer model chains (Pimm and Lawton 1977, 1978), and that this aspect of reduced stability might account for the rarity of longer chains. Subsequent work suggested that this result was an artefact caused by confounding the frequency of stabilizing density-dependent population regulation with the length of model food chains (Sterner *et al.* 1997). Another problem was that the inefficiency of energy transfer should predict that longer food chains should occur in more productive environments, but there was little empirical support for this pattern (Pimm and Lawton 1977; Pimm 1982; Post 2002; however, see Oksanen *et al.* 1981). If anything, longer food chains and more complex food webs seemed to occur in less productive situations, rather than in the most productive environments. This apparent inconsistency must be tempered by the realization that there were very few detailed descriptions of any food chains or webs at the time, and most of them were from aquatic systems (Cohen 1978). The relatively small number of highly resolved food webs that have since accumulated (Winemiller and Pianka 1990; Martinez 1991; Dunne *et al.* 2002) suggest that the empirical data used to motivate these theoretical studies were far from ideal.

Experiments designed to evaluate the roles of energy limitation or population dynamics in limiting food chain length take two forms: (1) manipulations of energy inputs with concordant measurements of resulting food chain lengths and (2) manipulations of food chain lengths with concordant measurements of population dynamics. In the first approach, if more energy allowed longer food chains to develop and persist, then energetics presumably played some role in setting food chain length. The basic theory underlying this idea has been developed by Fretwell (1977) and Oksanen *et al.* (1981). Assuming that efficiency of energy transfer between trophic levels is about 10%, energy availability would have to be manipulated by over an order of magnitude (e.g. at least 10-fold) to see the addition or loss of a top trophic level. It is essential to be able to clearly place the organisms involved on particular trophic levels – this means that relatively linear food chains without substantial omnivory are required. Very few experiments satisfy these requirements, but those that do suggest an important role for energy in determining food chain length.

Jenkins *et al.* (1992) tested the effects of productivity on the relatively simple food webs that develop in water-filled tree-holes in tropical Australia. They examined food chain development after experimentally reducing productivity over 2 orders of magnitude, a reduction that should result in the loss of at least one trophic level from the top of the food chain. Decreasing productivity reduced the number of coexisting species, the number of trophic links and maximum food chain length. There seemed to be clear evidence that energy played a role in limiting food chain length in this system.

Kaunzinger and Morin (1998) used a simple microbial system to test for effects of productivity on food chain length. The longest three-level food chains consisted of a basal level (the bacterium *Serratia marcescens*) consumed by a ciliated protist (*Colpidium striatum*), which was in turn consumed by a top predator (the ciliate *Didinium nasutum*). The system lacks omnivores, so the trophic position of each species was known without error. Productivity was manipulated by varying the nutrient concentration of growth medium consumed by the bacteria. Three-level food chains, those containing the top predator *Didinium*, persisted only at higher levels of productivity (Fig. 1.7). At lower productivity levels the third trophic level failed to persist, clearly supporting the role of energy transfer in limiting the length of simple linear food chains. Patterns of change in the abundance of species on each trophic level are also consistent with simple prey-dependent models of predator–prey interactions (e.g. Leibold 1996), but are not consistent with ratio-dependent models (e.g. Abrams and Ginzburg 2000).

There is scant evidence for comparable patterns in natural systems, perhaps because of the technical difficulties in unambiguously assigning species to trophic levels or measuring productivity. Post *et al.* (2000) failed to find a relationship between food chain length and productivity in a survey of natural lakes, but did find that larger lakes tended to support longer food chains. Their finding is superficially similar to the idea that larger predators located higher in the food chain should require larger home ranges (Slobodkin 1960), even disproportionally larger than one allometrically would expect from their size (Haskell *et al.* 2002). Post *et al.* relied on the indirect measurement of the trophic position of top predators using stable isotope ratios, rather than using knowledge of the structure of the entire food chain. Post *et al.* (2000) and Post (2002) have suggested that the dependence of food chain length on energy inputs shown by Jenkins *et al.* (1992) and

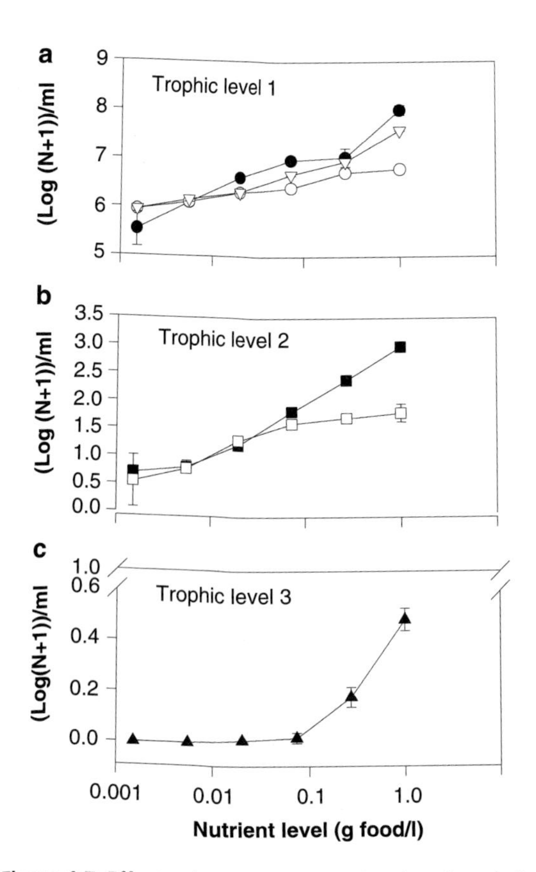

Figure 1.7 Effects of productivity on the abundance of species occupying basal (a; bacteria, trophic level 1), intermediate (b; *Colpidium*, trophic level 2) and top (c; *Didinium*, trophic level 3) levels in simple linear food chains. Three-level food chains persist only at higher levels of productivity, as shown by the failure of *Didinium* to persist at lower levels of productivity determined by nutrient levels in experimental microcosms. Solid and empty symbols in (b) show the abundance of the intermediate trophic level in food chains with and without the third trophic level, respectively. Reprinted with permission from Macmillan Publishers Ltd: *Nature* 395, 495–7, C.M.K. Kaunzinger and P. Morin, © 1998.

Kaunzinger and Morin (1998) may be a feature of relatively unproductive systems.

The other approach used to evaluate links between population dynamics and food chain length involves comparisons of the dynamics of species that occur in chains of differing length. If the models are correct, population dynamics should be more variable in longer chains, and that increased variation should lead to higher values of temporal variability for the same species embedded in longer food chains. Lawler and Morin (1993) found that the population dynamics of protists in relatively simple laboratory food chains become more variable with modest increases in food chain length. Comparisons of the temporal variability of populations of the same bacterivorous protists in short food chains in which the bacterivores were the top predators, and in food chains that are just one trophic level longer in which the bacterivores are intermediate species preyed on by another predatory protist, point to increased temporal variation in abundance in the majority of longer food chains. Increased temporal variation in abundance would be consistent with longer return times in longer food chains, as suggested by Pimm and Lawton (1977).

There is also reason to suspect that energy and population dynamics can interact in ways not directly considered by Pimm and Lawton (1977), as described in earlier models by Rosenzweig (1971) in the context of the so-called 'paradox of enrichment'. Rosenzweig found that a number of different predator–prey models became increasingly unstable as systems were made more productive – a consequence of increasing rates of increase or carrying capacities in the models. In this scenario, adding energy to a simple food chain might destabilize the system, and shorten the chain. Of course, it is possible that the addition of another trophic level to an energetically enhanced chain could offset the destabilizing effects of enrichment, though the findings of Pimm and Lawton (1977) might argue against this. However, addition of weakly interacting species to the food web has been suggested to confer increased stability on unstable systems (McCann *et al.* 1998), so some kinds of increased trophic complex may help to offset the predicted destabilizing effects of enrichment.

1.4.3 Unresolved issues

The relative contribution of foraging-based and stability-based mechanisms to the connectance of real food webs remains unresolved. Given the central importance of connectance for determining other structural features of food webs, this should be a priority research question about community topology.

It would be fascinating to see whether the addition of weak interactors can stabilize food chains, and whether those effects offset destabilizing effects of increased energy inputs, as some theories predict (McCann *et al.* 1998). Observations and models on soil food webs indeed suggest that this is the case (Neutel *et al.* 2002).

PART II

DYNAMICS

CHAPTER 2

Trophic dynamics of communities

Herman A. Verhoef and Han Olff

2.1 What types of dynamics can be distinguished?

Having examined the geometry and structure of the ecological networks in Chapter 1 we will move on to consider the dynamics of communities, and concentrate on the analysis of changes in the abundance of species in a multitrophic context. In 1927 Charles Elton stated that the structure of a community is determined by the net of feeding relations between trophic units, the *food web*. The topology of these feeding links (Chapter 1) naturally emerges from the *dynamics* of populations within ecological communities (May 1973; Neutel *et al.* 2002, 2007; Rooney *et al.* 2006), while the topology of the network in turn will affect the dynamics of the populations it contains (DeAngelis 1992).

One of the central goals in ecology is to discover why populations change over time. Much of the attention to this important subject is theoretical and the findings and consequent discussions are based on models outcome. However, empirical data on the different types of dynamics and the underlying mechanisms are increasingly reported. Four main patterns of population dynamics leading to coexistence (or long-term co-occurrence) of different species can be identified: coexistence at equilibrium, coexistence at alternate equilibria (with critical 'tipping points'), coexistence at stable limit cycles and coexistence at chaos.

2.1.1 Stable equilibria

Only a few studies address whether collections of multiple species show stable compositions corresponding to stable equilibria in mathematical models. Resource-based competition theory that makes such predictions for multiple species (Tilman 1982) seems not to hold for more than two species competing for two resources (Huisman and Weissing 1999, 2001). Also, the life span of the organisms is often too long compared with the length of the study, making the judgement of true coexistence across multiple generations problematic (Morin 1999). A good example is presented by Lawton and Gaston (1989). Despite natural perturbations affecting the community of about 20 herbivorous insects living on bracken fern, the relative abundances of the species and the taxonomic composition remained the same over a period of 7 years. The generation time of the respective populations is about 1 year. Similarly, in a 25-year-long study of large herbivore coexistence in a tropical savanna, Prins and Douglas-Hamilton (1990) found high community-level stability in species composition, despite fluctuations in abundance of individual species. A more recent study deals with a small microbial food web. Changes in the dynamics of a defined predator–prey system, consisting of a bacterivorous ciliate (*Tetrahymena pyriformis*) and two bacterial prey species, were triggered by changes in the dilution rates of a one-stage chemostat. The bacterial species preferred by the ciliate (*Pedobacter*) outcompeted *Brevundimonas*, the second bacterial species. At relatively high dilution rates *Brevundimonas* died off by the sixth day, whereas the remaining species existed in stable coexistence at equilibrium (Becks *et al.* 2005). Also in this experiment the generation time of the organisms involved was much shorter than the duration of the experiment.

2.1.2 Alternate equilibria

Evidence is accumulating that certain large-scale complex systems may have alternate equilibria and critical tipping points. Examples are the well-known trophic cascades in freshwater lakes (Carpenter and Kitchell 1993; Scheffer *et al.* 1993). Again, the criticism of studies that suggest the existence of alternate equilibria deals with the length of the study relative to that of the generation time of the organisms involved, and the physical characteristics of the different sites at which the species were studied (Connell and Sousa 1983). Also more recently the possible multiple stable states of a system have been stated to be difficult to be proven experimentally (Scheffer and Carpenter 2003; Schröder *et al.* 2005), despite several recent new suggestions (Kefi *et al.* 2007; van der Heide *et al.* 2007; Carpenter *et al.* 2008). The transition from one state to an alternate state is nowadays called a 'catastrophic regime shift', indicating the serious ecosystem-level implications of this phenomenon. The major problem is that ecological resilience cannot be measured in practice. Models can be used as indicators of ecological resilience (Carpenter *et al.* 2001), but the mechanisms of these transitions are often poorly known (Scheffer and Carpenter 2003). Recently, van Nes and Scheffer (2007) and Carpenter *et al.* (2008) have successfully explored various indicators of an upcoming catastrophic shift, such as the 'critical slowing down' known from physics.

2.1.3 Stable limit cycles

Apart from the predator–prey oscillations based on the Lotka–Volterra equations which are only neutrally stable, the Holling–Tanner model (Holling 1965; Tanner 1975) produces a range of dynamics. This model shows no tendency to return to the equilibrium point, but displays another form of predator–prey oscillations: stable limit cycles. In the above-mentioned microbial food web study, obvious stable limit cycles were established at low dilution rates of the chemostat systems (Becks *et al.* 2005). Maxima and minima for the predator and the two preys recurred during the whole period of the study.

2.1.4 Chaotic dynamics

The long-term persistence of complex food webs is not automatically linked to stability, and many mathematical models predict that species interactions can create chaos and species extinctions. Despite receiving an overwhelming amount of theoretical attention, experimental demonstrations of chaos are rare. Only a few single species systems (Costantino *et al.* 1997; Ellner and Turchin 2005), the already mentioned microbial food web study, in which at intermediate dilution rates of the chemostat systems chaotic dynamics were observed (Becks *et al.* 2005) as well as nitrifying bacteria in a wastewater bioreactor (Graham *et al.* 2007), show compelling evidence for chaos. Recently, it has been shown that in a long-term experiment with a plankton community, consisting of bacteria, several phytoplankton species, herbivorous and predatory zooplankton species, and detritivores, chaotic dynamics also appear (Beninca *et al.* 2008). The food web showed strong fluctuations in species abundances, attributed to different species interactions. We refer to this study later. As both the structure and the dynamics of a closed, local community are the result of the interactions among the constituting species, we need *population dynamic models* to represent them. Such interactions may be of different kinds, such as between predators and prey, between competitors for the same resource, and non-trophic interactions, e.g. through environmental modification (Olff *et al.* 2009). In the present chapter we will discuss the dynamics of *small food web modules* and those of *complex interactions*.

2.2 Dynamics of food web modules

Insights from specific 'few-species-interaction-configurations' or modules (Menge 1995; Holt 1997; Bascompte and Melian 2005) of consumer–resource interactions have much increased over the last few decades. For example, we know much more now about resource competition (Schoener 1974; Tilman 1982), mutualism (Oksanen 1988), apparent competition (Holt 1977), indirect mutualism (Vandermeer 1980; Ulanowicz 1997), intra-guild predation (Polis *et al.* 1989), positive interactions such as facilitation (Callaway 2007), positive feedbacks (DeAngelis *et al.* 1986), regulatory feedbacks (Bagdassarian *et al.* 2007), trophic cascades (Carpenter *et al.* 2008) and multiple stable state dynamics (Scheffer and Carpenter 2003). These may all be considered organizational forces that structure food webs, but they may not all be of equal importance. For example, Ulanowicz (1997) makes a strong case for the special importance of indirect mutualism as an organizational force in food webs, as the resulting feedback loops 'attract' resources towards them. Other authors, such as Tilman (1982), have emphasized the importance of competition as a key organizational force in ecological communities. Again others, such as Krebs *et al.* (1999) emphasize the importance of predator–prey interactions in structuring communities. Despite the insights gained into such specific processes, the question remains how such modules together organize into complex interaction webs, and how to address their relative importance.

Food web modules are characterized by the fact that they are small systems (two or three trophic levels) that possess explicit dynamics (Fig. 2.1a–e). With these simple 'building blocks', more realistic food webs can be 'built' (Fig. 2.1f), which in turn are subsets of the true complexity in trophic interactions found in real ecosystems. For example, Fig. 2.2 shows the network of trophic interactions as found on intertidal sand flats in the Wadden Sea, a soft-bottom intertidal ecosystem with complex trophic structure. This example shows how exploitative competition, food chains, apparent competition and intra-guild predation can operate simultaneously within the same ecosystem. In this, it should be realized that food web descriptions in terms of interaction topology and flows (as in Fig. 2.2) generally capture the long-term averages of organism densities and fluxes. The actual abundances may vary due to external drivers (such as varying weather conditions) and internal dynamics (e.g. limit cycles). To illustrate this point, Fig. 2.3 shows observations of the long-term population dynamics of some of the bivalve species shown in the food web of Fig. 2.2. In this case, winter

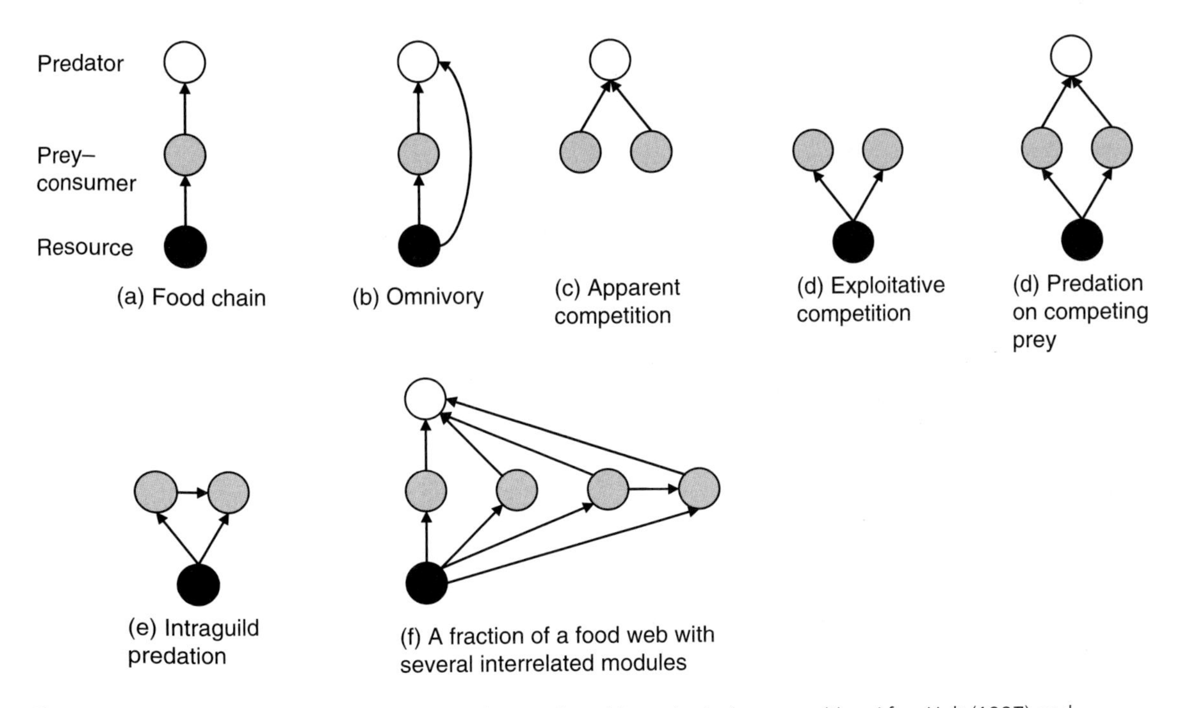

Figure 2.1 (a–f) Examples of trophic modules that are found in ecological communities. After Holt (1997) and Bascompte and Melian (2005).

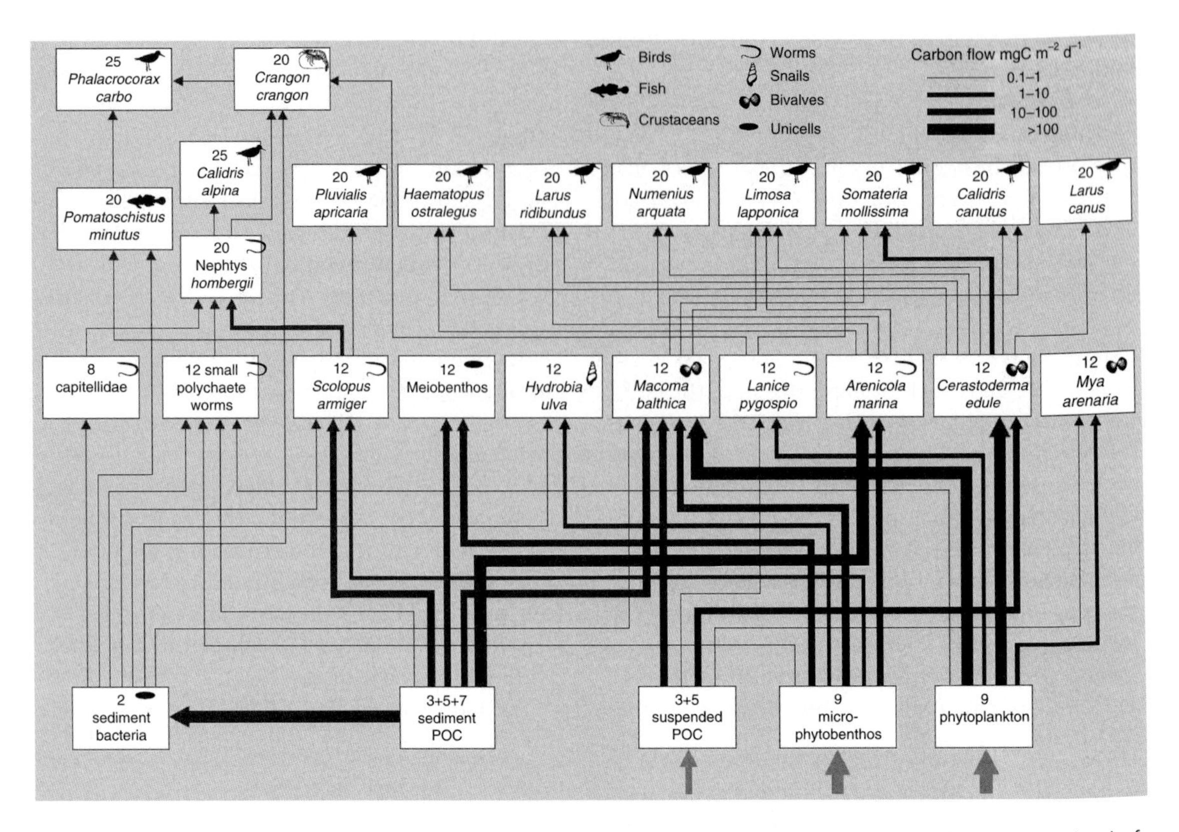

Figure 2.2 Example of a real food web, as observed on intertidal sand flats in Wadden Sea near the German island of Sylt, showing how the different types of modules from Fig. 2.1 can together form a complex network of interactions. POC, particulate organic carbon. Data from Baird *et al.* (2007).

temperatures are thought to control abundances through their effect on recruitment (Beukema *et al.* 2001).

Not all modules will be equally important in every ecosystem. For example, ecosystems that are dominated by many species within the same trophic level, such as diverse grasslands, may be strongly structured by exploitative competition. On the other hand, ecosystems such as the marine pelagic zone seem dominated by trophic chains. Bascompte and Melian (2005) recently compared the frequency of different types of modules across natural food webs. They found that apparent competition and intra-guild predation (Fig. 2.1) were generally overrepresented with respect to a suite of null models, while the level of omnivory varied highly across ecosystems.

The relative importance of external forcing versus internal dynamics as causes of dynamics in food webs under natural conditions is still hard to assess, despite a long history of research on the subject (Pimm 1982, 1991; Loreau and de Mazancourt 2008). Various research lines can be distinguished here, depending on their theoretical versus experimental nature, the complexity of the system under study and the level of control of variation in external conditions (exclusion or inclusion of forcing factors). Some theoretical studies have investigated the dynamics of species organized in simple modules (DeAngelis 1992). Several studies have been performed under experimentally controlled conditions, in which the influence of external variation on populations has been mostly eliminated, e.g. as in the study by Beninca

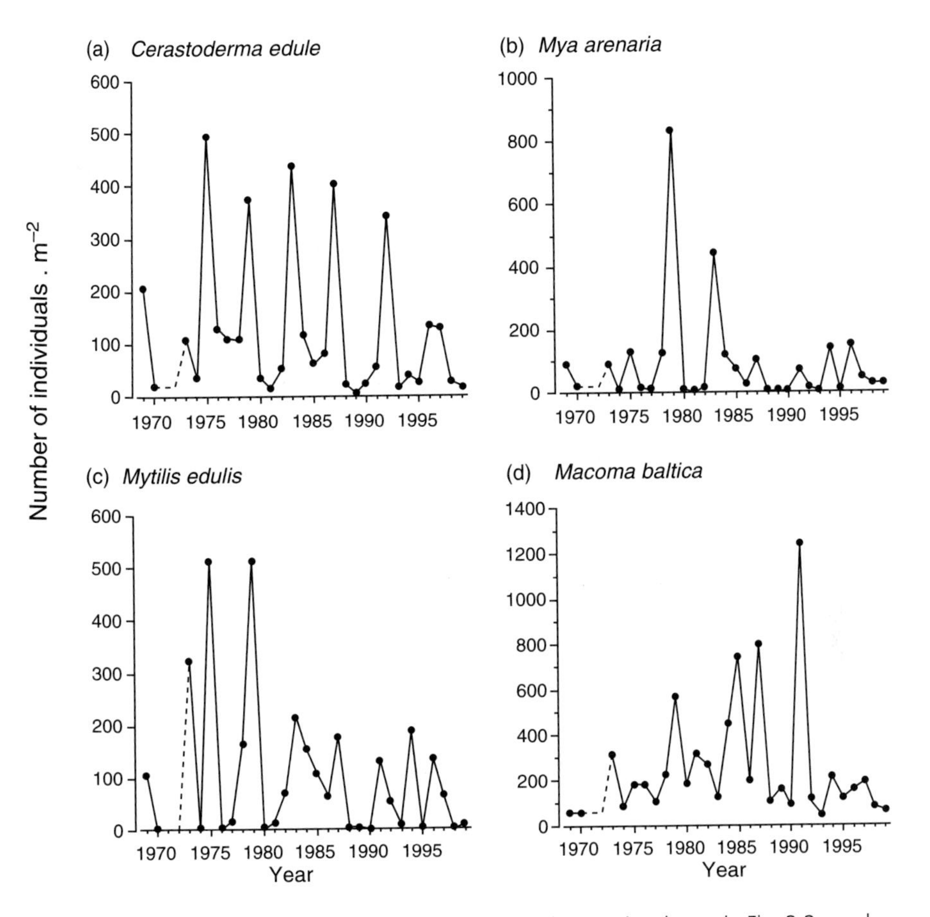

Figure 2.3 (a–d) Long-term population dynamics of some of the bivalve species shown in Fig. 2.2, as observed on the tidal flats of the western Wadden Sea in the Netherlands. Reproduced with permission from Beukema *et al.* (2001).

et al. (2008). Such studies are expected to mostly show dynamics that arise from the internal structure of food webs, and have been performed with both small modules and more complex webs. Other studies have explored the dynamics of interacting species populations under field conditions, which allows the assessment of the importance of external forcing factors. However, there is a necessary trade-off here.

2.3 Internal dynamics in food web modules or simple webs

In theory, consumer–resource interactions consist of two trophic levels that can fluctuate for a long time, but can also lead to unstable dynamics, with the precise type of dynamics depending on model formulation (Fig. 2.4). A good example of a study of a simple consumer–resource interaction is that of the limnetic crustacean zooplankton species *Daphnia* and its edible algal prey. McCauley *et al.* (1999) found large- and small-amplitude cycles in the same global environment, i.e. consumer–resource (predator–prey) and cohort (stage-structured) cycles (Fig. 2.5). In cohort cycles, demographic stages (usually thought to be the juvenile stage in *Daphnia*) are capable of strongly suppressing the other stages (adults) by competing for food. As the suppressing stage matures or dies, a pulse of reproduction or growth follows in the other stage, causing a cycling strongly out of phase (McCauley *et al.* 1999).

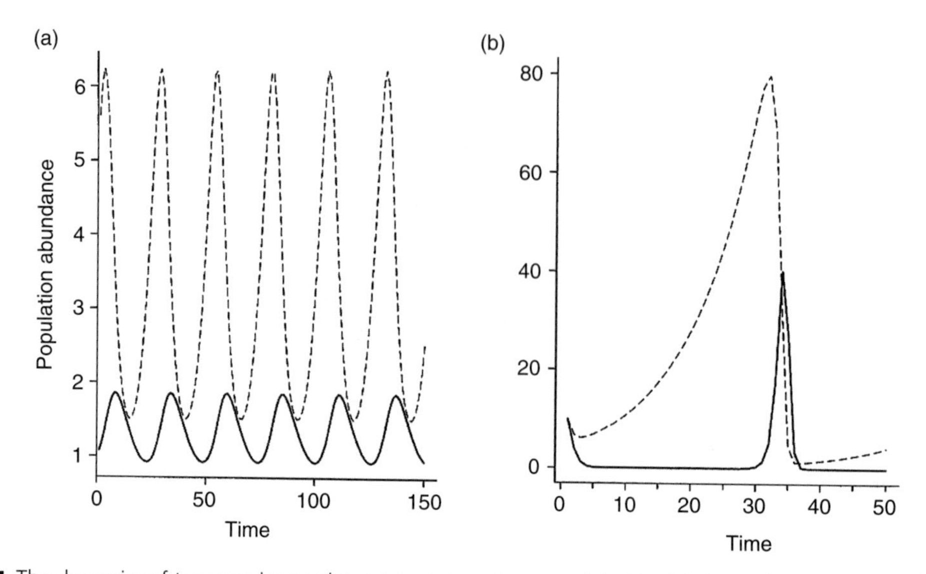

Figure 2.4 The dynamics of two-species predator–prey interactions predicted by (a) the Lotka–Volterra model (Lotka, 1926; Volterra, 1926) and (b) the Nicholson–Bailey model (Nicholson and Bailey, 1935). In both models, the dynamics are unstable. In the Lotka–Volterra model, the dynamics are neutral cycles with the period set determined by the model parameters and amplitude set by the initial conditions. Predators lag prey by one-quarter of a cycle. In the Nicholson–Bailey model, the dynamics show divergent oscillations with overexploitation by the predator, leading to the extinction of both prey and predators. Dashed lines, prey; solid lines, predators. Reproduced with permission from May and McLean (2007).

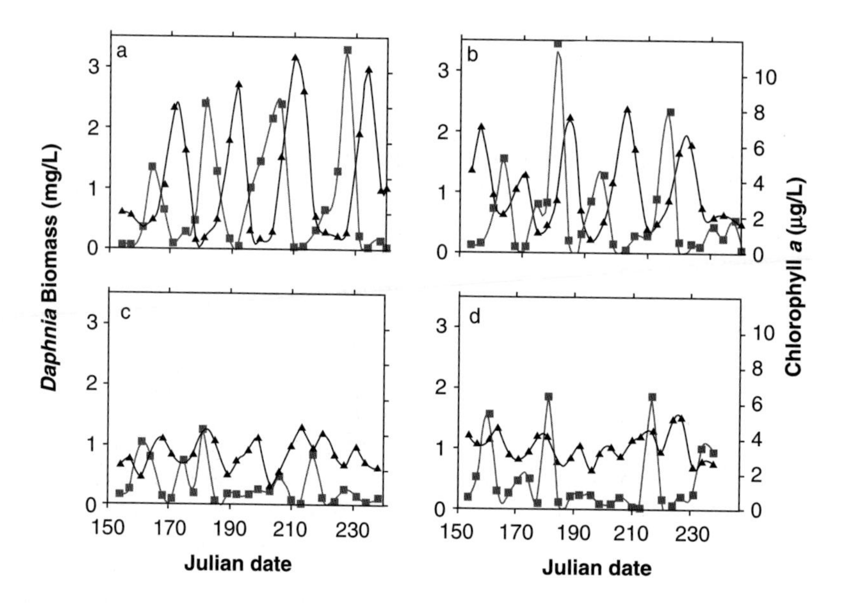

Figure 2.5 Large- and small-amplitude cycles in the same global environment. Population dynamics of *Daphnia* (triangles) and their edible algal prey (squares) in four nutrient-rich systems from one treatment. (a and b) Examples of large-amplitude predator–prey cycles. (c and d) Examples of small-amplitude stage-structured cycles. The initial biomass of the replicates is similar. *Daphnia* biomass is calculated from estimates of density and size structure, using length–weight relationships measured for the clone used in the experiment. Algal biomass is measured as chlorophyll *a* concentration. Reproduced from McCauley *et al.* (1999), with permission from *Nature*.

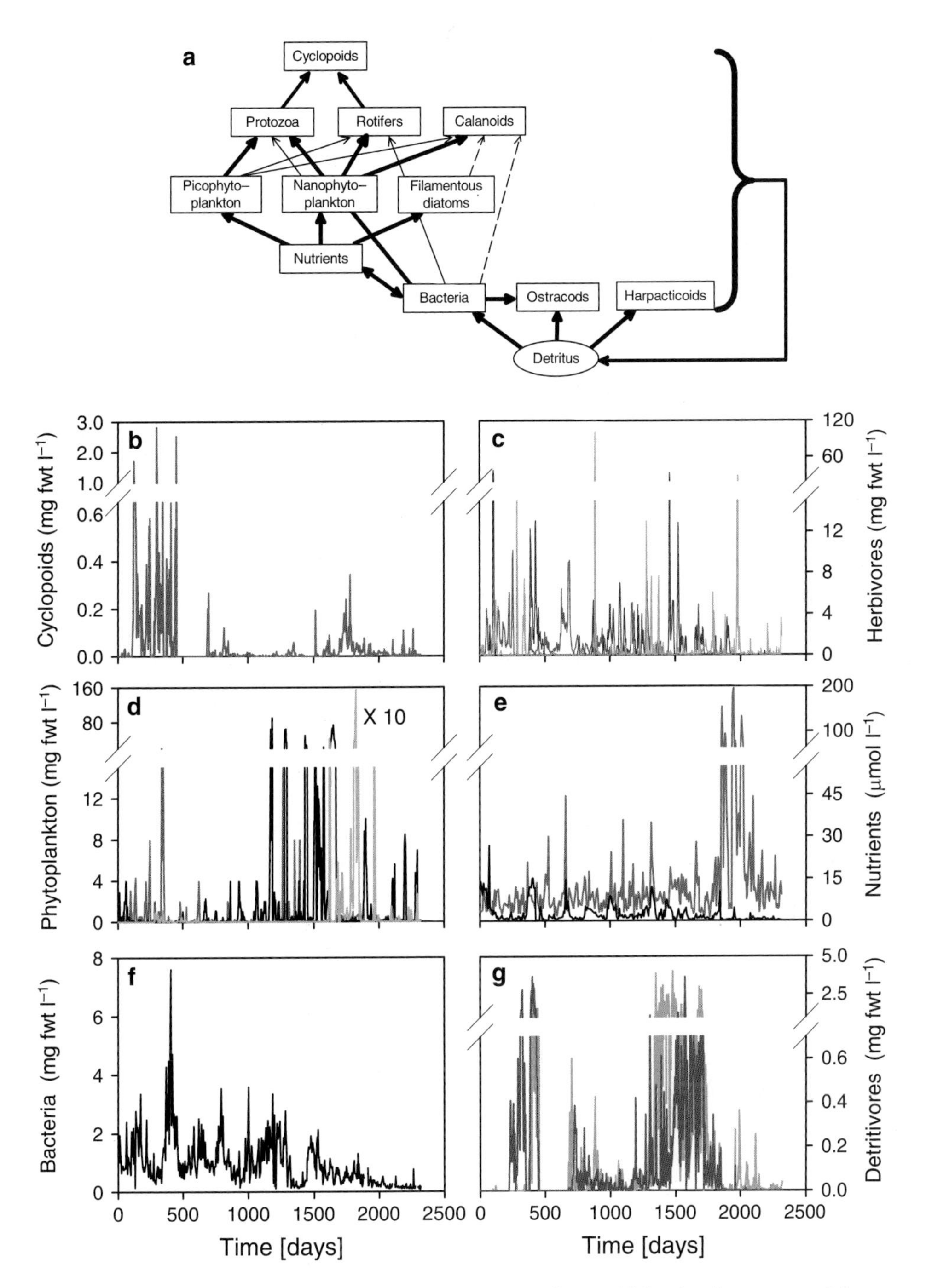

Figure 2.6 Description of a plankton community in a mesocosm experiment. (a) Food web structure of the mesocosm experiment. The thickness of the arrows gives a first indication of the food preferences of the species, as derived from general knowledge of their biology. (b–g) Time series of the functional groups in the food web (measured as freshweight biomass). (b) Cyclopoid copepods; (c) calanoid copepods (red), rotifers (blue) and protozoa (dark green); (d) picophytoplankton (black), nanophytoplankton (red) and filamentous diatoms (green); note that the diatom biomass should be magnified by 10; (e) dissolved inorganic nitrogen (red) and soluble reactive phosphorus (black); (f) heterotrophic bacteria; (g) harpacticoid copepods (violet) and ostracods (light blue). Reproduced with permission from Beninca *et al.* (2008). See plate 1.

This interaction between this grazer species and its prey can be thought of as an 'elemental oscillator', the basic building block for ecological communities (Vandermeer 1994). Leibold *et al.* (2005) working with a more complex food web, consisting of three grazer species (*Daphnia*, *Ceriodaphnia* and *Chydorus*) and edible algae, hypothesized that the dynamics in such complex food webs can be understood in terms of the simple subsets used by McCauley *et al.* (1999). The behaviour of these more complex food webs might be understood by thinking of coupled oscillators consisting of many such oscillators with interacting damping and amplifying harmonics (Hastings and Powell 1991). They also found both consumer–resource and cohort cycles. This indicates that interactions of zooplankton and algae in complex systems still consist of the same basic elements – in this case consumer–resource cycles and cohort cycles – and their dynamics can be understood from the dynamics of their component parts. In the study with a more complex marine web consisting of phytoplankton and zooplankton, Beninca *et al.* (2008) found strong chaotic fluctuations (Fig. 2.6). Species interactions in this food web are indicated as the driving forces. The persistence of this food web despite the great density fluctuations and unpredictability of the abundances is a rarely demonstrated phenomenon. It may also be more common than we think. The constant external conditions used in this study may be an artefact in itself, causing high productivity, leading to a situation that has been called the paradox of enrichment (Rosenzweig 1971). Many other important aspects of this study that distinguish it from real food webs under natural conditions are the absence of higher trophic levels and the exclusion of interactions between organisms and their abiotic environment, e.g. through local nutrient depletion or organism–sediment feedback. Even though isolated modules of species may exhibit chaotic dynamics, this may be highly dampened or even excluded by these effects in natural systems.

2.4 Dynamics enforced by external conditions

In addition to dynamics that arise internally within communities, species populations are also often subject to strong external forcing, e.g. where regional climatic conditions affect local air, water or soil temperature. Ecophysiological differences among species in ability to cope with these changes may result in species-specific responses (Karasov and Martinez del Rio 2007), thus leading to community dynamics under varying external conditions. This external forcing is the key 'point of entry' in studying effects of climate change on food webs, but also how toxic pollutants will affect trophic structure and ecosystem functioning. For example, ectotherms (at lower trophic levels) and endotherms (at higher trophic levels) are expected to respond very differently to short- or long-term temperature changes. Surprisingly, although there are good reasons to suspect its importance in natural populations, e.g. in the level of synchrony between species in long-term ecological monitoring (Bakker *et al.* 1996), environmental forcing has hardly received any attention in the study of consumer–resource interactions, food webs or other interaction webs. There is, however, some relevant theoretical work: a mechanistic-neutral model describing the dynamics of a community of equivalent species influenced by density dependence, environmental forcing and demographic stochasticity. The model shows that demographic stochasticity alone cannot oppose the synchronizing effect of density dependence and environmental forces (Loreau and de Mazancourt 2008). Vasseur and Fox (2007) have shown in their model food web study that the synchronization of dynamics is the result of environmental fluctuations. This synchrony promotes stability, because the maximum abundance of top predators is reduced by the synchronous decline in the density of consumers and synchronous increase in consumer density is changed by resource competition into synchronous decline. These authors conclude that future studies on food web dynamics should take into account the joint action of internal feedbacks and external forcing.

Another example concerns recovery after perturbation. The reaction of a system to a perturbation depends on the size of the 'basin of attraction'. The basin of attraction is a theoretical measure of the maximal perturbation that the system can absorb without shifting to another state and is often referred to as 'ecological resilience' (Peterson *et al.*

1998; Folke *et al.* 2004). Systems with a high ecological resilience can be seen as particularly stable systems, whereas systems with a low ecological resilience can be seen as unstable systems close to the tipping point into an alternate state. Recent studies have explored whether there are early warning signals for this shift into another state. The slow recovery from perturbations may be a possible indicator of an impending state shift (van Nes and Scheffer 2007).

2.5 Equilibrium biomass at different productivities

Important lessons can be learned from comparing the configuration of food webs (abundances of different species on different trophic levels) under different external conditions. A simple approach that has been used is to compare systems subject to different boundary conditions that impose constraints on their level of primary production. Such differences can be caused by variation in temperature or nutrient input into the system.

This approach can be used to study simple *food chains* consisting of a predator, a consumer and a resource species. This approach can also be used for simple systems with three levels, or for systems where all species at one trophic level are lumped into 'trophic species', under the assumption that the 'food chain module' in the system (see Fig. 2.1) strongly overrules the dynamics of the system with respect to other modules (whether this is true is, however, an open question). This approach to simplify systems leads to the concept of *trophic cascades*. Hairston *et al.* (1960) and Fretwell (1977) mention this trophic cascade phenomenon in their writings about population regulation, although they did not call the process by this particular name (Morin 1999). Paine (1980) used the term *trophic cascade* to describe how the top-down effects of predators could influence the abundances of species in lower trophic levels, a concept that was developed further by Oksanen *et al.* (1981) (Fig. 2.7a). If we increase the length of this *trophic food chain* by adding one additional trophic level (top carnivores) (Fig. 2.7b), *indirect mutualism* between non-adjacent trophic levels and a decrease or constancy in the abundance of 'odd' levels with increasing productivity appear.

Evidence for top-down trophic cascades is surprisingly scarce, and comes primarily from aquatic systems: stream communities (Power *et al.* 1985) and lakes (Carpenter and Kitchell 1993). There are few terrestrial examples: Emmons (1987), Terborgh

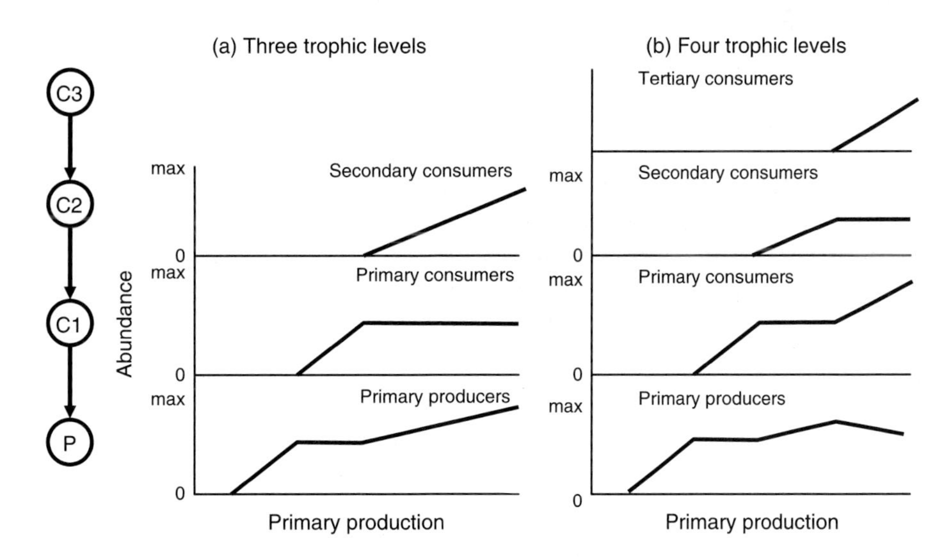

Figure 2.7 Equilibrium of three (a) and four (b) trophic levels along a gradient of primary productivity according to the ecosystem exploitation theory of Oksanen *et al.* (1981).

et al. (2006), Marquis and Whelan (1994) and Jefferies (1999). For example, Emmons and Terborgh *et al.* describe the disappearance of several common species of birds from Barro Colorado Island. The building of the Panama Canal created an island that was too small to sustain large predators, such as jaguars and pumas. Their extinction led to population increases of their prey species. These mesopredators fed on the eggs and young of ground-nesting birds and their increase in numbers was sufficient to wipe out many bird populations.

Another example concerns a terrestrial trophic cascade from a study by Marquis and Whelan (1994). They found strong effects of insectivorous birds foraging herbivorous insects on white oak trees. Birds significantly reduced the abundance of herbivorous insects on the oaks. Netting around some trees caused exclusion of birds from the insects, while other uncaged trees remained available to the birds. Oaks with birds and reduced herbivorous insects had less leaf damage from insects and subsequently had a higher biomass.

A slight modification of simple trophic cascades leads to intra-guild predation (IGP; Polis *et al.* 1989). This type of interaction can be seen as an extension of a simple predator–prey interaction: the predator also eats some of the consumers, but potentially competes with uneaten consumers as well for a common resource. Spiller and Schoener (1989) studied interactions between predatory *Anolis* lizards, predatory web-building spiders and their arthropod prey on small islands in the Bahamas. The interaction between lizards and spiders can be described as an IGP interaction, because lizards eat some spiders, but lizards also potentially compete with uneaten spiders for small arthropod prey. There is even an effect on the level of anti-herbivore defence of the dominant vegetation on the islands, *Conocarpus erectus* or buttonwood (Schoener 1988). On islands without lizards the leaves have trichomes to discourage insect attack; on islands with lizards, leaves are without them.

Summarizing we can state that in small food webs, oscillating consumer–resource interactions are not only predicted by models but also occur in natural systems. In chains of three or more levels trophic cascades are important, but experimental support is limited.

2.6 Dynamics of complex interactions

Food webs are conceptualized by their basic unit of interaction, consumption, and this basic process is oscillatory. When these basic units are connected, the conceptual framework becomes a system of *coupled oscillators.* This concept generated notable patterns in the theoretical literature as described by, for example, Vandermeer (2004). The conclusion that weak interactions can have strong effects on stabilizing ecosystems (McCann *et al.* 1998; Neutel *et al.* 2002) derives from this concept of coupled oscillators. According to May (1973), measures of interaction strength are the elements in a community matrix at equilibrium, which represent the direct effect of an individual of one species on the total population of another species at equilibrium. These results suggest that average interaction strength should be weak in species-rich, highly connected systems. The fundamental question is whether the configuration or distribution of interaction strengths within food webs is important for questions of community stability. de Ruiter *et al.* (1995) linked the differing approaches by deriving values of the matrices from empirical observations (Fig. 2.8). The data come from a terrestrial food web study, the Lovinkhoeve Experimental Farm (Integrated Management) in The Netherlands. This study indicates that the patterning of interaction strengths is essential for system stability. However, there is no direct correlation between interaction strength and stability. Weak interactions may be strong in terms of their stabilizing effects to the community. Further, it has been shown that long trophic loops contain relatively many weak links increasing food web stability because they reduce maximum loop weight, thus reducing the amount of intraspecific interaction needed for system stability (Neutel *et al.* 2002) (Fig. 2.9).

2.7 Conclusions

The types of dynamics leading to long-term co-occurrence of species in communities can be indicated as stable equilibria, alternate equilibria, stable limit cycles and chaotic dynamics. Although much of the attention to this subject is theoretical, empirical data are increasingly reported, although

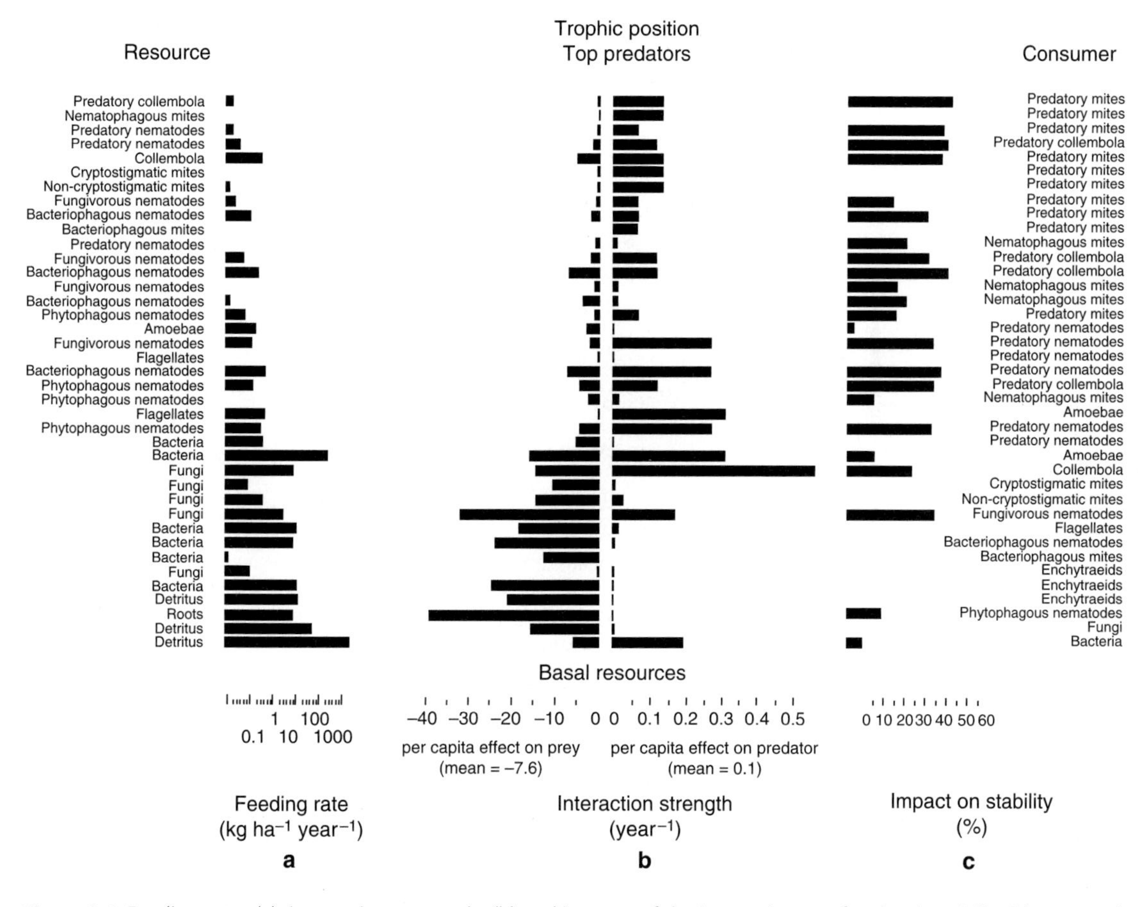

Figure 2.8 Feeding rates (a), interactions strengths (b) and impacts of the interactions on food web stability (c) arranged according to trophic position in the soil food web of arable fields with conventional agricultural practices at the Lovinkhoeve Experimental Farm in The Netherlands. From de Ruiter *et al.* (1995). Reprinted with permission from AAAS.

mostly still for experimental communities. For field studies, the life span of the organisms involved is often too long compared with the length of the study to unambiguously evaluate stability, but recent studies deal with that problem. Insights gained from the dynamics of specific food web modules have increased over the last few decades. But still the question remains how such modules are organized together into complex interaction webs, and how their interplay changes their dynamics as observed in isolated modules. The relative importance of external forcing versus internal dynamics for causing dynamics in food webs under natural conditions is still hard to assess. The study of the effects of climate change and toxic pollutants on food webs makes it necessary to concentrate in the future on the interplay of internal feedbacks and external forcing. Interesting lessons can be learned from the comparison of the configuration and dynamics of food webs under different external conditions, e.g. different temperatures that lead to different primary productivity. Where strong evidence for the existence of trophic cascades initially came from aquatic systems, terrestrial examples are now known. New studies on the dynamics of complex interaction webs have concentrated on the consequences of specific patterning of interaction strengths

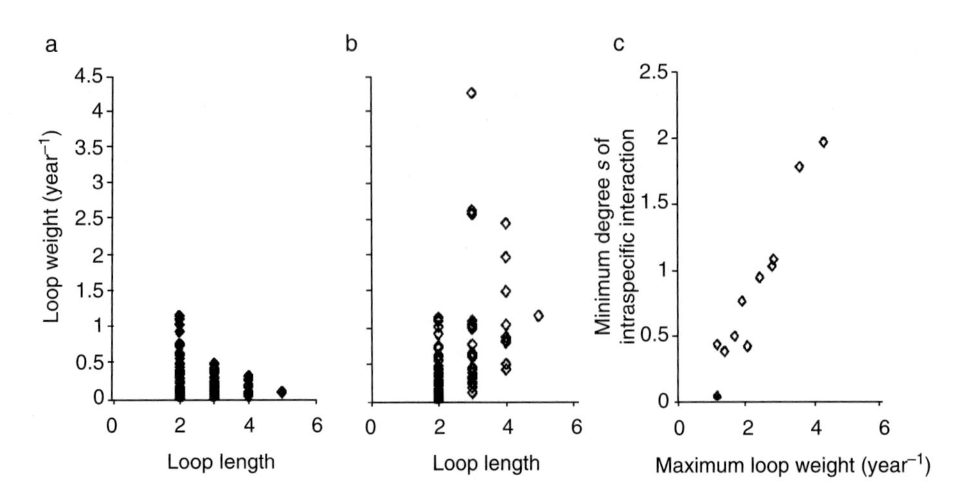

Figure 2.9 Loop length, loop weight and stability in the soil food web of the short grass prairie of the Central Plains Experimental Range and randomizations (20) of this matrix. (a) Loop weight versus loop length in the real matrix. (b) Loop weight versus loop length in a randomized matrix (a typical example). Long loops with a relatively small weight (those with many bottom-up effects) are not shown because they are not relevant for maximum loop weight. (c) Maximum loop weight and stability of the real matrix (solid diamond) and of 10 randomized matrices (open diamonds). Stability was measured as the value *s* that leads to a minimum level of intraspecific interaction strength needed for matrix stability. In a sensitivity analysis, it was found that variation in the parameter values within intervals between half and twice the observed value led to only a small variation in stability. From Neutel *et al.* (2002). Reprinted with permission from AAAS.

across the web for the stability of the overall system. Using classic stability analysis with Lotka–Volterra interaction terms, soil food webs with long trophic loops appear to contain relatively many weak links, which increases their stability. Future studies of more webs and with different modelling frameworks are required to show the generality of this phenomenon.

Acknowledgements

We thank Dick Visser for drawing Figure 2.2.

CHAPTER 3

Modelling the dynamics of complex food webs

Ulrich Brose and Jennifer A. Dunne

3.1 Introduction

The world is currently facing losses of biodiversity and habitats, species invasions, climate change, groundwater depletion and other anthropogenic perturbations that are resulting in the drastic reorganization of many ecosystems. In order to understand, predict and mitigate such reorganizations, which can severely affect ecosystem services (Daily 1997) that humans depend on such as water supply and purification and crop pollination, researchers need to broaden their focus from particular types of species to whole ecosystems. Ecological network research provides a very compelling framework for addressing the complexity of species interactions with each other and the environment. For example, the effects of any stressor – abiotic or biotic, natural or anthropogenic – on one population can cascade through ecological networks as a result of direct and indirect interactions, potentially affecting any other population in the same ecosystem. While much research has focused on direct effects between species or their populations, empirical and modelling studies have shown that effects due to indirect species interactions can be as important as direct effects in driving outcomes (Abrams *et al.* 1995; Menge 1997; Yodzis 2000). In fact, indirect effects can be stronger than the direct effects of a stressor, potentially greatly modifying the overall outcomes for population abundances in the face of extinctions (Ives and Cardinale 2004). Network analysis and modelling provide approaches for quantifying and assessing both direct and indirect effects. In general, determining the interplay among network structure, network dynamics and various aspects of stability such as persistence, robustness and resilience in complex 'real-world' networks is one of the greatest current challenges in the natural and social sciences, and it represents an exciting and dramatically expanding area of cross-disciplinary inquiry (Strogatz 2001).

3.2 Simple trophic interaction modules and population dynamics

While there are many types of interactions that species can have with each other, trophic (feeding) interactions are ubiquitous, are central to both ecological and evolutionary dynamics and are relatively easily observed, defined, quantified and modelled compared with other kinds of interactions. Within ecology, food web research, the study of networks of trophic interactions, represents a long tradition of both empirical and theoretical network analysis (Elton 1933; Lindeman 1942; MacArthur 1955; May 1973; Cohen *et al.* 1990; Pimm *et al.* 1991; see review by Egerton 2007). Ideally, food web research seeks to identify, analyse and model feeding interactions among whole communities of taxa including plants, bacteria, fungi, invertebrates and vertebrates, with feeding links representing transfers of biomass via various trophic interactions including detritivory, herbivory, predation, cannibalism and parasitism. A great deal of food web research, starting in the mid-1970s, has focused on various aspects of how such networks are structured (see review by Dunne 2006), with emerging strong empirical and model-based evidence that food webs from different habitats have similar characteristic topologies (Williams and Martinez 2000; Camacho *et al.* 2002; Dunne *et al.* 2002, 2004; Milo *et al.* 2002; Stouffer *et al.* 2005, 2006, 2007; see also Chapter 1).

However, food webs are inherently dynamical systems, since feeding interactions involve variable flows of biomass among species whose population densities are changing over time in response to direct and indirect interactions. Because it is difficult to compile detailed, long-term empirical datasets for dynamics of two or more interacting species, much research on species interaction dynamics relies on modelling. Many modelling studies of trophic dynamics have used analytically tractable approaches to explore predator–prey or parasite–host interactions (Yodzis and Innes 1992; Weitz and Levin 2006) or small modules of interacting taxa (McCann *et al.* 1998; Fussmann and Heber 2002), generally ignoring all but the most simple of possible network structures. In natural ecosystems, such interaction dyads or modules are embedded in diverse, complex food webs, where many additional taxa and their direct and indirect effects can play important roles for both the stability of focal species and the stability of the broader community. Therefore, it is critically important to ask whether knowledge about population dynamics obtained in small interaction modules (Yodzis and Innes 1992; McCann and Yodzis 1994; McCann *et al.* 1998; Fussmann and Heber 2002; Weitz and Levin 2006) applies to population dynamics in more realistically complex food webs.

Analyses of such modules suggest that every additional feeding link between the species may change the population dynamics dramatically. For instance, the dynamics of three populations in a tri-trophic food chain (Fig. 3.1a) can be stabilized or destabilized by an additional link from the top species to the basal species (McCann and Hastings 1997; Vandermeer 2006). By convention, this module (Fig. 3.1b) is termed an omnivory module (or intra-guild predation module). Depending on the relative strength of the links this module represents an intermediate stage between a tri-trophic food chain (in which the top species does not consume the basal species) and an exploitative competition module (in which the top species does not consume the intermediate species). Now, omnivory as the stage between these extremes can stabilize population dynamics by either eliminating chaotic dynamics or bounding the minima of the population densities away from zero (McCann and Hastings 1997). However, opposite results can be obtained with different parameters for population traits (Vandermeer 2006). Generally, omnivory can stabilize the population dynamics if the tri-trophic food chain and the exploitative competition module at the extremes of the gradient are unstable, whereas omnivory should destabilize the system if the modules at the extremes of the gradient are stable (Vandermeer 2006). Furthermore, the consequences of omnivory in a comparable simple experimental system are highly dependent on nutrient enrichment, since coexistence of both consumers is restricted to intermediate nutrient saturations (Diehl and Feissel 2001). Extending the size of the food web modules beyond three populations, Fussmann and Heber (2002) demonstrated that the frequency of chaotic dynamics increases with the number of trophic levels, but decreases with other structural properties that cause higher food web complexity. Interestingly, these results indicate that population stability might increase or decrease with food web complexity, depending on which process dominates in a particular food web. These studies have emphasized the critically important roles of network complexity and the distribution of interaction strengths across feeding links in determining

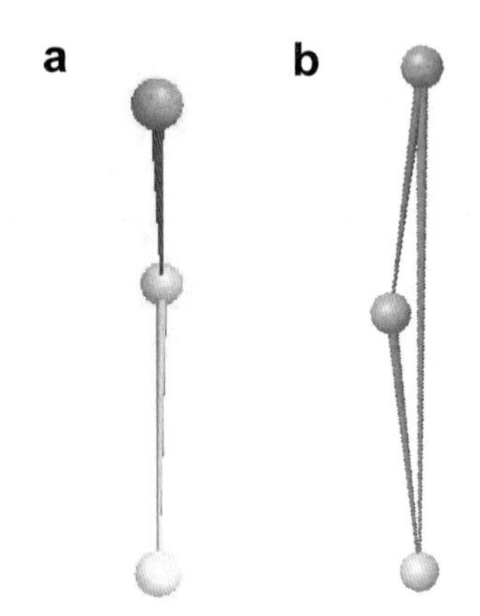

Figure 3.1 Structure of food web modules. (a) Tri-trophic food chain; (b) omnivory or intra-guild predation module.

population dynamics. However, it remains to be seen (1) whether our understanding of population dynamics in small modules can be scaled up to complex food webs and (2) how global characteristics of complex food webs such as connectance (a measure of link richness – the probability that any two species will interact with each other; see Chapter 1) affect population dynamics. In the next section, we describe an approach using keystone species that addresses the first question.

3.3 Scaling up keystone effects in complex food webs

In keystone species modules, a keystone consumer of a competitively dominant basal species facilitates the coexistence of competitively subordinate basal species (Fig. 3.2). When the keystone species is experimentally removed or goes locally extinct, competition can lead to extinction of the subordinate basal species. Such facilitation of basal species co-

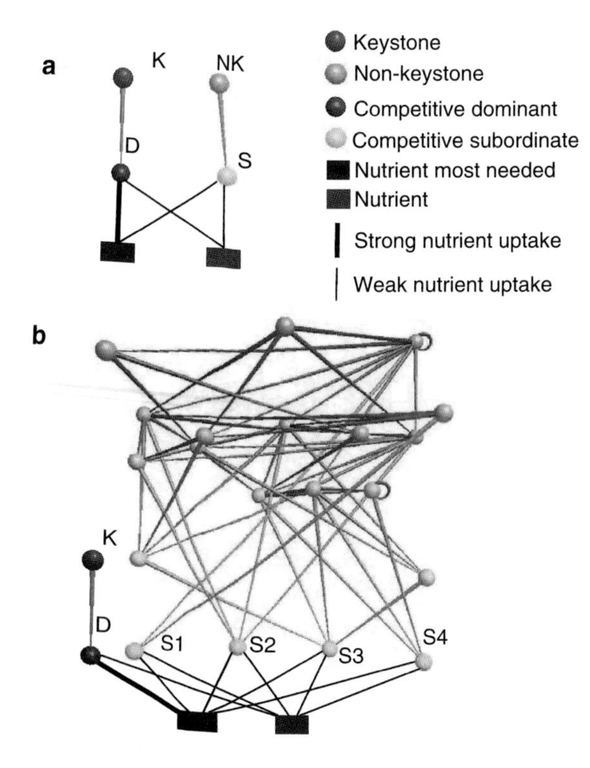

Figure 3.2 Keystone consumer in (a) a small module and (b) a complex food web. See plate 2.

existence by a keystone consumer was first described for the starfish *Pisaster ochraceus* that preferentially consumes the competitively dominant mussel *Mytilius californianus*, and thereby facilitates the coexistence of a diverse community of basal species that are competitively subordinate to the mussel (Paine 1966, 1974). Thus, the starfish has a positive effect on the biomass density of most species in the intertidal food web (i.e. the biomass density of most species is higher when the starfish is present than when it is absent). This facilitation by *Pisaster* was termed a 'keystone effect'. Similar keystone effects of other consumer species have subsequently been documented for many other ecosystems (Power *et al.* 1996), which suggests a broad generality of this phenomenon. However, the strength of the keystone effect varies dramatically between years and sites within ecosystems (Paine 1980; Menge *et al.* 1994; Berlow 1999). Analyses of keystone modules have shown that keystone effects vary substantially with the presence or absence of peripheral (non-keystone) species and links (Brose *et al.* 2005). This suggests that keystone effects in complex food webs (Fig. 3.2b) might be highly context dependent.

Systematic simulation analyses of complex food webs have revealed surprisingly simple determinants of keystone effects (Brose *et al.* 2005). Generally, distant effects of the global network structure or effects of populations that are more than two degrees (trophic links) separated from the keystone module are buffered by the network structure. Most likely, the multiple pathways between populations in complex food webs are characterized by effects of different signs that cancel each other out. Thus, effects between two species over pathways longer than two degrees often cancel each other out, and only effects over one or two degrees of separation that dominate in food webs (Williams *et al.* 2002) are not balanced by other interaction pathways of opposite sign and could systematically vary the strength of the keystone effect. These results suggest that effects within keystone modules that are embedded in complex food webs are affected by other populations within a 'local interaction sphere' of influence, which includes effects of non-keystone species that are separated by one or two links from the keystone module (Brose *et al.*

2005). If these results generalize to other types of modules such as omnivory modules, population dynamics in natural food webs may be determined by local interaction spheres that are larger than the classic modules, but smaller than entire complex food webs.

3.4 Diversity/complexity–stability relationships

Much of the research on food web dynamics in model systems with more than two taxa has been orientated around the classic (May 1972) and enduring (McCann 2000) debate on how diversity and complexity of communities affect food web stability. In the first half of the 20th century, many ecologists believed that natural communities develop into stable systems through successional dynamics. Aspects of this belief developed into the notion that complex communities are more stable than simple ones (Odum 1953; MacArthur 1955; Elton 1958; Hutchinson 1959). Some popular examples included the vulnerability of agricultural monocultures to calamities in contrast to the apparent stability of diverse tropical rainforests, and the higher frequency of invasions in simple island communities compared with more complex mainland communities. It was thought that a community consisting of species with multiple consumers would have fewer invasions and pest outbreaks than communities of species with fewer consumers. This was stated in a general theoretical way by MacArthur (1955), who hypothesized that 'a large number of paths through each species is necessary to reduce the effects of overpopulation of one species'. MacArthur concluded that 'stability increases as the number of links increases' and that stability is easier to achieve in more diverse assemblages of species, thus linking community stability with both increased trophic links and increased numbers of species. Other types of theoretical considerations emerged to support the positive complexity–stability relationship. For example, Elton (1958) argued that simple predator–prey models reveal their *lack* of stability in the oscillatory behaviour they exhibit, although he failed to compare them with multispecies models (May 1973). The notion that 'diversity and complexity beget stability', which already had great intuitive appeal as well as the weight of history behind it, was thus accorded a gloss of theoretical rigor (e.g. a 'formal proof' according to Hutchinson 1959), and took on the patina of conventional wisdom by the late 1950s.

The concept that complexity implies stability, as a theoretical generality, was explicitly and rigorously challenged by the analytical work of May (1972, 1973), a physicist by training and ecologist by inclination. He used local stability analyses of randomly assembled community matrices to mathematically demonstrate that network stability decreases with complexity. In particular, he found that more diverse systems, compared with less diverse systems, will tend to sharply transition from stable to unstable behaviour as the number of species, the connectance *or* the average interaction strength increase beyond a critical value. May's analytical results and his conclusion that 'in general mathematical models of multispecies communities, complexity tends to beget instability' (May 2001, p. 74) turned earlier ecological 'intuition' on its head and instigated a dramatic shift and refocusing of theoretical ecology. His results left many empirical ecologists wondering how the astonishing diversity and complexity they observed in natural communities could persist, even though May himself insisted there was no paradox. May framed a central challenge for ecological research this way: 'In short, there is no comfortable theorem assuring that increasing diversity and complexity beget enhanced community stability; rather, as a mathematical generality, the opposite is true. The task, therefore, is to elucidate the devious strategies which make for stability in enduring natural systems' (May 2001, p. 174).

3.5 Stability of complex food webs: community matrices

Despite methodological criticism of May's methodology (e.g. Cohen and Newman 1984) much subsequent work related to food webs was devoted to finding network structures, species' strategies and dynamical characteristics that were consistent with May's theorem or that would allow complex communities to be stable or persist. May (1972,

1973) used random matrices of interactions for his analyses, which yield random network structures ('who eats whom') and random distributions of traits and interaction strengths ('how much is eaten') across the populations and links in the networks. In natural food webs, however, neither the matrices of interactions nor the distributions of interaction strengths are random. Systematic patterns of low assimilation efficiencies, strong self-regulation (negative effects of a species on its own biomass) or donor control of biomasses can cause positive complexity–stability relationships (DeAngelis 1975). Subsequent analyses demonstrated that non-random empirical network structures of natural food webs are more dynamically stable than random networks (Yodzis 1981), suggesting that natural food webs possess a topology that increases the stability of the population dynamics.

Extending the approach of adding empirical realism to stability analyses, de Ruiter *et al.* (1995) parameterized May's general community matrix model with empirical food web structures and interaction strengths among the species. In their empirical data, they found a pattern of strong top-down effects of consumers on their resources at lower trophic levels in food webs and strong bottom-up effects of resources on their consumers at higher trophic levels. Adding empirical interaction strength patterns to the community matrices increased their local stability in comparison with matrices with random interaction strength values (de Ruiter *et al.* 1995). Most importantly, these results demonstrated that natural food web structures as well as the distribution of interaction strengths within those structures contribute to an increased local stability of the corresponding community matrices. Neutel *et al.* (2002) explained this finding with results that showed that weak interactions are concentrated in long loops. In their analysis, a loop is a pathway of interactions from a certain species through the web back to the same species, without visiting other species more than once. They defined loop weight as the geometric mean of the interaction strengths in the loop and showed that loop weight decreases with loop length. Again, when applied to the community matrix, this empirically documented pattern of interaction strength distributions increased its local stability in comparison with random networks (Neutel *et al.* 2002). Together, these studies demonstrate that characteristics of the distribution of links and interaction strengths within natural food webs account for their stability.

3.6 Stability of complex food webs: bioenergetic dynamics

More recent theoretical studies have extended a numerical integration approach of ordinary differential equations to complex food web models (Williams and Martinez 2004; Martinez *et al.* 2006). In this approach, the structure of the complex networks is defined by a set of simple topological models: random, cascade, niche or nested-hierarchy model food webs (Cohen *et al.* 1990; Williams and Martinez 2000; Cattin *et al.* 2004). The dynamics follow a bioenergetic model (Yodzis and Innes 1992) that defines ordinary differential equations of changes in biomass densities for each population. Numerical integration of these differential equations yields time series of the biomass evolution of each species, which allows exploration of population stability and species persistence. Similar to results from community matrix models, non-random network structure increases an aspect stability in these bioenergetic dynamics models of complex food webs: the overall persistence of species (Martinez *et al.* 2006).

One key parameter of population dynamic models is the functional response describing the *per capita* (per unit biomass of the predator) consumption rate of a predator depending on the prey biomass density. Generally the *per capita* consumption rate is zero at zero prey density and then increases with increasing prey density. According to the shape of this increase classical functional response models are characterized as (1) linear or type I, (2) hyperbolic or type II, or (3) sigmoid or type III functional responses (Fig. 3.3). The linear functional response was used in classic population dynamics studies (Lotka 1925; Volterra 1926), but lacks a biologically necessary saturation in consumption rate at high prey density. A generalized non-linear saturating functional response model (Real 1977) is

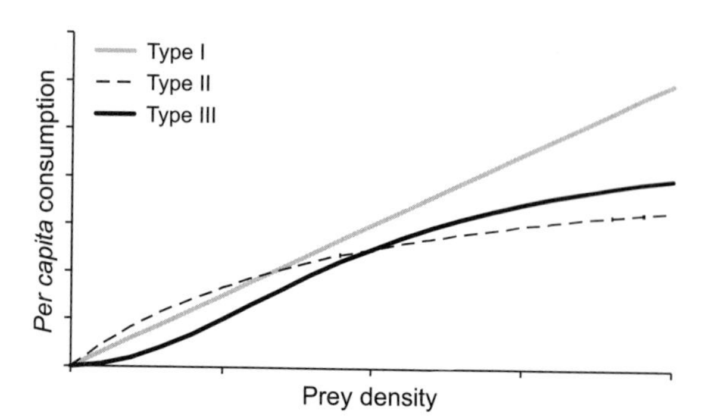

Figure 3.3 Functional responses: *per capita* consumption rates of consumers depending on prey density.

$$F(B) = \frac{cB^h}{1 + cT_h B^h} \quad (3.1)$$

where F is the predator's consumption rate, B is prey biomass density, T_h is the handling time that includes the time a predator needs to catch, kill and ingest a unit biomass of the prey, c is a constant that describes the increase in the attack rate, a, with the prey abundance ($a \sim cB^{h-1}$), and h is the Hill coefficient that varies between 1 (type II functional response) and 2 (type III functional response). Varying the Hill exponent between 1 and 2 gradually converts a type II into a type III functional response. Williams and Martinez (2004) showed that in food webs with a type II functional response many populations are unstable and prone to extinction. Slight increases in the Hill exponent can have dramatic effects on stabilizing the dynamics of particular species, as well as overall species persistence (Williams and Martinez 2004). Additionally, Hill exponents slightly higher than 1 buffer predator–prey modules and complex food webs against the destabilizing effects of nutrient enrichment (Rall *et al.* 2008). One important stabilizing feature of this variation in functional response is the slight relaxation of consumption at low resource densities, which leads to accelerating consumption rates with increasing prey density. This yields a strong top-down pressure, which controls the prey population to low equilibrium densities (Oaten and Murdoch 1975). The relaxation of feeding at low resource density is evocative of a variety of well-documented ecological mechanisms including prey switching and refuge seeking. These types of trophic and non-trophic behaviours allow rare or low-biomass resource species to persist in both natural and model ecosystems, increasing overall community persistence.

In another approach to integrating complex structure and dynamics, and in contrast to most prior dynamical studies, Kondoh (2003, 2006) allowed consumer preferences for resources to adaptively vary. This adaptation process increases the predator preferences for prey of above-average density and decreases preferences for prey of below-average density. This adaptive foraging model yields positive complexity–stability relationships in complex food webs if (1) the fraction of adaptive foragers and (2) the speed of adaptation are sufficiently high, and (3) the number of basal species does not vary with food web complexity (Kondoh 2003, 2006). Interestingly, the process of decreasing preferences for prey of low density causes relaxation of feeding at low prey density, which is mechanistically similar to type III functional responses.

3.7 Stability of complex food webs: allometric bioenergetic dynamics

Most species' traits, T, follow close allometric power-law relationships with their body mass, M:

$$T = aM^b \quad (3.2)$$

where a and b are constants (often b is approximately equal to ¾) and T can represent the biological rates of respiration, biomass growth or maximum consumption (Brown *et al.* 2004). These relationships can be used to parameterize population dynamic models, thus collapsing parts of the multidimensional parameter space into a body-mass axis. Moreover, natural food webs have a distinct body-mass structure of invertebrate and vertebrate predators being on geometric average roughly 10 and 100 times, respectively, larger than their prey (Brose *et al.* 2006a). Implementing theoretically and empirically supported allometric relationships in population dynamic models in complex food webs yields population dynamics models that are constrained by the body-mass structure of the food webs (Brose *et al.* 2006b). In these allometric models, consumer–resource body-mass ratios define the body mass of a consumer relative to the average body mass of its resources.

Brose *et al.* (2006b) varied the consumer–resource body-mass ratios in allometric food web models systematically between 10^{-2} and 10^{6}, creating a gradient of food webs with predators that are 100 times smaller than their prey to food webs of predators that are 10^{6} times larger than their prey (Fig. 3.4). The persistence of species in the food webs (i.e. the fraction of the initial populations that persisted during the simulations) increased with increasing body-mass ratios. Persistence is low when predators are smaller than or equal in size to their prey, but persistence increases steeply with increasing body-mass ratios (Fig. 3.4a). This increase saturates at body-mass ratios of 10 and 100 for invertebrate and vertebrate predators, respectively (Brose *et al.* 2006b), which is highly consistent with the geometric average body-mass ratios found in natural food webs (Brose *et al.* 2006a). Moreover, persistence decreases with the number of populations in the food web at low body-mass ratios, whereas it exhibits a slight increase in persistence at high body-mass ratios (Brose *et al.* 2006b).

This result confirms classic food web stability analyses (May 1972) showing negative diversity–stability relationships in random food webs, in which species are on average equally sized. But it also confirms the earlier notion of empirical ecologists (Odum 1953; MacArthur 1955; Elton 1958), who assumed positive diversity–stability relationships in natural food webs, in which species are on average 10–100 times larger than their prey. Consideration of the body-mass structure of food webs may thus reconcile lingering gaps between the perspectives of theoretical and empirical ecologists on diversity–stability relationships. It is interesting to note that, in this allometric modelling framework, a different measure of stability, the mean coefficient of variation of the species population biomasses in time in persistent webs ('population stability'), *decreases* with increasing body-size ratios until inflection points are reached that show the lowest stability, and then increases again beyond those points (Brose *et al.* 2006b). Those inflection points also correspond to the empirically observed body-size ratios (Brose *et al.* 2006a; Fig 3.4b). Thus, at intermediate body-size ratios high species persistence is coupled with low population stability. Interestingly, this demonstrated that an aspect of increased stability of the

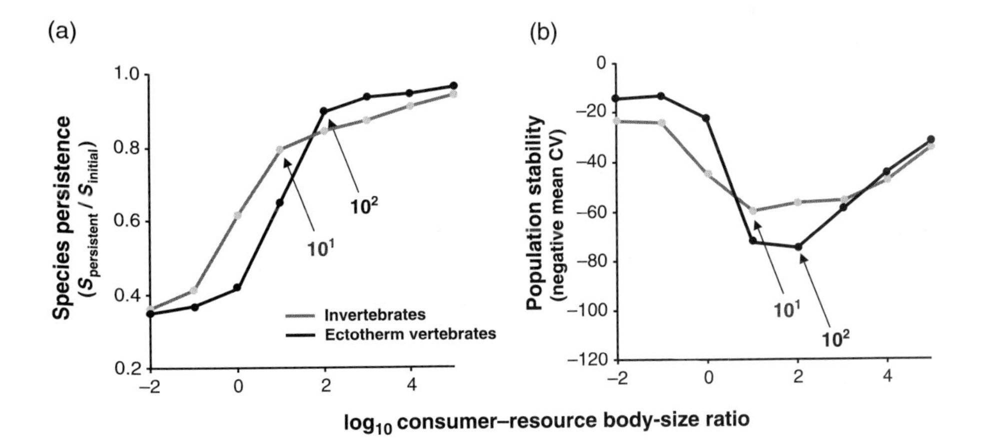

Figure 3.4 Impact of consumer–resource body-size ratios on (a) species persistence and (b) population stability depending on the species metabolic types. The inflection points for shifts to high-persistence dynamics are indicated by arrows for both curves, and those inflection points correspond to empirically observed consumer–resource body-size ratios for invertebrate dominated webs (10^1, consumers are on average 10 times larger than their resources) and ectotherm vertebrate dominated webs (10^2, consumers are on average 100 times larger than their resources). *S*, species richness; CV, coefficient of variation. Figure adapted from Brose *et al.* (2006b).

whole system (species persistence) is linked to an aspect of decreased stability of components of that system (population stability).

3.8 Future directions

Our knowledge about the structure and dynamical constraints of complex food webs has greatly improved over the last decade. The framework of dynamic food web models offers a great possibility to study the consequences of ongoing abiotic and biotic effects on natural ecosystems while including indirect effects between species. Future applications of this approach will need to further address how mechanistic knowledge gained in simple food web modules can be scaled up to predict patterns and processes in complex food webs. Local interaction spheres identified to influence the dynamics of keystone modules may help in scaling up dynamics to networks. Unravelling the factors that determine population dynamics of these local interaction spheres will be an important scientific challenge.

Implementing allometric scaling relationships in models of complex food webs has helped in understanding the processes that lead to particular distributions of interactions and network stability. Further integrations of metabolic allometry with food web research is likely to elucidate much of the constraints on structure and dynamics of complex food webs. Here, it remains particularly important to understand the allometric scaling of other model parameters such as the functional responses. Moreover, incorporating other types of consumer–resource interactions such as parasite–host links into basic food web models remains a challenge that needs to be addressed. Most likely, the framework of dynamic food web models described in this chapter will be an important backbone for future integrations of these processes into theoretical community ecology.

CHAPTER 4

Community assembly dynamics in space

Tadashi Fukami

4.1 Introduction

Species live in a complex web of interactions in the ecological community. What effects do species exert on one another, and how strongly? If species interactions are mostly strong, how do species cope with one another and coexist in the same community? In other words, what level of species diversity and what patterns of species composition should we expect to see if species interactions strongly affect community structure? These are some of the fundamental questions that community ecologists seek to answer (Morin 1999).

Much remains unknown to fully answer these questions, and one major challenge is that species interactions can bring about two contrasting types of community dynamics (Fig. 4.1). In theory, strong interactions can make communities either deterministic or historically contingent (Samuels and Drake 1997; Belyea and Lancaster 1999; Chase 2003; Fukami *et al.* 2005). When deterministic, the effect of species interactions on community structure is determined by environmental conditions. On the other hand, when historically contingent, community structure diverges among localities as

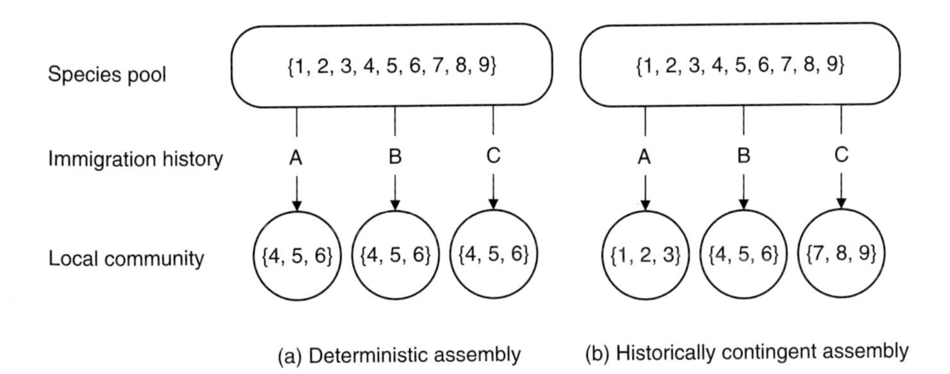

Figure 4.1 (a) Deterministic and (b) historically contingent community assembly. Numbers represent hypothetical species, sets of numbers in brackets represent the species composition of local communities, arrows from the species pool to local communities represent species immigration, and alphabets represent different immigration histories. Deterministic community assembly refers to situations in which different patches converge to the same species composition regardless of immigration history as long as the communities initially share the same environmental conditions. Historically contingent community assembly refers to situations in which different patches diverge to contain different sets of species if immigration history differs between them, even if the communities initially share the same environmental conditions. Specific species compositions in the figure are arbitrary. Modified from Fukami (2008). See also Chase (2003).

a result of stochastic variation in the history of species arrivals, even under identical environmental conditions and an identical regional species pool.

It is difficult to determine which of these two scenarios happens in natural communities. One main reason is simply that the immigration history of most communities is unknown. This problem is apparent in observational studies of community assembly. First popularized by Connor and Simberloff (1979) in response to Diamond (1975), these studies use statistical methods called null models to compare observed community structures with what would be expected if species interactions did not exert significant effects on structure (Gotelli 2001). The null models are a useful tool for detecting effects of species interactions, but only when used with caution. An incorrect assumption sometimes made when using null models is that strong interactions should always lead to community structures that are significantly different from null expectations. Strong interactions, when combined with variable immigration history, can result in historically contingent community development, which can produce apparently random community structure. As pointed out by Wilbur and Alford (1985) and Drake (1991), species interactions, even when strong, do not necessarily create community structures that are distinguishable from null expectations that are based on deterministic effects of species interactions. This limitation arises largely because most null-model studies use data taken at only one point in time. Temporal changes in community structure, let alone the history of species immigration, are usually not considered, simply because such data are rarely available.

Is it possible at all, then, to deepen our understanding of species interactions and community structure without historical information on species immigration? Studies have recently begun to evaluate possible conditions that make community assembly deterministic or historically contingent. For example, it has been suggested that the rate of nutrient supply determines the extent of historical contingency (Chase 2003; Steiner and Leibold 2004). If this is true, then we should be able to calculate at least how predictable community structure will be, based on nutrient supply rate. These studies indicate a potentially promising way in which we can deepen our understanding of community structure without knowing immigration history. Building on this framework, this chapter will consider the spatial scale of community assembly dynamics as a potentially important yet relatively overlooked factor that may critically determine the likelihood of deterministic versus historically contingent community assembly. My aim here is not to provide a comprehensive review of community assembly research. I will instead use the results of several recent studies to highlight ideas that I believe are worthy of further exploration.

4.2 Determinism and historical contingency in community assembly

Before considering spatial issues relating to community assembly, I would first like to clarify what is meant by determinism and historical contingency. In this chapter, I define community assembly as the construction and maintenance of local communities through sequential arrival of potential colonists from an external species pool (Drake 1991; Warren *et al.* 2003). As Warren *et al.* (2003) pointed out, 'viewed in this way, community assembly emphasizes changes in the community state rather than embracing all evidence for pattern in community structure, the broader context in which the term assembly is sometimes used'.

While community assembly can be historically contingent or deterministic in the absence of species interactions, the focus of this chapter will be comparison of the two scenarios in their presence. Community assembly starts with a disturbance, such as a fire, flood or hurricane. Because space, nutrients and other resources are often abundant in the recently disturbed area, competition and other interspecific interactions are unlikely to exert strong effects on community structure at this stage. Also, of the potential colonists that can immigrate into the disturbed area, only some will have reached the new patch thus far, and which species have arrived can be a matter of chance (e.g. Walker *et al.* 2006). In this sense, communities are historically contingent, but not as a joint consequence of immigration history and species interactions. Once more time has passed since disturbance, most potential colonizers may have arrived, even though species interactions have not yet started to affect community structure (e.g. Mouquet *et al.* 2003). In other words, most species expected to be found there are present. In this sense, community structure is deterministic, but because this determinism does not involve species interactions, it is not what I would like to focus on here, either.

Given more time after disturbance, species interactions will start to influence community composition more strongly as each species increases in abundance in the patch. These species interactions can make community structure become either deterministic or historically contingent. It is these two contrasting outcomes that are the focus of this chapter. According to the deterministic view, the environmental conditions under which community assembly happens determine which of the species from the regional pool will remain in the community as a consequence of species interactions. In this case, immigration history does not influence the final species composition of the community. Such communities are said to follow deterministic 'assembly rules' (Weiher and Keddy 1995; Belyea and Lancaster 1999). This idea is rooted in Clements's (1916) climax concept of succession. More recently, deterministic assembly rules have been indicated to drive community assembly not just through immigration, but also through evolutionary diversification (Losos *et al.* 1998; Gillespie 2004).

In contrast, if communities are historically contingent, environmental conditions do not determine a single climax community. Instead, even if two communities are originally under the same environmental conditions, they may contain different sets of species if they have different immigration histories. Lewontin (1969) is often cited as the first author to articulate this idea. Here there is more than one final stable state (called alternative stable states, multiple stable points, multiple stable equilibria, etc.; see Schröder *et al.* 2005) that communities may approach through assembly; once a community assumes a stable state, it cannot move to another stable state unless heavily disturbed. This phenomenon is caused by 'priority effects', in which early-arriving species affect, either negatively or positively, the performance of species that arrive late in terms of population growth (see Almany 2003 and references therein). A simple example of priority effects involves pre-emptive competition, in which species that arrive early make resources unavailable, by virtue of being there first, to other later-arriving species that need those resources to survive and grow (e.g. MacArthur 1972; Sale 1977; Tilman 1988). However, priority effects need not involve only competition, and can happen via predation (e.g. Barkai and McQuaid 1988; Holt and Polis 1997), environmental modification (e.g. Peterson 1984; Knowlton 2004) and other types of species interactions. Recently, experiments have shown that not only community assembly over ecological time, but evolutionary assembly through diversification can also be historically contingent (Fukami *et al.* 2007).

Ever since Lewontin's early writings (1969), much emphasis has been placed on alternative stable states in studying historically contingent assembly. Historical contingency should be considered from a broader perspective, however. There are two ways that communities can be historically contingent even when there is only one final stable state to which communities tend over time.

First, communities can exist in what is called a permanent endcycle. Morton and Law (1997) suggested that there are theoretically two types of final states that communities reach. One is called a permanent endpoint, and the other a permanent endcycle. Permanent endpoints consist of subsets of species from the species pool that are resistant to invasion by any species that are not members of the endpoint. When ecologists refer to alternative stable states, they are in many cases referring to alternative permanent endpoints. In contrast, a permanent endcycle is 'the union of the sets of species that occur in a cyclic or more complex sequence of communities' (Morton and Law 1997). Each set of species in a permanent endcycle can be invaded by at least one of the other species in the endcycle, but cannot be invaded by any species not in the endcycle (Fig. 4.2). Communities in a permanent endcycle are contingent on immigration history, because species composition at a given point in time depends on the sequence of species invasion as the communities go through the endcycle (Lockwood *et al.* 1997; Fukami 2004b; Steiner and Leibold 2004; Van Nes *et al.* 2007). This is true even with just

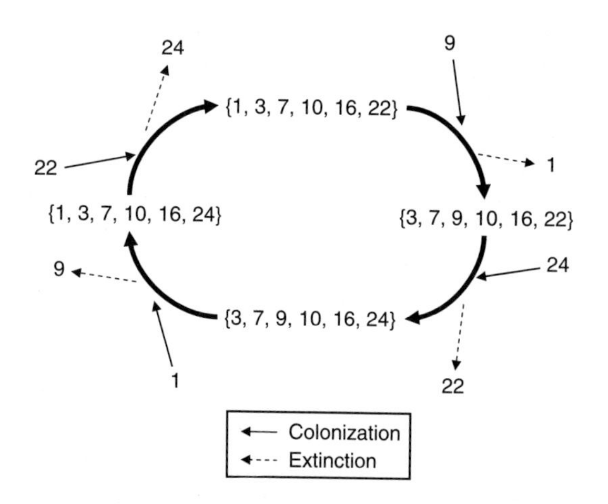

Figure 4.2 An example of a permanent endcycle. Numbers represent hypothetical species in a computer simulation. In this simulation, species 1–12 are autotrophs and species 13–24 are heterotrophs. Sets of numbers in brackets represent species composition of a local community, thick arrows represent temporal changes in species composition, thin arrows represent species colonization, and dotted arrows represent local species extinction. In the example shown here, the species pool consists of 24 species, but only those that participate in the endcycle are shown. Modified from Morton and Law (1997) and Fukami (2008).

one final permanent endcycle, in the absence of alternative final states.

Second, communities can also exhibit alternative long-term transient dynamics. It can take a long time, relative to the generation times of the species involved, for a community to reach a stable state. In such cases, communities will be historically contingent for a long time on their way to a stable state if they follow alternative successional trajectories (Fukami 2004a). For example, suppose that species A competitively excludes species B regardless of immigration history. Even so, species B can remain dominant for a long time if it arrives before species A, and before competitive exclusion eventually occurs. This phenomenon is particularly likely when dispersal ability and competitive ability are similar among species (Sale 1977; Knowlton 2004; Fukami *et al.* 2007; Van Geest *et al.* 2007). For the rest of this chapter, I will consider permanent endcycles and long-term transients as well as alternative stable states in discussing community assembly.

4.3 Community assembly and spatial scale

Having clarified what is meant by determinism and historical contingency, I would now like to develop the main thesis of this chapter, namely that explicit consideration of spatial scale should help us to better understand the conditions in which community assembly is deterministic and those in which it is historically contingent. Drawing on recent theoretical and empirical studies, I will focus on three factors relating to spatial scale: (1) patch size, (2) patch isolation and (3) environmental heterogeneity. In discussing these, it will become clear that it is the relative spatial scale of all of these factors simultaneously considered that brings us the closest to a full understanding of community assembly dynamics.

4.3.1 Patch size

The local patch is the scale at which community assembly occurs (Fig. 4.1). Recent research has suggested that the size of local patches can affect the degree of historical contingency in community assembly. In their pioneering work, Petraitis and Latham (1999) proposed that historical contingency leading to alternative stable states occurs only when patch size exceeds a threshold value. When a newly created patch is too small, the species dominant in and around the patch before disturbance quickly colonize it from adjacent areas and continue to dominate. In this sense, the fate of community assembly in the patch is deterministic. In contrast, when the patch is large, species that are not dominant in adjacent areas may immigrate from a certain distance away and subsequently become abundant before adjacent dominant species take over the patch. In this situation, the history of species immigration can influence community membership. Thus, this is a historically contingent assembly.

Petratis and Latham's (1999) idea is mainly derived from their work on rocky intertidal communities in the New England region of North America, where each patch appears to be in either of two states, algal-dominated or mussel-dominated. It was suggested that, for a disturbance such as ice scour to cause a patch to move from algal-dominated to mussel-dominated or vice versa, patch size needed to be sufficiently large to prevent nearby dominants from always driving community assembly. It should be noted here, though, that there is some debate about whether these really represent two alternative stable states (Bertness *et al.* 2004).

Fukami (2004a) proposed a hypothesis that seems contradictory to the Petraitis and Latham (1999) hypothesis. Experiments showed that community assembly was historically contingent to a greater extent in smaller rather than larger patches. Microbial communities were assembled in the laboratory by introducing 16 species of freshwater protists and rotifers in four different orders in each of four different microcosm sizes. The results showed that species diversity was affected more by immigration history in smaller microcosms. This was explained as follows. Given the same initial population size, early arriving species can achieve high population density more quickly in smaller patches. Consequently, resource availability and other conditions in patches are more greatly altered by early immigrants in smaller patches, which then has a greater effect on late-arriving species in smaller patches (see also Orrock and Fletcher 2005).

The apparent contradiction between Petraitis and Latham (1999) and Fukami (2004a) stems partly from different assumptions made about the source of immigrants. Petraitis and Latham assume that immigration rates, particularly of species that are dominant near the patches, are higher for smaller patches. Immigration history itself is, then, more deterministic there, resulting in more deterministic assembly. On the other hand, Fukami assumes that immigration rate and history do not vary with patch size. In this situation, the inverse relationship between patch size and the rate of increase in population density causes larger patches to be more deterministic.

Which assumption is more realistic? The answer depends partly on the environment around the patch. Petraitis and Latham's assumption would be more realistic if patches are surrounded by areas that provide immigrants. Besides rocky intertidal patches, forest gaps (e.g. Hubbell 2001) are possible examples. On the other hand, Fukami's assumption may be more realistic if the source of immigrants is distant from the patches. Examples may include entire islands acting as patches that undergo community assembly after an island-wide volcanic eruption (e.g. Thornton 1996) and entire ponds acting as patches that undergo assembly after drought and subsequent refilling of water (e.g. Chase 2007).

But even in patches distant from the species pool, immigration rates may vary with patch size. Just as darts are more likely to hit a larger dartboard, species may be more likely to arrive at a larger patch. Under this target size effect (Lomolino 1990), larger patches receive more individuals, consequently reducing the between-patch difference in the population density of early-arriving species. The effect of patch size on historical contingency suggested by Fukami (2004a) may not be as strong then. It is also possible, however, that slower immigration rates in smaller patches make more time available for early immigrants to alter the environment before other species arrive. This can strengthen priority effects in smaller patches relative to larger ones, making the difference in the extent of historical contingency between small and large patches more pronounced. The relative importance of these two opposing ways in which patch size affects historical contingency requires further investigation.

Clearly, the effects of patch size on community assembly are complex. In particular, it has become clear that considering the effect of patch size necessitates consideration of the areas surrounding the patches as well. The following sections will explore surrounding areas a little further. I will first consider the degree of patch isolation and then the spatial scale at which environmental heterogeneity is observed relative to the scale of patches.

4.3.2 Patch isolation

Several studies suggest that community assembly is more sensitive to immigration history when the patch is located farther from the species pool. Robinson and Edgemon (1988) assembled microbial communities by introducing phytoplankton species into aquatic microcosms in three different orders at three different rates. Results showed that the effect of introduction order on species composition was greater when communities were assembled with lower immigration rates. Because immigration rate is generally expected to be lower when the distance from the species pool is greater (MacArthur and Wilson 1967), Robinson and Edgemon's results suggest that community assembly is historically contingent to a greater extent when the patch is more isolated.

In a related study, Lockwood *et al.* (1997) conducted computer simulation of community assembly using Lotka–Volterra equations modelling competition and predation within patches. Using two immigration rates, they found that immigration history influenced species composition under both immigration rates, but that the type of effect differed between the two rates. When immigration rate is low, different immigration histories lead communities to alternative stable states, whereas when immigration rate is high, permanent endcycles occur. The likely reason for this difference has to do with whether the assembling communities approach an equilibrium between immigration events. Low immigration rate allows for this, eventually resulting in a stable state of species composition. In contrast, high immigration rate prevents the community reaching any possible equilibrium between immigration events. Thus, high immigration rate maintains species composition in a transient state of change, resulting in permanent endcycles.

Fukami (2004b) used a similar Lotka–Volterra model to find that community assembly resulted in permanent endcycles regardless of immigration rate, but that the number of species involved in permanent endcycles was greater when immigration rate is low. As a result, immigration history has a greater effect on species composition when immigration rate is lower (see also Schreiber and Rittenhouse 2004).

These studies all assume that the species pool that provides immigrants exists externally, such that patch community dynamics do not affect the species pool (Fig. 4.3b). The model of community assembly based on this assumption is typically referred to as the mainland-island model. An alternative model has been termed the metacommunity model, which describes a collection of multiple local patches each undergoing community assembly through occasional dispersal of species between the patches (Wilson 1992; Leibold *et al.* 2004; see Chapter 5). In metacommunities, the species pool is internal instead of external, and local patches serve as the source of immigrants (Fig. 4.3a). In terms of patch isolation, when patches are more isolated from one another, the rate of *internal* dispersal is lower (Fig. 4.3a), whereas when patches are more isolated from the species pool, the rate of *external* dispersal is lower (Fig. 4.3b).

Computer simulations show that higher internal dispersal (or how isolated patches are to one another) could make community assembly more deterministic (Fukami 2005). This theoretical result is consistent with findings from empirical studies (e.g. Chase 2003; Cadotte 2006). However, Fukami (2005) also showed that whether this effect of

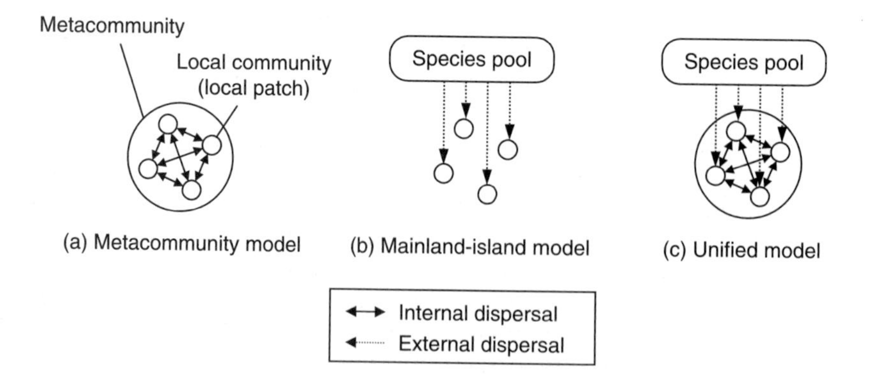

Figure 4.3 (a) Metacommunity model, (b) mainland-island model and (c) unified model of community assembly. Arrows represent dispersal between patches within the metacommunity (referred to as internal dispersal). Dashed arrows represent dispersal from the external species pool (referred to as external dispersal). Modified from Fukami (2005).

internal dispersal occurs depended on the rate of external dispersal. Specifically, frequent internal dispersal reduces the extent of historical contingency if external dispersal is not frequent, but internal dispersal does not affect historical contingency if external dispersal is frequent. Therefore, the two dispersal types can reciprocally provide the context in which each affects species diversity. These results indicate that in order to understand historical contingency in community assembly, it is important, though rarely done, to distinguish internal and external dispersal and to know the relative frequency of the two types of dispersal (Fig. 4.3c).

4.3.3 Scale of environmental heterogeneity

Many studies, including those discussed above, assume that local patches share identical environmental conditions. They also assume that the region within which local patches are embedded is homogeneous across space. Clearly these assumptions are not met in many ecological landscapes. An interesting question then is how the scale at which environmental heterogeneity is observed may influence the degree of historical contingency in community assembly.

A study by Shurin *et al.* (2004) is relevant here. They used a mathematical model to study conditions for coexistence of two competing species at a regional scale. The region modelled consists of multiple patches that vary in resource supply ratio. In the absence of variation among patches in resource supply ratio, one of the two species competitively excludes the other. When patches vary in the ratio, historical contingency occurs in terms of which species occupies a given patch. Specifically, in patches where resource supply ratio is intermediate, the species that arrives first prevents the other from colonizing that patch. In other patches where the ratio takes more extreme values, one or the other species dominates. These patches serve as a species pool that provides immigrants to patches of intermediate environmental conditions for historically contingent community assembly to be realized there (though it involves only two species in the model). Historically contingent assembly occurs only when there is an external species pool that is not influenced by the patches in which historical contingency is observed.

In terms of the spatial scale of environmental heterogeneity, the results of Shurin *et al.* (2004) can be interpreted as follows. Historically contingent assembly occurs when environmental conditions are sufficiently heterogeneous across patches (Fig. 4.4d) rather than within patches (Fig. 4.4c). Thus, it is the scale of environmental heterogeneity relative to the patches in question, rather than its absolute scale independent of patch size, that affects the degree of historical contingency in community assembly.

The experiment conducted by Drake (1991) provides additional insight into environmental heterogeneity and community assembly. Similar in design to Robinson and Edgemon (1988) and Fukami (2004a), Drake (1991) assembled aquatic microbial microcosms through sequential introductions of species in various orders using two different sizes of microcosms. The results showed that, in small patches, the same species dominated the assembled community regardless of introduction order, whereas, in large patches, species introduced early dominated over those introduced late. These results

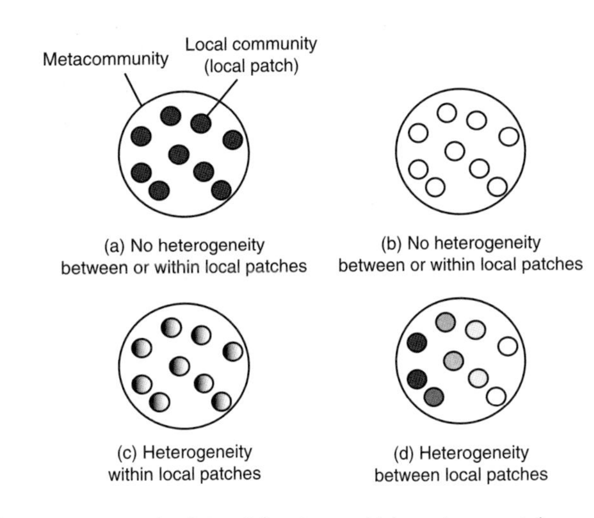

Figure 4.4 (a–d) Spatial scale at which environmental heterogeneity is observed. Shading indicates variation in environmental conditions (e.g. rate of nutrient supply). Heterogeneity is drawn arbitrarily as a gradient. Modified from Fukami (2008).

appear to contradict those of Fukami (2004a) discussed above, while being congruent with those of Petraitis and Latham (1999). But neither seems to be the case. Drake (1991) invoked differences in environmental heterogeneity between small and large patches to explain his results, whereas neither Fukami (2004a) nor Petraitis and Latham (1999) explicitly considered environmental heterogeneity. Drake postulated that environmental heterogeneity increased with patch size (light availability was more variable in larger microcosms owing to increased depth; depth was standardized across patch size in Fukami (2004a)), and that variation among species in their competitive ability was small when environmental heterogeneity was great. Communities are thought to be more sensitive to historical contingency when species are competitively more similar (e.g. MacArthur 1972; Hubbell 2001). If this applies to Drake's microcosms, then it explains smaller historical contingency in smaller patches. Drake's (1991) explanation would need to be tested to be rigorously validated, but the suggested potential relationship between patch size, environmental heterogeneity, competitive relationship and historical contingency remains novel to this day.

4.3.4 Synthesis

In summary, I have considered patch size, patch isolation and the spatial scale of environmental heterogeneity as three spatial factors influencing the degree of determinism and historical contingency in community assembly. These factors do not affect community assembly independently of one another. Instead, their scale and consequently their role in community assembly are determined relative to those of the others. Through consideration of these three factors, several conditions for historical contingency have emerged. Specifically, community assembly is hypothesized to be historically contingent to a greater extent when (1) immigration rate is lower, (2) immigration history is more variable and (3) the species pool that provides immigrants to local patches undergoing assembly exists more independently of the community dynamics within the patches.

4.4 Community assembly and species traits

Ultimately, consequences of patch size, patch isolation and environmental heterogeneity for community assembly depend on the spatial scale of species movement (Cadotte and Fukami 2005). For this reason, it is important to know the dispersal ability of the species involved in community assembly in question, in order to address the determinism versus historical contingency question. Furthermore, the degree of variation in dispersal ability among species can also influence historical contingency in community assembly. This is because the more similar species are in dispersal ability, the more stochastic immigration history is expected to be, which can then lead to less deterministic assembly. Smaller variation in competitive ability should also result in less deterministic assembly, as priority effects act stronger between competitively more similar species.

Furthermore, dispersal ability and competitive ability are thought to sometimes show a trade-off, such that species that are good dispersers are poor competitors, and vice versa (e.g. Petraitis *et al.* 1989; Cadotte 2007). In terms of succession, this means that early-successional species are competitively inferior to late-successional species (Petraitis *et al.* 1989). This trade-off, too, may influence historical effects in community assembly. For example, assembly may be more deterministic when the species show a clearer trade-off between these two traits. This is because, under a clear trade-off, community assembly is expected to progress predictably to eventually end with a predictable set of late-successional competitive species dominating the community.

Of course, dispersal ability and competitive ability are just a few of many traits that characterize species. There has recently been a renewed interest in explaining community dynamics from species traits (e.g. Fukami *et al.* 2005; McGill *et al.* 2006; Ackerly and Cornwell 2007). Other traits that can influence community assembly include disturbance tolerance, intrinsic rate of growth and predator avoidance. Including these traits in a framework for community assembly should enhance our predictive power. For example, even when there is a clear trade-off between dispersal ability and competitive ability, community assembly can be historically contingent in the presence of predators. Recent experimental work suggests that the timing of predator arrival at the local patch can influence the final structure of prey communities under a competition–colonization trade-off (Olito and Fukami 2009).

Dispersal ability and other traits may ultimately be determined by the spatial scale of patches that the species have experienced over evolutionary time (Denslow 1980). Patch sizes that species have experienced in the past and those of the present are not necessarily the same. This is particularly true in the presence of anthropogenic disturbance, habitat fragmentation and exotic species introduction. Anthropogenic disturbance can be evolutionarily novel; habitat fragmentation can create new kinds of patch size and isolation; and exotic species can differ from native species in the spatial scale of patches that they have adapted to, consequently differing in the way native and exotic species perceive spatial scale. How do these anthropogenic changes in the scale of community assembly affect historical contingency in assembly? We currently know little to answer this question. A better understanding of the role of scale in community assembly may contribute to advancing not only community ecology as a basic science, but also solving applied issues regarding the community-level impacts of species invasions.

4.5 Conclusions and prospects

I have discussed how the spatial scale at which community assembly occurs may influence the degree to which community assembly dynamics are deterministic versus historically contingent. As spatial factors, I have focused on patch size, patch isolation and the scale at which environmental conditions vary. In combination, these factors are proposed to jointly affect three elements of community assembly dynamics: the rate of immigration to local communities, the degree to which the species pool is external to local community dynamics and the extent of variation in immigration history between local communities. I have argued that these three elements will in turn determine the extent of historical contingency and determinism in community assembly. Additionally, I have briefly pointed out that the spatial scale of community assembly is defined relative to dispersal ability of species involved. But dispersal ability is often not independent of other traits such as competitive ability, disturbance tolerance and predator avoidance. Explicit consideration of these traits should lead to a better understanding of the conditions for contingent versus deterministic assembly.

As discussed in the introduction, much of community assembly research has traditionally relied on null-model approaches using observational data. This is because experimental assembly of natural communities is difficult in most situations owing to the large spatial and temporal scales involved in this type of work. However, direct experimental manipulation of immigration history is necessary in order to rigorously evaluate historical effects in community assembly (Schröder *et al.* 2005). For this reason, I expect that experiments will become increasingly important in community assembly research. Experiments have so far been limited mainly to those with microorganisms in the laboratory owing to their logistical advantages, but we will also need to do more field experiments to ensure that the concepts we develop are firmly placed in natural context. Though difficult, field experiments are feasible by, for example, incorporating experimental research into ecological restoration projects (e.g. Fukami *et al.* 2005; Weiher 2007).

In addition, research on community assembly has mainly considered systems in which environmental conditions do not vary considerably except when pulse disturbance events initiate a new round of community assembly. However, environmental conditions can of course fluctuate greatly in many systems. How do temporal fluctuations affect the role of spatial scale in determining the degree of historical contingency in community structure? Does the temporal scale of environmental fluctuations relative to that of community assembly affect the extent of historical contingency? These questions remain unanswered. My focus here has been spatial scale, but temporal scale should also be explored further in future research in relation to community assembly, over both ecological and evolutionary time.

This chapter has largely consisted of exploration of ideas rather than evaluation of data. We do not yet have sufficient data to draw general conclusions as to how often or to what extent natural communities are governed by historical contingency. If it turns out in the future that many communities are indeed highly sensitive to historical effects, then one may question whether community ecology can be called a science in the first place. The answer could be no if science was defined as discovering general patterns in nature and explaining these patterns within a predictive framework. In fact, we do know that clear general patterns are rarely observed in community structure. Historical contingency may well be a main reason behind the absence of such patterns. Nonetheless, like other authors (e.g. Long and Karel 2002; Chase 2003), I believe a good understanding of the conditions for determinism versus historical contingency will contribute to building a predictive theory of community ecology. Here I have sought to provide a first step in this endeavour, with a special focus on the spatial scale of community assembly dynamics. Many of the ideas presented here are only exploratory, and some may prove wrong. Even so, it is my hope that they serve to stimulate further research on the dynamics of community assembly.

PART III

SPACE AND TIME

CHAPTER 5

Increasing spatio-temporal scales: metacommunity ecology

Jonathan M. Chase and Janne Bengtsson

5.1 Introduction

Community ecology is a dynamic and rapidly changing field. While community ecology in the 1980s and 1990s focused primarily on how local environmental conditions, and species interactions within those localities, influenced patterns of coexistence and relative species abundances at the local scale, more recent years have seen an increasing recognition (or rediscovery) of the important role of space and time (Ricklefs 1987, 2004; Ricklefs and Schluter 1993; Hubbell 2001; Leibold *et al.* 2004; Holyoak *et al.* 2005). We refer to the implicit recognition (and study) of the important role of spatio-temporal dynamics as metacommunity ecology.

By analogy with a metapopulation, which refers to a series of populations interconnected by dispersal among patches, a metacommunity refers to a collection of communities of potentially interacting species that are interconnected by dispersal (Leibold *et al.* 2004; Holyoak *et al.* 2005). Accordingly, the simplest model of a metacommunity is one of several non-interacting species each existing in co-occurring metapopulations. In practice, however, the symmetry between metapopulations and metacommunities is loose at best. Although population biologists often make great efforts to define the scope of a metapopulation, the scales under consideration can be highly variable within the context of metacommunities. Spatial scales can range from small-scale environmental gradients or patches in which species colonize or go extinct, to large-scale biogeographic studies across provinces and continents. Furthermore, temporal scales under consideration within the metacommunity context range from seasonal or yearly fluctuations to eons of global climate change and phylogenetic inertia. Individual species in a metacommunity respond to scale differentially. Finally, among species in a metacommunity, there are often large differences in movements and dispersal, and rates of reproduction and mortality, which depend on several factors, including behaviour, body size and trophic level. Thus, rather than trying to define the range of a metacommunity, we instead focus on the concept of metacommunities as incorporating aspects of community ecology at larger spatio-temporal scales. These can range from as small a scale as considering the variation of plant species that occur on different aspects of a hillside slope to as large a scale as considering the variation in plant species composition across biogeographic provinces. When setting out to test a given metacommunity model, it is important to recognize the constraints on those models, and to address their predictions at appropriate spatio-temporal scales. It would, for example, be rather silly to test the predictive ability of a 'mass effects' metacommunity model, which inherently assumes spatially heterogeneous environments (e.g. Mouquet and Loreau 2003), at a scale so small that the landscape is essentially homogeneous.

Throughout this chapter, we will intermingle discussions of more historical perspectives of metacommunity ecology into the contemporary perspective. We do this because we feel that in order to have a deep understanding of the contemporary views of metacommunity ecology, it is necessary to know how the ideas have evolved. In addition, when possible, we discuss theory and empirical work hand-in-hand, which we feel is a more appropriate way to discuss this information, rather than keeping them separate. We argue that metacommunity ecology will most rapidly advance with an intimate, rather than superficial, connection between theory and data.

As a first order of business, it is necessary to recognize that, although metacommunity ecology as a field has exploded in the past decade or so (Tilman 1997; Leibold *et al.* 2004; Holyoak *et al.* 2005), its historical roots are much older. For example, early experimentalists testing Lotka–Volterra competition and predator–prey models quickly recognized that it was exceedingly difficult for competitive or predator–prey species pairs to coexist without some sort of spatio-temporal variation that was missing from the simple models (Gause 1934; Park 1948, 1954; Huffaker 1958). Insect ecologists have long recognized the importance of large-scale processes and dispersal in population dynamics, and aspects of metapopulation ecology were inherent in the writings of Andrewartha and Birch (1954). Similarly, during the 'renaissance' of community ecology – the 1960s and 1970s – spatio-temporal perspectives were quite common (e.g. MacArthur and Levins 1964; MacArthur and Wilson 1967; Levins and Culver 1971; MacArthur 1972; Horn and MacArthur 1972; Levin 1974; Slatkin 1974).

Probably as a consequence of the increase in statistical and experimental rigor that arose from a series of critiques (e.g. Strong *et al.* 1984), community ecology studies in the 1980s and early 1990s were primarily aimed at local-scale processes, such as the mechanisms that influence species coexistence and relative abundances. The recent recognition of the importance of spatial processes probably has several simultaneous origins. First, while the focus on experiments and local-scale mechanisms offers much to community ecology, it is not able to account for many of the patterns that are observed in nature, at both local and larger scales (e.g. Ricklefs 2004). Second, Hubbell's (2001) 'neutral' theory served to catalyse the field by espousing the controversial perspective that local interactions are irrelevant at larger scales, and that a perspective based solely on stochastic processes of colonization and extinction (and speciation) could approximate natural patterns (see Chave 2004; Alonso *et al.* 2006; Holyoak and Loreau 2006). Third, in the context of applied community ecology, spatio-temporal perspectives are often necessary to understand how communities are degraded, and how they can be restored (see Chapter 9). Fourth, statistical analyses, including spatial, multivariate and computational analyses, have become more sophisticated and powerful (e.g. Clark *et al.* 2007), and, when applied in combination with theoretical and experimental information, can provide a deeper understanding of both the patterns and underlying processes of community structure than could have been gained previously.

5.2 The varied theoretical perspectives on metacommunities

Recent syntheses have suggested that metacommunity theory can be roughly categorized into four general conceptual frameworks: Neutral, patch dynamics, species sorting and mass effects (Leibold *et al.* 2004; Holyoak *et al.* 2005). The four perspectives are not exclusive. For example, mass effects and patch dynamics can be viewed as occurring along a gradient of organism movement intensity and habitat heterogeneity. The relative importance of species sorting, patch dynamics and mass effects in metacommunities is a primary issue when examining the influence of local and regional processes on patterns of community composition (see Chapter 9). Furthermore, the stochastic processes inherent in the neutral theory are one component of more complex spatio-temporal niche models (Chesson 2000; Adler *et al.* 2007).

There has been considerable attention directed towards testing empirical data from a variety of systems against the predictions of metacommunity models based on neutral versus niche differences (e.g. Chave 2004; McGill *et al.* 2006). Unfortunately, patterns from one type of data are generally not able to unambiguously separate the different model predictions (Chase 2005; Chase *et al.* 2005) (Table 5.1). Although data on the relative abundances of species in a community are often used

Table 5.1 Summary of predictions from the four metacommunity modelling frameworks (modified from Chase *et al.* 2005)

	Model prediction			
Effect	**Neutral**	**Patch dynamics**	**Species sorting**	**Mass effects**
Overall local diversity	Extinction and colonization balance	Extinction and colonization balance	Depends on species interactions	Depends on species interactions and colonization/ extinction
Overall regional diversity	Extinction and speciation balance	Depends on competition–colonization trade-off	Same as above and degree of habitat heterogeneity	Same as above and degree of habitat heterogeneity
Relative species abundance	Zero-sum multinomial (skewed towards rare species)	Variable depending on level of migration and degree of interaction	Variable depending on environmental conditions	Variable depending on level of migration
Dispersal effects on local diversity	Increase	First increase, then decrease (hump-shaped)	No effect	First increase, then decrease (hump-shaped)
Dispersal effects on regional diversity	Decrease	Decrease	No effect	Decrease
Dispersal effects on β-diversity	Decrease	Global: no effect Local: decrease	No effect	Global: decrease Local: decrease
Local disturbance	Return immediately	Unpredictable	Return immediately	Return following succession
Regional disturbance	Random walk	Return following succession	Return immediately	Return following succession
Temporal variation: local	Variable	Variable	Static unless environment changes	Static unless environment changes
Temporal local variation: regional	Variable	Static unless environment changes	Same as above	Same as above

to compare the different model predictions (e.g. McGill *et al.* 2006), several models can predict the same pattern of relative abundances as the neutral model (reviewed in Chase *et al.* 2005). In addition, statistically distinguishing between different models using empirical species-abundance distributions is not always straightforward (McGill *et al.* 2007). This indicates the limitations of using one pattern, especially relative abundance patterns, to discern among the validity of the assumptions underlying the different perspectives.

The four models make different assumptions regarding how species respond to environmental and spatial gradients, and thus make different predictions regarding patterns of community structure along those gradients. We do not go into detail regarding the theoretical reasoning behind each of the predicted patterns here, but instead simply review them in Table 5.1 (for more detail, see Chase *et al.* 2005). We start with the simplest model, and then move to models using increasingly complex assumptions.

5.2.1 Neutral

Although Hubbell's (2001) construct is the most widely recognized neutral model, there are actually many related 'neutral' theoretical constructs that make no specific assumptions about species traits or their responses to the environment (reviewed in Chave 2004). The neutral model assumes that species are neutral with respect to their interspecific interactions as well as the underlying environment. This means that the numbers of individuals and species that occur in any given locality result from purely stochastic processes (e.g. colonization and extinction). MacArthur and Wilson's (1967) Equilibrium Theory of Island Biogeography is probably the most widely recognized neutral model, and similar concepts form the basis of Hubbell's (2001) neutral theory. In the MacArthur and Wilson theory, the number of species in a habitat is solely determined by the balance between the colonization rate of species from a species pool (usually mainland) and the local extinction rates of species. Hubbell's (2001) model expands on the MacArthur and Wilson theory in two important ways. First, the stochastic processes of colonization and extinction at the patch level from the MacArthur and Wilson theory are transferred to the individual level so that they denote birth–death processes, allowing it to make predictions of species' relative abundances. Second, rather than defining a species pool (e.g. from a mainland), Hubbell allows the species pool to result from speciation, which is also derived from stochastic processes. While specific details of the neutral model and its assumptions have been heavily debated (e.g. Ricklefs 2003; Etienne *et al.* 2007), neither Hubbell (2001) nor others who have developed neutral theories (e.g. Chave 2004) truly believe that the assumptions of the neutral model can fully capture the true nature of these communities. Instead, the neutral models emphasize how far one can go by making the simplest assumption that communities are structured by stochastic processes only.

5.2.2 Patch dynamics

This perspective is also borne out of stochastic processes of colonization and extinction, but in this case at the patch level. In Levins' (1969) original metapopulation model, patches are assumed to be similar, and species can colonize and go extinct from a patch at defined rates. However, patch dynamics models of metacommunities go beyond the neutral model by assuming that species may display trade-offs in their relative abilities to colonize and compete, in turn allowing coexistence under some circumstances. There are many ways to depict these sorts of colonization–competition and other trade-offs, each of which gives slightly different sets of predictions (Levins and Culver 1971; Hastings 1980; Tilman 1994; Hanski and Gyllenberg 1997; Yu and Wilson 2001; Calcagno *et al.* 2006).

5.2.3 Species sorting

This framework leaves behind the stochastic processes of colonization and extinction inherent in the two frameworks described above, and instead focuses on deterministic processes that result from differential responses of species to heterogeneous environments; i.e. niche differences. Here, a species will occur in a locality if the abiotic and biotic environments are favourable. Trade-offs in tolerance to different environmental conditions are typically invoked here to suggest that species that are favoured under some environmental conditions will be disfavoured under others (Tilman 1982; Chase and Leibold 2003).

5.2.4 Mass effects

This framework combines aspects of patch dynamics and species sorting by assuming that species have differential responses to environmental conditions and differential dispersal, colonization and extinction rates (e.g. Mouquet and Loreau 2003; Amarasekare *et al.* 2004). Dispersal is assumed to be high enough to influence local dynamics. The coexistence and relative abundances of the species will depend on rates of dispersal and extinction, as well as on the source–sink relationship between the different habitats for each species. The existence of source patches is crucial for a species to persist in a dispersal-driven metacommunity.

5.3 Metacommunity theory: resolving MacArthur's paradox

Each of the four metacommunity perspectives provides important insights, and they should not be viewed as strict alternatives but rather as different ways to view problems of coexistence and diversity. 'MacArthur's paradox' (Schoener 1989) refers to the fact that Robert MacArthur played a significant role in the simultaneous development of metacommunity models based on niche differences (MacArthur and Levins 1964) and those based on more neutral perspectives (MacArthur and Wilson 1967). Here, we emphasize that, rather than viewing these perspectives as dichotomous alternatives, they illustrate different ends of a continuum, and that each approach is valid for problems at different scales and for different questions. For example, when the focus is at a relatively coarse scale, it may be an acceptable first approximation to assume that all species are equal. This could explain the reasonable success that simple 'neutral' models like the MacArthur and Wilson theory and their derivatives have had for addressing questions at coarser scales (Simberloff 1974; Whittaker and Fernández-Palacios 2007), despite the abstraction of much ecological detail that obviously is important for local coexistence. When the focus is local, or when the questions being addressed are fine-scaled, these more abstract models often need to be expanded to focus specifically on niche differences among species and how those differences influence patterns of community structure (e.g. Chase and Leibold 2003).

Any theoretical model of metacommunities – be it based on neutral, patch dynamic, species sorting or mass effect/dispersal processes – can only be a cartoon of the processes that are actually occurring in that metacommunity. We argue that, although it is instructive to examine patterns observed in natural systems or from experiments in the context of these model predictions (e.g. Table 5.1), this should not be done to test the validity of one model over the other. Instead, we echo several recent syntheses that have explicitly recognized that, in any metacommunity, a variety of processes, including those related to neutral models (e.g. stochasticity) and those related to niche models (e.g. determinism), are operating simultaneously (Tilman 2004; Chase 2005, 2007; Gravel *et al.* 2006; Leibold and McPeek 2006; Adler *et al.* 2007; Clark *et al.* 2007). The challenge for metacommunity ecologists is to disentangle the relative importance of the different processes, and, most importantly, to identify features of a given habitat type that would make one or the other set of processes exert a stronger influence on community patterns.

5.4 As easy as α, β, γ: the importance of scale

Why is it necessary to explicitly consider the role of space in a metacommunity? If spatial processes are not important, then the factors that influence the number and relative abundance of coexisting species at any given locality should simply 'add up' to describe those features at the regional scale. If, on the other hand, there is something inherent in the way that space differentially influences patterns of local and regional diversity, then scale (and thus space) matters (e.g. Whittaker *et al.* 2001; Chase 2003).

This problem is illustrated by explicitly considering how scale influences patterns of diversity. At local (small) spatial scales, species diversity (= species richness) is the number of species counted in the area defined; known as α-diversity. Because there is generally variation in the types of species (species composition) observed from one locality to the next, the turnover of species composition among sites is known as β-diversity. Finally, regional diversity, known as γ-diversity, can be derived by either an additive ($\gamma = \alpha + \beta$ (e.g. Lande 1996)) or multiplicative ($\gamma = \alpha\beta$ (Whittaker 1972)) partitioning. This emphasizes how understanding patterns of β-diversity is critical to understanding the scalar relationship between local and regional diversity, which is the domain of metacommunity ecology. β-Diversity can emerge from both deterministic and stochastic processes. First, if localities differ environmentally, high β-diversity can emerge from deterministic processes that favour different species in different environments, i.e. species-sorting processes (e.g. Whittaker 1972; Chase and Leibold 2003). Second, even when localities have identical environmental conditions, high β-diversity can emerge from stochastic processes (e.g. local extinctions because of demographic stochasticity; Hubbell 2001), or dispersal limitation and priority effects, often connected to a patch dynamics perspective (Chave and Leigh 2002; Condit *et al.* 2002; Chase 2003).

Scale-dependent diversity relationships have been explicitly observed along gradients of disturbance (Chase 2003, 2007; Ostman *et al.* 2006) and gradients of productivity (Mittelbach *et al.* 2001; Chase and Leibold 2002; Chase 2003). Scale dependence will occur whenever responses of α- and γ-diversity do not scale linearly, i.e. β-diversity varies along the gradient. For example, in a meta-analysis of the well-studied relationship between productivity and species diversity, Mittelbach *et al.* (2001) found that hump-shaped productivity–diversity relationships tend to emerge more frequently at local scales, whereas monotonically increasing productivity–diversity relationships tended to emerge more frequently at regional scales. Chase and Leibold (2002) showed this scale dependence in surveys of invertebrates and amphibians from small fishless ponds in Southwestern Michigan, USA. When considered on a 'per-pond' basis (local scale), they found a hump-shaped relationship between productivity and diversity, whereas when the same data were considered on a 'per-watershed' basis (regional scale), a monotonically increasing relationship between productivity and diversity was observed (Fig. 5.1a).

Such scale dependence is not universal (e.g. Chase and Ryberg 2004; Harrison *et al.* 2006a). For example, when comparing the ponds discussed above with regions where ponds were physically closer and dispersal was more likely among ponds, Chase and Ryberg (2004) found that β-diversity was lower in regions where dispersal was more likely among ponds, and no such scale dependence emerged (Fig. 5.1b). This result could have been due to higher

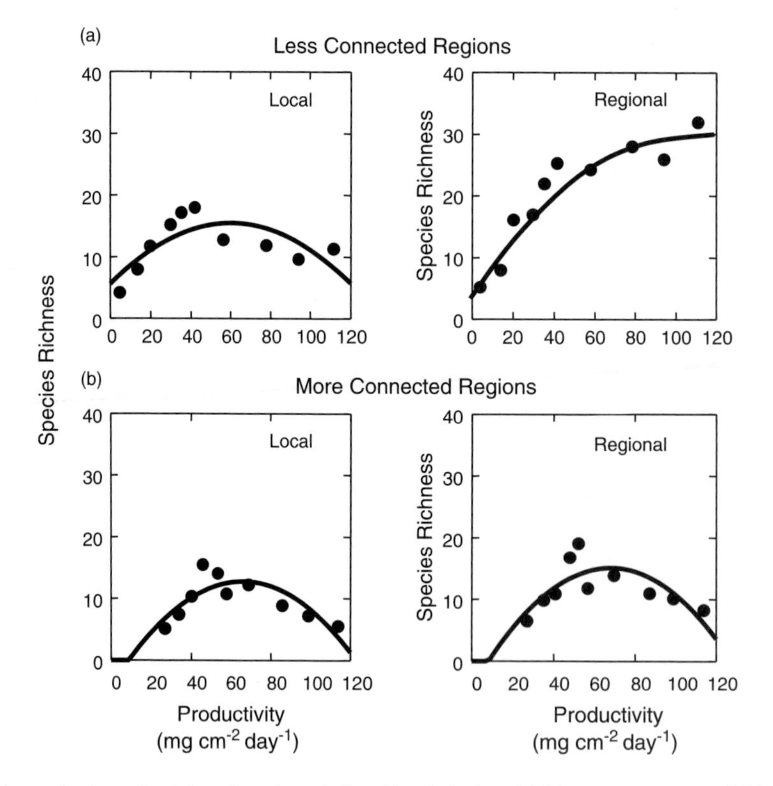

Figure 5.1 Scale-dependent productivity–diversity relationships in isolated (a) but not connected (b) watersheds. Redrawn from Chase and Ryberg (2004).

dispersal rates among the more closely aligned metacommunities, resulting in a reduced degree of stochasticity and higher similarity in community composition. In Effects of dispersal rates on local community structure (below), we discuss how variation in the proximity of habitats (which is related to the rates of dispersal) within a metacommunity might alter patterns of α-, β- and γ-diversity.

5.5 Species–area relationships and metacommunity structure

Although it is conceptually quite useful, partitioning diversity into α, β and γ components is not always straightforward. Theoretically, one can imagine localities as patches surrounded by inhospitable matrix (e.g. Mouquet and Loreau 2003; Amarasekare *et al.* 2004). In some metacommunities with patchy structure this assumption is largely fulfilled, and several such systems have been used as empirical models of metacommunities. Examples include water-filled inquilines such as pitcher-plant leaves (e.g. Kneitel and Miller 2003; Gotelli and Ellison 2006), ponds and rockpools (e.g. Bengtsson 1989, 1991; Chase 2003), islands (e.g. Simberloff and Wilson 1969; Diamond 1975; Schoener *et al.* 2002), host plants for insects (e.g. Zabel and Tscharntke 1998; Kreuss and Tscharnkte 2000), moss patches on rocks (Gilbert *et al.* 1998) and rocky outcrops (e.g. Harrison 1997, 1999; Ostman *et al.* 2007). However, in many terrestrial (forests, grasslands) and aquatic (oceans, large lakes) landscapes, the distinction between where one 'locality' ends and another begins is ambiguous at best (Loreau 2000). In these continuous landscapes, diversity might be scaled or partitioned more objectively using species–area relationships (SARs).

The positive SAR is one of the oldest and best-supported patterns in ecology (Lomolino 2000). SARs can be calculated in two ways: (1) using incrementally larger areas in which smaller areas are nested or (2) using data on richness and area from distinct separate patches or islands (e.g. Rosenzweig 1995). The identity of mechanisms driving variation in the shape of SARs remains at the forefront of research on metacommunity ecology (e.g. Drakare *et al.* 2006). The function most frequently used to describe the SAR is a power law:

$$S = CA^z$$

where S is species richness and A is area. When log-transformed, we get the familiar linearized SAR,

$$\log S = \log C + z \ \log A$$

where and C and z are curve-fitting parameters depicting the intercept (C) and rate of increase (z) of species with area.

Of primary interest here is z, the slope of the log-transformed SAR, indicating the steepness of the increase in the number of species with increasing area. Although somewhat more complex, variation in the parameter z is related to variation in β-diversity (e.g. Connor and McCoy 1979; Rosenzweig 1995; Drakare *et al.* 2006). For example, SARs are often calculated using a 'nested' design (Scheiner 2003), by adding together the area of similar sized smaller patches or sampling units into a larger area. In this case, if the smaller patches are highly divergent in their community composition (higher β-diversity), the increase in species richness with increasing area – the z-value – will be higher than if the smaller patches are more similar to one another (lower β- diversity). Thus, variation in z among sites can provide important information regarding the partitioning of diversity across scales.

Several factors can create variation in the slope (z) of log-transformed SARs. Even when sampling methodology is controlled, z varies among communities (reviewed in Drakare *et al.* 2006). Some of this variation can be attributed to factors such as organism size or trophic position (Holt *et al.* 1999; Drakare *et al.* 2006). However, variation in z can also emerge from differences in metacommunity structure. For example, z is generally higher on islands than on continents (Rosenzweig 1995; Whittaker and Fernández-Palacios 2007). This might result because islands have lower rates of dispersal, and thus higher inter-island differentiation than similar sized areas of continents that have higher rates of dispersal (see Effects of dispersal rates on local communities, below). In addition, in a meta-analysis of SARs, Drakare *et al.* (2006) found generally higher z (and thus higher β-diversity) at lower latitudes (Fig. 5.2). This indicates the possibility that metacommunity

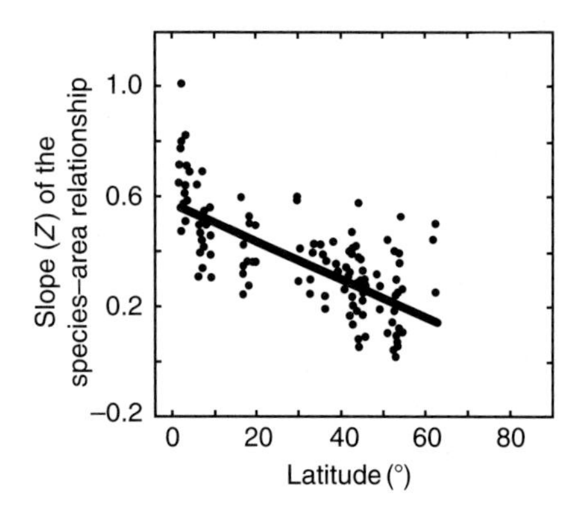

Figure 5.2 Decline in the slope of the species area curve (z) with decreasing latitude. Redrawn from Drakare *et al.* (2006).

assembly processes are different between temperate and tropical areas. Similarly, Ryberg and Chase (2007) showed both theoretically and empirically how the presence of top predators can reduce *z*. Other community stressors with effects similar to predators (e.g. increasing local extinction rates) should have similar effects.

5.6 Effects of dispersal rates on local communities

As noted above, variation in dispersal rates is a primary factor that can influence metacommunity structure and the partitioning of diversity. It is well known that islands that are closer to the mainland, and thus have higher rates of immigration, have higher diversity than comparable islands farther from the mainland, which receive lower rates of dispersal (e.g. MacArthur and Wilson 1967). Further, in a conservation context, experimental data are emerging that show higher levels of diversity in habitat patches that are connected by habitat corridors than in habitat patches without such corridors (Gilbert *et al.* 1998; Damschen *et al.* 2006).

There will often be a limit to the positive effect of dispersal on local diversity, such that diversity can show an asymptotic (MacArthur and Wilson 1967) or hump-shaped (Mouquet and Loreau 2003) relationship with increasing dispersal rates. Using meta-analysis, Cadotte (2006) compared standardized dispersal rates with their influence on local species diversity among animals and plants. Among animals, there was a positive, but potentially asymptotic or hump-shaped relationship between the enhancement of local diversity relative to the control with increasing dispersal rates, whereas there was generally a positive effect of dispersal on local diversity (most treatment effects relative to the control are greater than zero), but this did not seem to vary with the rates of enhanced dispersal.

Higher dispersal through habitat connections does not always increase local diversity. For example, Hoyle and Gilbert (2004) used an experimental landscape of moss and its microarthropod inhabitants similar to that used by Gilbert *et al.* (1998), but found no positive effects of habitat corridors on species diversity during a year with favourable weather conditions. A possible explanation for this difference is that, during drought years, rescue effects via corridors may be more important in preventing local extinctions. Similarly, Kneitel and Miller (2003) found strong effects of dispersal on the diversity of a protist community in the absence of predatory mosquitoes, but much weaker dispersal effects in the presence of mosquitoes.

While dispersal's effects on diversity have been well-studied at local scales (e.g. reviewed in Cadotte 2006), its effects on β-diversity are less studied. However, both theoretical and empirical results show that rates of dispersal often result in lower β-diversity (and lower *z*), homogenizing communities, and potentially ameliorating or reversing the local-scale positive effects of dispersal, when diversity is measured at the regional scale (Harrison 1997, 1999; Amarasekare 2000; Forbes and Chase 2002; Chase 2003; Mouquet and Loreau 2003; Chase and Ryberg 2004; Fukami 2004; Cadotte and Fukami 2005). Finally, the influence of dispersal on β-diversity and scale-dependent patterns of diversity will also depend on the mechanisms that allow species to coexist both locally and regionally. Among experimental communities treehole-dwelling protists, Ostman *et al.* (2006) found that dispersal homogenized local communities under benign conditions (reducing β-diversity) and thus reduced overall γ-diversity. However, when drought disturbance was imposed on those communities, dispersal allowed species to recolonize habitats where they were driven extinct, increasing overall β- and γ-diversity.

5.7 Local–regional richness relationships

It is commonly observed that the relationship between γ-diversity and α-diversity – called local–regional richness (LRR) relationships – is positive (e.g. Ricklefs 1987, 2004; Cornell 1993; Rosenzweig and Ziv 1999; Shurin and Srivastava 2005). That is, as the number of species in the regional pool increases, so does the number of species that coexist in any given locality. Originally, the shape of this relationship was thought to be able to discern whether communities were saturated with species (in which case the LRR should be asymptotic) or whether they were unsaturated (in which case the LRR should be linear) (e.g. Terborgh and Faaborg 1980; Cornell 1993). However, more recent analyses have suggested that LRR relationships can be linear even when communities are saturated with species, or asymptotic even when communities are not saturated (reviewed in Shurin and Srivastava 2005).

Despite the potential problems with inferring causation from LRR patterns, they can provide valuable information on metacommunity structure. Using a dataset of several *Daphnia* species co-occurring in Baltic Sea rockpools (Bengtsson 1989, 1991), combined with a modelling approach, Hugueny *et al.* (2007) were able to discern between the patch dynamic (Levins' metapopulation) and source sink models. Since these models have different underlying colonization and competition structures, Hugeney *et al.* were able to gain a deeper understanding of the processes structuring this community. Additionally, multivariate analyses exploring the effects of environmental conditions on patterns of diversity allow a more direct comparison of how these factors directly and indirectly influence local and regional richness when the LRR is explicitly considered (Harrison *et al.* 2006b).

5.8 A synthesis of metacommunity models

In any metacommunity, a variety of processes, including those related to neutral models (e.g. stochasticity) and those related to niche models (e.g. determinism), are operating simultaneously (Chase 2005, 2007; Gravel *et al.* 2006; Leibold and McPeek 2006; Adler *et al.* 2007; Clark *et al.* 2007). In analogy to population genetics, where genetic drift is only part of the equation and is tempered by the importance of natural selection, in metacommunity ecology, the ecological drift invoked by neutral theory (Hubbell 2001) is only part of the equation, and can be tempered by deterministic factors, which Chase (2007) referred to as 'niche selection'.

When niche selection is strong, such as in low productivity or highly disturbed communities, the stochasticity associated with ecological drift should be lower than when niche selection is weaker (Booth and Larson 1999; Chase 2003, 2007). Indeed, in an experimental study on the assembly of small pond communities, Chase (2007) found that patterns of β-diversity (site-to-site differences in species composition) were not different from what would be expected from purely stochastic (neutral) assembly when ponds were permanent. However, when drought was experimentally imposed on one-half of those ponds, community structure became more similar between ponds, and patterns of β-diversity were better predicted by expectations based on models that incorporate species niches (Chase 2007).

Another way to depict the same idea is to consider the dual mechanisms that can influence coexistence among species in a metacommunity. Coexistence results from a balance between stabilizing and equalizing forces (Chesson 2000; Adler *et al.* 2007). Equalizing factors serve to make species responses to variation in the environment similar, enabling them to persist in the same patches, while stabilizing factors serve to make species responses to the environment more different, i.e. niche differentiation and partitioning (e.g. Chesson 2000). The equalizing factors represent the components of species coexistence that are neutral, whereas the stabilizing factors represent the components of species coexistence that are niche related (Adler *et al.* 2007). It is easy to see here how both niche and neutral perspectives can (and indeed must) coexist in the same conceptual space, since the neutral theory is simply a special case of a more complete theoretical construct.

5.9 Adding food web interactions into the equation

All of the models discussed above consider species interactions only within a single trophic level. Although many of the empirical studies above implicitly considered both predator and prey species in their analyses, few explicitly considered the role that food web interactions might play. There is increasing evidence that not only do predators play an important role in structuring communities (reviewed in Chase *et al.* 2002), but that they are expected to respond differentially to spatial factors relative to prey communities, thus indirectly altering metacommunity structure (Holt *et al.* 1999; Holt and Hoopes 2005).

Predators often are more sensitive to habitat area than prey species (reviewed in Holt and Hoopes 2005; Ryall and Fahrig 2006), primarily because predators typically need larger areas in order to maintain viable populations. Smaller habitat patches are often less likely to maintain large predators, and thus can have higher diversity and abundance of prey species than larger habitat patches (e.g. Terborgh *et al.* 2001; Gotelli and Ellison 2006; Ostman *et al.* 2007). Further, because predators are more likely to be present in larger habitats, and because predators can strongly influence prey community structure, this can have a strong influence on species-area curves (Holt *et al.* 1999; Holt and Hoopes 2005). For example, on rocky outcrops (glades) in the Missouri Ozark mountains, Ostman *et al.* (2007) found no relationship between habitat area and insect species richness. However, when the occurrence of the voracious insectivorous lizard *Crotophytus collaris*, primarily only found on larger glades, was statistically controlled, significant species-area relationships emerged, albeit with different slopes with and without the predatory lizards (Fig. 5.3).

Habitat isolation can also disproportionately influence predators, and thus alter prey metacommunity structure (reviewed in Holt and Hoopes 2005). Several empirical studies have shown that more isolated habitats have lower rates of predation (and parasitism), which in turn can alter the structure of the entire food web (Zabel and Tscharntke 1998; Kruess and Tscharntke 2000; Watts and Didham 2006; Shulman and Chase 2007). In pond-dwelling invertebrate communities, Shulman and Chase (2007) showed that satellite ponds that were more isolated from the source pond had lower predator–prey species richness ratios relative to ponds that were closer to the source pond. This resulted primarily because predator richness was highest near the source pond, but decreased with distance, allowing prey species richness to increase farther from the source.

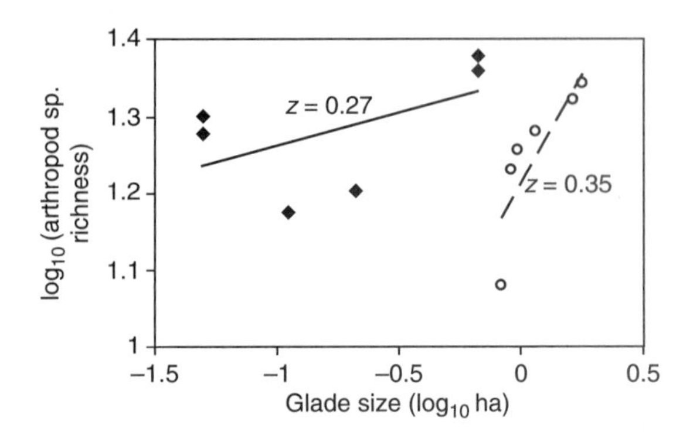

Figure 5.3 Effects of patch area on insect species richness in Ozark glades with and without insectivorous lizard predators. Redrawn from Ostman *et al.* (2007).

However, predators may also have larger home areas and be able to utilize a large area of a set of patches that appear more isolated to prey species (Oksanen 1990; Van de Koppel *et al.* 2005). This may result in different effects of spatial structure on predator–prey interactions, as it basically decouples predators from single prey populations and may lead to local overexploitation (e.g. van de Koppel *et al.* 2005). Hence predators may integrate dynamics over several local ecosystems, which may be similar or of varying quality (see below).

5.10 Cross-ecosystem boundaries

There is a growing body of evidence suggesting that resource subsidies and predators (often termed bottom-up and top-down regulation, respectively) frequently cross traditionally defined ecosystem boundaries (e.g. Polis *et al.* 2004). Such fluxes of resources and consumers are not generally considered under traditional metacommunity ecology, since the ecosystems being connected often transcend those types to which ecologists typically restrict themselves. Grassland ecologists rarely study woodlands. Marine ecologists rarely study freshwater. Aquatic ecologists rarely study terrestrial habitats. Above-ground ecologists seldom study soil ecology. However, many of the organisms that ecologists study readily traverse these boundaries seasonally, or across their life span, and do so either through behavioural means, such as migration, through life-history shifts (sometimes called ontogenetic niche shifts; Werner and Gilliam 1984), or, like plants, just by growing with roots in the soil and green parts above ground, linking the two subsystems of terrestrial ecology.

Organisms frequently move between ecosystem types daily or seasonally for foraging, predator avoidance, breeding, overwintering, etc. For example, many birds, ungulates and marine mammals are strong interactors (as both predators and potential competitors) that can migrate over very long distances. The interactions of these organisms in one local community can be the result of processes that occurred in a very different ecosystem a very long distance away.

Many species undergo life-history switches which take them across different types of ecosystems. For example, many pelagic marine fishes spend their juvenile periods in estuaries, coastal wetlands or freshwater streams. Additionally, some species undergo much more dramatic ontogenetic niche shifts, and in doing so can shift ecosystem types (e.g. aquatic to terrestrial) and trophic roles (e.g. herbivore to carnivore) (reviewed in Werner and Gilliam 1984). These include a large number of insects (e.g. *Diptera, Odonata, Megaloptera*) and amphibians that have aquatic larval and terrestrial adult stages. In a recent study emphasizing the important connections across ecosystem boundaries, Knight *et al.* (2005) found cascading effects of fish in ponds to terrestrial plants through their influence of dragonflies with ontogenetic niche shifts. Specifically, dragonflies are strongly limited by predatory fish in ponds, which translates into fewer adult dragonflies near ponds with fish relative to near ponds without fish. In turn, adult

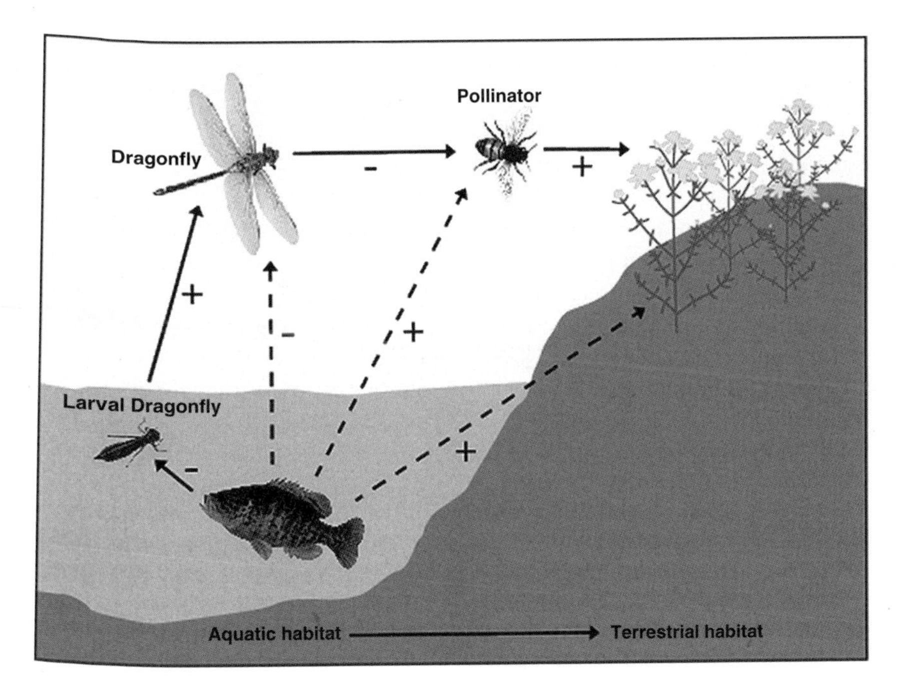

Figure 5.4 Schematic of the effects of fish in ponds on the adjacent terrestrial food web. Redrawn from Knight *et al.* (2005).

dragonflies are voracious predators on a variety of terrestrial insects, including species that are important pollinators of terrestrial plants. Thus, fish, by reducing the number of larval and consequently adult dragonflies near ponds, facilitated higher levels of pollination to plants in the surrounding terrestrial community (Fig. 5.4).

The environmental conditions and interactions that organisms face in one part of the landscape can strongly influence their interactions in the other. When these sorts of movements are frequent and important, we need to expand our view of the metacommunity. In doing so, we will need to develop a more general theory that can explicitly account for cross-ecosystem flows of organisms. This more general theory will have to consider interactions among organisms that move between ecosystem types in a landscape context. An organism will be constrained by the availability of the ecosystem type that is most limiting to the population. For example, for anadramous fishes, the availability of pelagic habitat for the adult stage will have little influence on abundance if breeding and juvenile habitats (e.g. streams) are degraded (Allendorf *et al.* 1997). For many amphibians, the quality of the pond in which larvae will develop is largely irrelevant in the absence of adequate terrestrial habitat (Semlitsch 1998). We can consider the availability of ecosystem types in a similar manner to the way we consider resources in more traditional consumer–resource models (Chase and Leibold 2003). Specifically, if one ecosystem type that is needed for an organism to persist is in limited supply, it is that availability that will govern the dynamics of the population, and thus the interactions of that organism in the other ecosystem types that it uses.

5.11 Conclusions

We have provided a broad overview of metacommunity ecology. By our definition, metacommunity ecology encompasses a large number of spatio-temporal processes that occur above the level of the local community, and thus our treatment is a bit broader than previous synthetic treatments of metacommunities (Leibold *et al.* 2004; Holyoak *et al.* 2005). In our view, metacommunity ecology encompasses the 'mesoscale' (*sensu* Holt 1993) of community ecology, above the level of the local community, but below that of large-scale biogeographic studies. Thus, there is a rather large spatial scale and span of questions that can be addressed under the metacommunity umbrella. These include: (1) patterns of metacommunity structure, including the relationships between α-, β- and γ-diversity; (2) patterns of interspecific interactions, relative abundances and coexistence in local communities that are linked through dispersal; (3) SARs and their variation; (4) interactions among species across traditional ecosystem boundaries.

Metacommunity ecology is a rapidly growing field, but one in which considerable new insights, synthesis, theories and empirical studies are needed. We have discussed a number of recent advances in metacommunity ecology, including diversity partitioning at different spatial scales, the interaction between stochastic and deterministic factors, food web interactions and cross-ecosystem boundaries. Additionally, although beyond the scope of the current chapter, evolutionary and historical factors can also play a significant role in the structuring of metacommunities, and this aspect is just beginning to be considered (e.g. Urban and Skelly 2006; Kraft *et al.* 2007). Thus, there is considerably more work needed in these and other areas to gain a better understanding of the processes that maintain community diversity and structure at different spatial scales. In addition, this understanding will be critical in order to develop the concepts and tools necessary to manage, conserve and restore ecosystems in this increasingly human-dominated world (see Chapter 9).

Acknowledgements

We thank Jens Åström, Wade Ryberg, Herman Verhoef and Peter Morin for comments and discussions that helped us to improve this chapter.

CHAPTER 6

Spatio-temporal structure in soil communities and ecosystem processes

Matty P. Berg

6.1 Introduction

For some time ecologists have been interested in the relationship between the structure of communities and the consequences this has for the rate of ecosystem processes, i.e. the structure–function debate (Vanni 2002; Hättenschwiler *et al.* 2005), and the role of community architecture as a stabilizing mechanism of communities (McCann *et al.* 2005). A wealth of evidence exists that links the dynamics of populations to rates of ecosystem processes (Jones and Lawton 1995; Wardle 2002 and examples therein). The role of invertebrates often relates to their differential utilization of resources or scarce elements in the habitat, which can result in interspecific functional dissimilarity (Heemsbergen *et al.* 2004): the more different species are in the functions they fulfil in ecosystems, the more important they become. However, the enormous diversity in organisms in many communities makes the evaluation of their role in ecosystems a Herculean task. To reduce this diversity into smaller components a food web approach has been adopted. In this way the impact of community structure on ecosystem processes and stability and visa versa can be analysed (DeAngelis 1992).

Although community structure has been shown to be of great importance in regulating ecosystem processes, the question remains how important spatio-temporal variability in community structure is for the regulation of processes, and if we need this kind of detail to understand underlying mechanisms (Berg and Bengtsson 2007). In an attempt to answer this question I will use a terrestrial example, the organic horizon of a coniferous forest soil, (1) to quantify community variability in time and across space and (2) to assess possible consequences of community variability for an important soil process, the degradation of organic matter and the subsequent flow of energy and nutrients through soil. First, I will briefly introduce soil communities, how feeding links between species in the community can be visualized in connectedness food webs, and how organic matter is structuring these food webs. Second, I will show that soil communities are highly variable in time and across space, much more variable than currently is appreciated in many empirical and theoretical community and food web studies, and elucidate the importance of including community variability in our studies to improve our mechanistic understanding of how soil organisms regulate soil processes. Finally, in order to understand the mechanisms behind the interplay between community structure and soil organic matter quality, I will present a conceptual scheme in which the impact of vertical stratification of soil communities on degradation is further elucidated.

6.2 Soil communities, detrital food webs and soil processes

Soil ecologists have long been fascinated by the enormous diversity of life in soil. There is a tremendous variety of microbes and fauna in soil, with much variation in body size, feeding specialization, life history strategies, spatial distribution and responses to abiotic factors. Although their coexistence is still an enigma, patterns and determinants in soil biodiversity are emerging (Wardle 2002; Bardgett *et al.* 2005). Factors that seem to regulate the temporal and spatial patterning of soil biota relate primarily to the heterogeneity of their environment, especially variability in resource quantity and quality and microclimate. This heterogeneity provides possibilities for resource and habitat specialization, or niche partitioning. The importance of detritus for species composition and abundance is derived from the observation that soil communities are bottom-up or donor controlled, which implies that resource supply, i.e. detritus, determines the structure of the community (Pimm 1982; Wardle 2002). Detritus is the main resource for fungi and bacteria and detritivores, such as earthworms, isopods, millipedes and enchytraeids. Microbes are fed upon by a number of fungivores, such as springtails, mites and some types of nematodes, whereas bacteria are fed upon by various protozoan groups, such as amoebae, flagellates and ciliates and other types of nematodes. Fungivores and bacterivores, in turn, are predated by predaceous mites, spiders and carabid and staphylinid beetles (Fig. 6.1). When detritus is added to the soil there is an increase in microbial biomass, which is often associated with an increase in soil fauna biomass (Wardle 2002 and references therein).

The coexistence of so many species also raises the question of whether all these species are of importance for the functioning of soil ecosystems. Many experiments show that ecosystem processes depend greatly on the diversity of the community in terms of functional characteristics of the species present and the abundance and distribution of these species over time and across space (Loreau *et al.* 2002; Hättenschwiler *et al.* 2005; Hooper *et al.* 2005). Collectively, decomposer organisms are

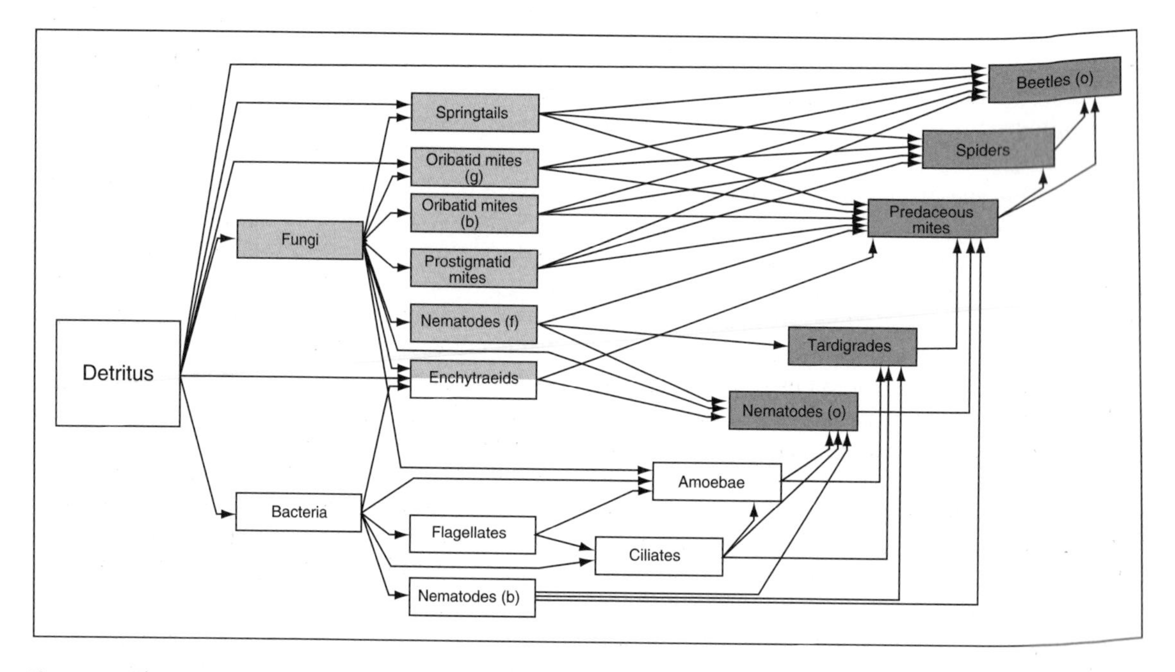

Figure 6.1 The connectedness food web of a coniferous forest soil. The boxes symbolize the functional groups of organisms and the vectors indicate the direction of the flow of C and nutrients through the web. The white boxes indicate the bacteria-dominated energy and nutrient flow and the light grey boxes indicate the fungi-dominated flow. Both flows merge on the level of intermediate and top predators, here in dark grey. The high acidity of the soil prevents the occurrence of earthworms. g, grazers; b, browsers; f, fungivore; b, bacterivore; o, omnivore. After Berg *et al.* (2001).

essential for the mineralization of carbon and nutrients that are bound to dead organic matter and the provision of resources for primary production. To quantify the functional significance of soil organisms, for instance for nutrient mineralization, species have been assembled in guilds or functional groups (Fig. 6.1). The rationale behind collecting species in functional groups, based on similar food types, predators, metabolic efficiencies and location in the soil profile, is that they may affect ecosystem processes in a similar way (Moore *et al.* 1988). Interestingly, the assumption that these shared characteristics of species reflect similar functions is seldom tested. The flow of carbon and nutrients through soil largely depends on trophic interactions between these functional groups, and these trophic relationships can be visualized in so-called connectedness food webs. A connectedness food web shows the architecture of trophic relationships within soil communities; for example, in Fig. 6.1 a connectedness food web is shown for a pine forest soil in the temperate region (Berg *et al.* 2001). This type of food web indicates only the most dominant feeding relationships and does not necessarily depict all physiologically possible trophic interactions between functional groups. Often, only those trophic interactions are included for which we have proof that they matter for carbon and nutrient flows. A food web, therefore, is not more than a reflection of a real community. It does not necessarily depict the real organization of the soil community, because other types than feeding interactions are excluded, such as facilitation or mutualism. More examples of connectedness food webs are given by Hunt *et al.* (1987; shortgrass prairie), de Ruiter *et al.* (1993; agricultural field) and Schröter *et al.* (2003; coniferous forests). Given the importance of detritus, both quantitatively and qualitatively, for the structure of soil communities, we need a closer look at the composition of soil organic matter, the process of organic matter decay and its spatio-temporal distribution in soil.

6.3 Soil organic matter

Detritus, or soil organic matter, is a collection of various organic compounds, such as plant remains, dead animals and metabolic by-products of microbial degradation. Input of detritus onto the forest floor is dominated by leaves, flower heads, seeds, twigs and bark and is often characterized by a seasonal pattern. Although plant species vary in their timing of leaf litter abscission over a year, most temperate tree species shed their litter in autumn. In coniferous forests needle fall is more or less continuous over a year, but also peaks in autumn. On some occasions the input of plant remains other than leaves can be high. After a severe storm large amounts of branches are blown from the trees and unusually heavy rains cause mast seeding and pulses of primary production (Ostfeld and Keesing 2000). Annual fluctuations in the supply of detritus to soil organisms may be a reason why many species of soil biota show strong seasonal patterns in activity or abundance (Wardle 2002).

In mixed forests where tree species that produce litters of different qualities coexist, a patchy horizontal spatial distribution in detritus quality may occur. Patchy spatial distributions of soil biota communities may be correlated with spatial variability in the amount of litter produced. For example, in the Amazon forest *Dicorynia quianensis* produces litter with a high content of polyphenolic complexes, whereas *Qualea* spp. produce litter with low levels of polyphenols, but high aluminium levels (Wardle and Lavelle 1997). Earthworms are absent from the litter of *D. quianensis*, whereas in litter which accumulates under *Qualea* endogenic earthworms are found, with a high abundance near the tree trunk where litter input is highest. More evidence that the spatial positioning of plants is a key determinant of the spatial patterning in soil biota can be found in Wardle (2002).

The organic compounds that make up detritus differ greatly in physical attributes and chemical complexity and, subsequently, in degradability. The process of organic matter decay is subdivided into two main phases. In the initial phase the readily decomposable organic compounds, mainly polysaccharides and non-lignified carbohydrates, are degraded. In a subsequent phase the more recalcitrant compounds, such as lignified carbohydrates, lignin and organic intermediates, are broken down

(Berg and Matzner 1997). The net effect is a relative increase in recalcitrant substrates as decomposition proceeds, while the amounts of un-decomposable material accumulate with depth in the organic profile. As nitrogen is usually the limiting nutrient in the initial phase of litter decomposition, the C:N ratio of organic matter is a general index of the quality of litter (Gadisch and Giller 1997; Berg and Laskowski 2006). However, to understand the mechanisms that regulate the processes of decomposition, the type of carbon in organic matter, the concentration of other nutrients than nitrogen and, especially, the composition of various secondary plant compounds are also important. For instance, litters with high lignin content, and that are rich in polyphenols, have a low degradation rate. Lignin is difficult to decompose by microbial enzymes (Shah and Nerud 2002) and can mask cell wall polysaccharides from degradation (Chesson 1997). Polyphenols, condensed tannins, terpenes and surface waxes strongly reduce the activity of microbes and the palatability of plant residues for detritivores (Hättenschwiler and Vitousek 2000).

During degradation, the morphology, biochemistry and physical properties of organic matter are continuously modified. These physicochemical transformations result in a decline in substrate quality for microbes and detritivores and are accompanied by succession of soil biota (Ponge 1991; Dilly and Irmler 1998). Over time, this older organic matter is covered by newly formed detritus with a higher quality, and this stratification in substrate quality results in shifts in the vertical distribution of soil biota (Faber 1991; Ponge 1991; Berg *et al.* 1998a). In the absence of earthworms detritus is not mixed through the soil and a stratified organic horizon is formed, with a subsequent litter, fragmented litter and humus layer on top of parent material. Although in the presence of earthworms organic matter is mixed through the soil, there still exists a vertical stratification in organic matter quality and quantity, but on a larger vertical spatial scale.

These examples imply that the distribution of soil organic matter and soil organisms is often strongly interlinked. To evaluate the importance of organic matter for community composition and to assess whether variability in community composition is linked to variability in soil organic matter quality, I present the results from an experiment that simultaneously describes variability in organic matter degradation and community structure over time and across horizontal and vertical space. This study was performed in a first-generation, 40-year-old Scots pine, *Pinus sylvestris*, forest, in The Netherlands, planted on a heather field on an inland dune of a former sand-drift area (Berg *et al.* 2001). First, I shall describe the variability in community composition over time, then introduce some factors that can explain temporal variability in community composition and finally indicate possible consequences for food web studies.

6.4 Variability in time in soil communities

Although it has been repeatedly pointed out that temporal and spatial heterogeneity in soil is crucial for understanding the distribution of soil organisms and how biodiversity affects key ecosystem processes (Bengtsson 1994; Ettema and Wardle 2002; Wardle 2002), surprisingly little is known about temporal and spatial variability in soil communities and detrital food webs (Bengtsson and Berg 2005). Variability in community composition has most often been studied in aquatic ecosystems. These studies found a considerable temporal and spatial variability in the composition of the food web of a freshwater pond (Warren 1989), an intermittent stream (Closs and Lake 1994) and a tidal freshwater river (Findlay *et al.* 1996). Schoenly and Cohen (1991) analysed a set of aquatic and terrestrial food webs for temporal variability in community composition. Very few species in the collection of food webs occurred on each sampling occasion, and most of the species were found only once. To examine the temporal and spatial variability in the structure of a soil community I performed a stratified litterbag experiment (for details of the design, see Fig. 6.2). Compositional variability, expressed using the Bray–Curtis (BC) similarity index, was measured over 2.5 years, for three organic horizons, namely litter, fragmented litter and humus (for details, see Fig. 6.3). From litter to humus the quality of organic matter for soil biota strongly declines. As far as I know, this is the first time that temporal and

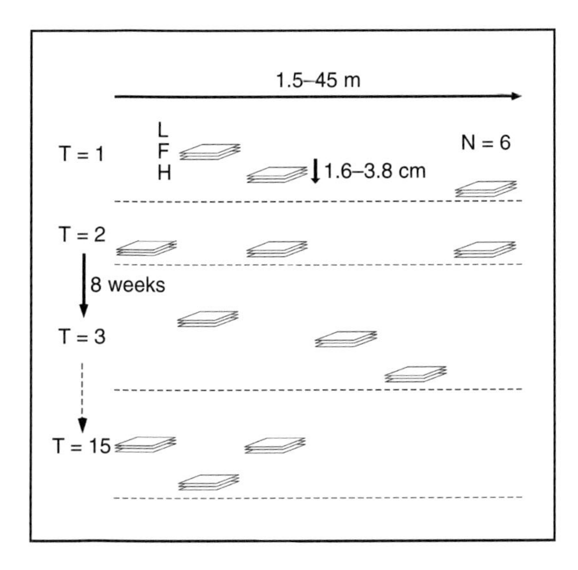

Figure 6.2 Stratified litterbag sets to measure organic matter and functional group dynamics over time and across space. The litterbags in a set, sized 10 × 10 cm each, were filled with homogenized pine litter (L; mesh size 3 mm), fragmented pine litter (F; mesh size 1.3 mm) or pine humus (H; mesh size 1 mm on top, 0.2 mm at bottom) and were 1.0 cm, 2.2 cm and 2.2 cm thick, respectively. Each litterbag set was 5.4 cm thick and reflected the thickness and detrital composition of the local organic soil horizon in which the sets were introduced. The 180 litterbag sets (12 sets, 15 samplings), each consisting of three litterbags filled with the three successive horizons, were placed randomly within a (40 × 50 m) area, and sampled at 8 week intervals over a period of 2.5 years. The distance between litterbag sets ranged between 1.5 m and 45 m. On each sampling occasion six litterbag sets were used to extract soil meso- and macrofauna and to analyse mass loss and chemical composition of organic matter. The remaining six sets were used to extract the microflora and microfauna. See Berg and Bengtsson (2007) for details about extraction methods of soil organism and biomass calculation procedures.

spatial variability in soil community composition has been simultaneously analysed.

The composition of the soil community varies more over time in the litter horizon (L) than in the underlying fragmented litter (F) and humus horizons (H) ($BC_L = 0.56 \pm 0.057$, $BC_F = 0.67 \pm 0.030$, $BC_H = 0.68 \pm 0.032$). The observed higher average variability in litter is caused by a significant decrease in similarity between samples when the time interval between sampling increases (decrease in BC_L from 0.73 to 0.39 over 2.5 years). Only the litter horizon shows a significant increase in variability over time. This difference in compositional variability between horizons corresponds with the mass loss of organic matter after 2.5 years, which is significantly higher in litter (mass loss L = 44.2% of initial mass, versus 4.2% and 2.8% for F and H, respectively). Moreover, mass loss in litter is correlated with change in chemical properties of litter, especially C:N ratios (Berg *et al.* 1998b). This strongly suggests that increase in compositional variability in litter over time is linked to changes in the quantity and/or quality of organic matter. Organic matter turnover is mentioned as one of the most important parts of environmental variability in forests (Bengtsson 1994).

Within-year variability in community composition is larger than between-year variability in community composition, but only in litter. Samples taken a year apart have significant lower community variability than samples obtained with half-year intervals. This confirms the findings of Bengtsson and Berg (2005), who showed for a managed pine forest and a virgin spruce forest that within-year variability in the composition of animal functional groups can be as large as between-year variability. The seasonal periodic oscillation in community variability coincides with a similar seasonal pattern in soil temperature and moisture values (Berg and Verhoef 1998). The annual amplitudes of soil temperature and water content are greater in litter than in underlying horizons. Owing to their short generation times many soil organisms can react relatively fast to short-term changes in the environment. This can be of the order of some days for organisms at the base of the food web to several months for animals at the top of the food web (Hunt *et al.* 1987). The high within-year variability in the litter community compared with deeper horizons can be explained by its higher resource quality. However, within-year variability in community composition is not easily explained by changes in organic matter quality, although input of detritus onto soil shows a seasonal pattern too. In an additional set of litterbags, filled with freshly fallen litter and replaced in the field every 8 weeks to obtain a constant litter quality over time, similar temporal trends in community

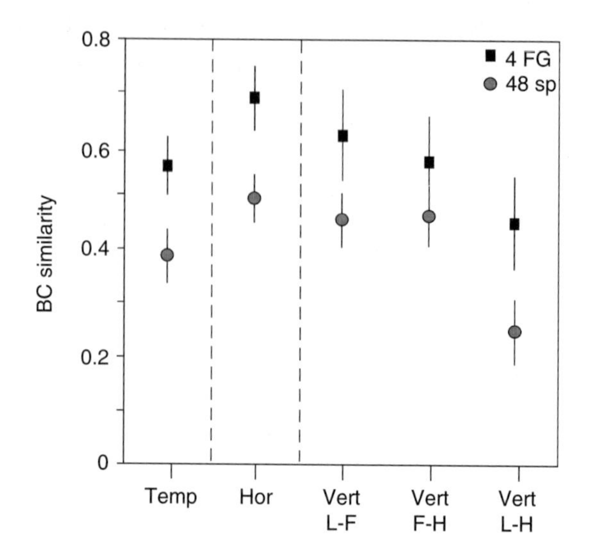

Figure 6.3 Temporal variability, and spatial variability in the micro-arthropod species composition (circles, 48 species combined) and the micro-arthropod functional group composition (squares, four functional groups combined (springtails, oribatid browser mites, oribatid grazer mites and predaceous mites)) in which the species were lumped, for a period of 2 years. The spiders and omnivorous beetles (no species determined) and prostigmatid mites (dominated for 95% by one species) were excluded from these analyses. The observed values for functional groups (squares) are similar to the values observed for the whole community. Compositional similarity clearly declines when the taxonomical level of resolution (from functional group to species) increases. The Bray–Curtis (BC) similarity index (Legendre and Legendre 1998) is used as a measure of community variability. Mean values of BC similarity are shown for the litter horizon over time (Temp), the litter horizon in horizontal space (Hor) and between the adjacent litter and fragmented litter horizons (Vert) and fragmented litter and humus horizon (Vert) and between the non-adjacent litter and humus horizons (Vert) in vertical space. Bars indicate the 95% confidence intervals. Lower BC-similarity values indicate higher variability. The mean biomass values from the six litterbags in each horizon collected at the same time were taken to calculate temporal similarity. Hence, the measures of temporal similarity are made on the spatial scale of the whole study plot. For analyses of horizontal spatial similarity, pair-wise similarities were calculated between the micro-arthropod functional group or species composition in each of the six litterbag sets sampled at each time. Each horizon was examined separately. For vertical similarity on the plot scale, the average abundances in each horizon of the six litterbag sets was used to calculate similarity between composition in adjacent (L–F and F–H) and non-adjacent (L–H) horizons. For more detail about analyses and for additional analyses in community composition at other levels of taxonomic resolution, see Berg and Bengtsson (2007).

variability are observed to those in litter that degraded over time.

The relatively high short-term variability of soil fauna communities implies that soil food webs, at least in seemingly stable and homogeneous temperate forests, do not have a fixed structure but are instead dynamic. Therefore, conclusions from community and food web analyses based on yearly averages should be treated with caution, especially in ecosystems with a fast turnover of populations and basic resources. For example, connectedness food webs, as shown in Fig. 6.1, are a template for simulation models that calculate the amount of C and N processed by a functional group or the whole food web. Using one of these models we derived the mineralization rates for C and N from the trophic interactions among groups of organisms, based on the annual biomass of the functional groups depicted in Fig. 6.1. The values of C and N mineralization rates indicated by the model were compared with observed C and N losses from organic matter in litterbags. The food web model underestimated C mineralization in litter by a factor of 2, whereas the model predictions for fragmented litter and humus were closer to the actual C mineralization. The simulated N mineralization rates resembled the observed rates in litter, but for fragmented litter and humus the simulated values accounted for only approximately 40% and 20% of the actual N mineralization (Berg *et al.* 2001). However, when we based our model predictions on the sampling biomasses, not averaged over time, and averaged the derived mineralization rates over 1 year, the model significantly overestimated mineralization rates (data not published). Sensitivity analyses of this food web model have shown that small changes in the biomass of particular functional groups, for example basal organisms or predators, may have a marked and disproportionate effect on the mineralization of nutrients (Hunt *et al.* 1987; de Ruiter *et al.* 1993). The discrepancy between model predictions, when based on temporal variability in functional group biomass versus their annual biomass, might relate to the existence of two distinct energy channels within detrital food webs (Rooney *et al.* 2006). One energy channel is based on bacteria, the other on fungi, and these two channels differ in both productivity and turnover rate. Size differences between fungi and fungivores, on the one side, and bacteria and bacterivores, on the other, result in dissimilar population dynamics, with faster turnover in the bacteria-dominated energy channel compared with the fungi-dominated channel. Averaging of time in these compartmentalized food web models may account for deviation in C and N flows found between observed values and model predictions. It may be more informative to take temporal variability in food web structure into account when examining C and N dynamics, especially when effects of variability are non-linear. Moreover, identification of which animals are of prime importance for flow of C and N might change when the organization of the food web is not fixed but dynamic. For a better understanding of how food web structure affects food web stability and nutrient fluxes through food webs we may need to include more temporal details.

6.5 Variability across horizontal space in soil communities

Soil organisms are not homogeneously distributed across space, both horizontally and vertically. A limitation of studies on spatial variability is that they have been done on individual organism groups rather than on whole communities. In our example, the horizontal spatial variability in community composition is rather low and does not differ between horizons ($BC_L = 0.69 \pm 0.062$, $BC_F = 0.72 \pm 0.046$ and $BC_H = 0.70 \pm 0.047$). Similarly, a low variability in mass loss of organic matter across space is found for all three subsequent horizons. Moreover, a horizontal spatial structure in community composition and mass loss of detritus is not observed, which means that the level of variability does not depend on the distance between sampling points. These observations emphasize that organic matter turnover is an important factor explaining variability in community composition, and in horizontal space.

The observed low variability in community composition is not expected, as the horizontal variability measures are based on single sets of litterbags rather than on averages of six litterbags (see Fig. 6.3 for details), and horizontal variability in plants, microbes and animals has been shown to be substantial (Saetre and Bååth 2000; Ettema and Wardle 2002; Laverman *et al.* 2002). A likely explanation is that the study site lacks ecosystem engineers, such as earthworms, because of the acid, non-calcareous soil type. Soil engineers can create small-scale variability in the environment by forming casts, redistributing organic matter and altering soil texture (for an overview, see Lavelle and Spain 2001). Plants have been shown to increase spatial variability (Klironomos *et al.* 1999; Wardle 2002). The possibility of vegetation inducing variability was greatly reduced because the litterbags were incubated at a more or less constant distance from equidistantly dispersed trees, in a forest with homogeneous grass–moss undergrowth. The relatively low spatial variability and absence of horizontal spatial structure may thus be indicative of the absence of factors that create variation, especially given the short duration of the study – only 2.5 years. In older, more mature forests with a

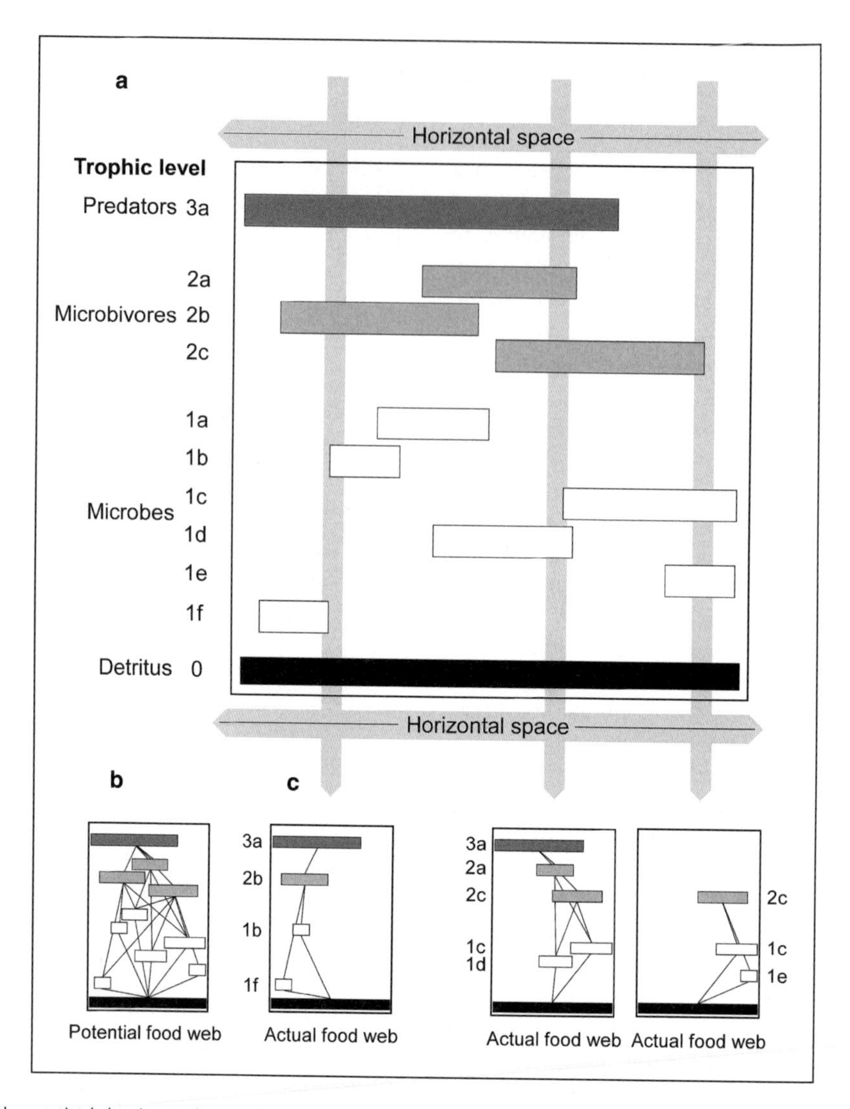

Figure 6.4 A theoretical, horizontal spatially explicit connectedness food web of three trophic groups (1, microbes; 2, microbivores; 3, predators) with 10 species (indicated with a letter). (a) Distribution of the 10 species (boxes) in horizontal space, at a specific time. The length of the boxes indicates the amount of horizontal space (home range) occupied by a species. (b) The potential connectedness food web showing ecologically possible feeding interactions (lines) between the 10 species (boxes). (c) Three samples at different points in space (vertical grey arrows) result in three local or actual food webs. The actual food web contains a subset of the species and feeding interactions of the potentially possible food web. Adapted from Brose *et al.* (2005).

high plant diversity and patchy distribution of decaying logs a higher horizontal spatial variability in detritus quality will be found than observed in this plantation. When soil engineers are also present horizontal spatial variability in community composition will increase, owing to their direct and indirect effects on organic matter quantity and quality.

Nevertheless, horizontal spatial variability in community composition is present, and this implies that connectedness food webs can be structured into compartments in which interactions between species are localized. Compartmentalization might strongly influence food web stability properties (Hall and Raffaelli 1997; but see Morin 1999). This aspect of spatial scale has so far been largely ignored in empirical and theoretical food web studies. In most studies that focus on horizontal space trophic interactions are aggregated over broader spatial scales, the scale of a forest or even a landscape, and not the scale of the organisms (Brose *et al.* 2005). This results in potential food web architectures that depict all physiological possible feeding links. In reality, when sampling on a small spatial scale, not all potential feeding links are realized, because not every species occurs everywhere (Fig. 6.4). The larger the species, the larger its home range, but the lower its numerical abundance and frequency in a local community (Cohen *et al.* 2003; Jennings and Mackinson 2003). Similarly, not all small-sized organisms occur locally, owing to environmental heterogeneity. In Fig. 6.4, three samples at different locations yield three different local connectedness food webs. The top predator is not always present, i.e. the right-hand example in Fig. 6.4c, owing to its low abundance and large home range. The spatial dimensions of resource use of the top predator links the other two local webs (Holt 1996; Pokarzhevskii *et al.* 2003; Hedlund *et al.* 2004). The more local webs are sampled and combined, the more the potential food web is approached (Brose *et al.* 2005). Many local versus one potential food web and coupling of local webs by predators are a challenge for future food web research, especially from a modelling point of view. However, the first attempts to deal with these aspects of spatial variability have already been made (Kondoh 2003, 2005; Teng and McCann 2004).

6.6 Variability across vertical space in soil communities is high

The distribution of groups of soil organisms often shows a distinct vertical spatial pattern (Faber 1991; Ponge 1991; Berg *et al.* 1998a). Likewise, in our forest example community composition has a distinct vertical pattern. The community is significantly more variable between non-adjacent organic horizons, the litter–humus comparison ($BC_{L-H} = 0.47 \pm 0.088$), than between adjacent horizons, the litter–fragmented litter comparison ($BC_{L-F} = 0.62 \pm 0.084$) and fragmented litter–humus comparison ($BC_{F-H} = 0.57 \pm 0.085$). Moreover, vertical spatial variability in the community composition of non-adjacent horizons is greater than both temporal and horizontal variability (Fig. 6.3). Thus, a high proportion of the variation in community structure in soils is likely to be attributed to different soil horizons, even if these horizons are only a few centimetres apart.

Soil temperature and soil moisture content, organic matter texture, quantity and quality are factors that differ between organic horizons. Differences in soil temperature and moisture content between successive horizons are generally small (Berg and Verhoef 1998), whereas significant modifications in organic matter morphology (from needle to amorphous colloids) are accompanied by a decrease in particle size, organic matter quality and, subsequently, degradation rates. Decay rates of organic matter decline from litter to humus (mass loss/year: L = 17.6% versus 1.7% and 1.1% for F and H, respectively). These morphological and chemical transitions, which are more pronounced the further the horizons are apart, may cause considerable variation in the composition of soil communities. When ecosystem engineers are present, the scale of vertical gradient in organic matter texture, quantity and quality can change from a few centimetres, in coniferous forests, to decimetres or even metres in other ecosystems, owing to mixing of detritus through the soil by earthworms.

The observation of vertical structure in community composition implies that, even more so than for horizontal space, food webs can be structured into compartments in which interactions are localized (Fig. 6.5). Most soil food web studies, however, do not consider vertical space, and feeding links in connectedness food webs are aggregated over the whole organic layer, or over the depth of soil core samples. However, on the scale of the organisms, not all physiologically possible trophic interactions are realized, as not all species are everywhere. For instance, the small particle size of humus prevents large-bodied animals, such as spiders, occurring in the humus horizon (Berg *et al.* 1998a), whereas the decline in resource quality with depth strongly affects succession, hence species composition, of microbes (Kendrick and Burges

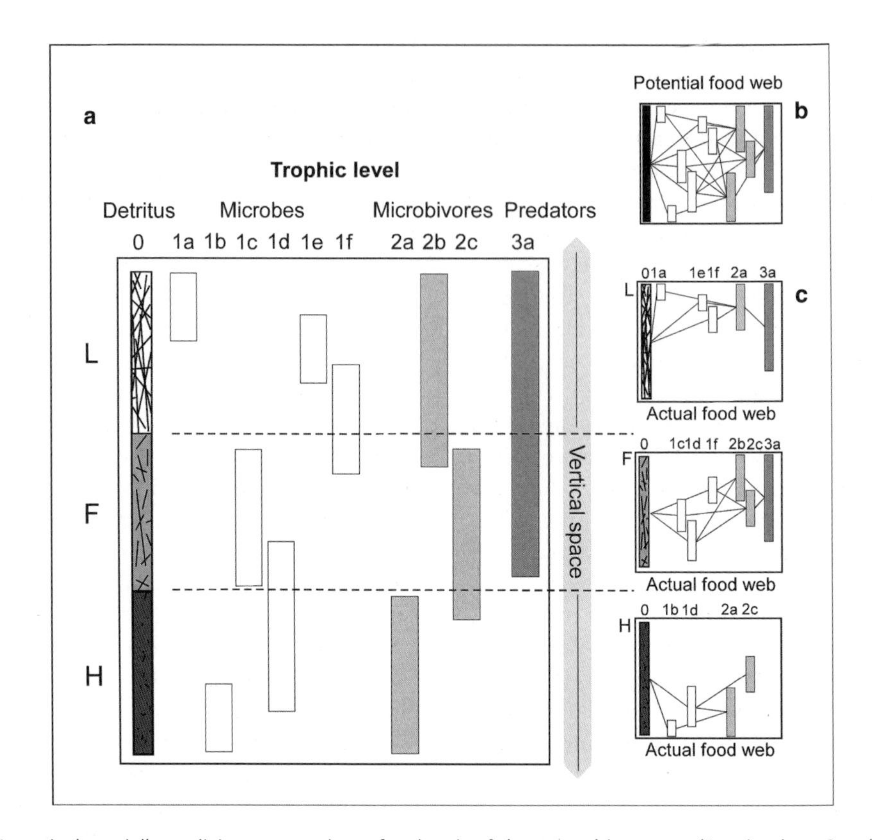

Figure 6.5 A vertical spatially explicit connectedness food web of three trophic groups (1, microbes; 2, microbivores; 3, predators) with 10 species (indicated with a letter). (a) Distribution of the 10 species (boxes) in vertical space. The length of the boxes indicates the amount of vertical space occupied by a species. The dashed lines separate the organic horizons. (b) The potential connectedness food web showing ecologically possible feeding interactions (lines) between the 10 species (boxes). (c) Three samples from subsequent organic horizons (L, litter; F, fragmented litter; H, humus) result in three local or actual food webs. The actual food web contains a subset of the species and feeding interactions of the potentially possible food web.

1962) and fauna (Dilly and Irmler 1998). As a result, successive organic horizons yield different local connectedness food webs (Fig. 6.5). Small-sized organisms, such as microbes and Protozoa, are highly abundant, but their occurrence is vertically localized because of adaptation to particular organic matter qualities, in combination with low dispersal ability. Fungivores, like springtails and mites, and small predaceous mites may couple local food webs by vertical migration (Setälä and Aarnio 2002), often induced by fluctuations in temperature or soil moisture content (Briones *et al.* 1997), or spatial dimensions in resource use. From a food web model perspective the variance in local vertical food web structure and the structural differences between local food webs versus one aggregated food web is as challenging as for horizontal space.

Berg *et al.* (2001) have shown, in one of the few food web studies in which vertical space is explicitly taken into account, that the rates of C and N mineralization may strongly depend on vertical stratification of functional groups and are related to differences in food web composition between soil compartments. The observed significant decrease in C and N losses from litter to humus (50 mg C/g/year in L to 1.5 mg C/g/year in H; 0.9 mg N/g/year in L to 0.05 mg N/g/year in H) correlates strongly with model simulations of C and N mineralization rates for the three successive horizons (Pearson correlation coefficient for C, $r = 0.999$; for N, $r = 0.996$). Moreover, total food web biomass did not explain the observed decrease in decay rates of detritus with depth. This indicates the importance of indirect contributions of soil fauna, via enhancing the mineralization rates of microbes, sometimes exceeding the direct contribution of fauna to C and N mineralization rates. The results imply that the regulation of ecosystem processes is inextricably linked to the structure of the soil community, and that accounting for vertical variation in food web structure seems essential for understanding energy and nutrient flows in soils. Empirical evidence that micro-stratification of soil organisms can have specific effects on ecosystem processes is given by Briones and Ineson (2002). Using radiocarbon techniques, they found that enchytraeids in a blanket bog assimilate carbon components of predominantly 5–10 years old. Vertical movement of enchytraeids due to abiotic factors does not affect this as they show similar ^{14}C enrichment values at various depths. In response to warming, however, worms are forced deeper into the soil, where they feed on older carbon components. Similarly, the vertical stratification of fungivores determines their impact on ecosystem processes (Faber 1991). Surface-living, epigeic species affect the colonization of fresh leaf litter, and potentially enhance the immobilization of nutrients. Hemiedaphic species enhance the net mineralization and nutrient mobilization in fragmented litter. Euedaphic species living in the humus horizon have the potential to affect plant growth by interference with mycorrhizal establishment or nutrient uptake by the roots. Therefore, aggregating organic horizons, with their own species compositions and specific impact on processes, is an oversimplification that obscures our understanding of the mechanisms behind many soil processes.

6.7 Spatio-temporal scales of community studies

There exists a disparity between the short temporal scale, days to weeks, and small spatial scale, centimetres to metres, at which most soil organisms operate and the long temporal scales, years to decade, and large spatial scales, plots to landscape, adopted in many studies. In community and food web studies we have to use the scales at which organisms operate and include their basal resources to fully understand their role in communities and ecosystems. Time and space are, however, the same side of the coin. They act simultaneously on species in the soil community over time by biochemical and physical changes in resources and across space by burrowing organic matter owing to input of fresh litter. Spatio-temporal modifications in detrital characteristics as degradation progresses are perceived differently by bacteria and fungi (Bosatta and Ågren 1991). Bacteria have shorter generation times than fungi (Rooney *et al.* 2006), and react differently to disturbances (Orwin *et al.* 2006). Hence, modification in detritus affects the fungi- or bacteria-based energy or nutrient channels of the food web in a different way, and this may contribute to the observed high within-year variability in community composition within organic horizons. Moreover, bacteria gain importance from litter to humus, because they benefit from the increase in resource surface owing to a decline in particle size (Fig. 6.6). Fungi penetrate whole leaves, a trait that bacteria lack, and do not profit from the decrease in particle size. A decline in particle size and alteration in chemical composition with soil depth and time may explain the often observed succession of soil organisms when degradation proceeds, and result in specific local communities in subsequent organic horizons (Fig. 6.6).

Resource-based modification in food web structure in turn feeds back to organic matter degradation, because food web composition strongly affects the rate of C and N mineralization. Trophic interactions within local communities can result in either immobilization of nutrients, in litter, or mobilization of nutrients, in fragmented litter and humus. This shift in the role soil organisms play in the decomposition processes warrants the subdivision of the organic horizon in its successive layers, each representing a different phase in the process of decomposition, each with a specific soil community composition (Fig. 6.6). However, spatio-temporal changes in resources are not described adequately

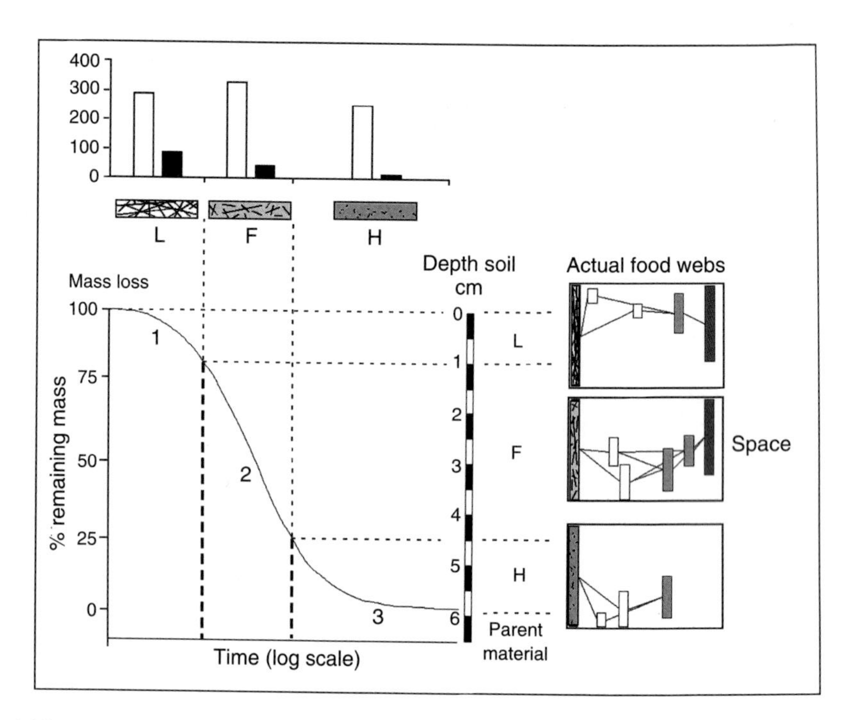

Figure 6.6 Model for mass loss and rate-regulating factors during degradation of detritus (left; after Berg and Laskowski 2006). Phase I curve (equals litter), mass loss is due to loss of water-soluble substances and non-lignified carbohydrates. Mass loss is stimulated by high levels of N, P and S. Trophic interactions result in nutrient immobilization. Phase II curve (equals fragmented litter), mass loss is due to loss of lignified carbohydrates, lignin and lignin-like compounds. Mass loss is stimulated by Mn and suppressed by high levels of N. Trophic interactions result in nutrient mobilization. Phase 3 curve (equals humus), mass loss reaches its limit value; concentrations of lignin and lignin-like compounds are about constant. Trophic interactions result in nutrient mobilization. Over time, which corresponds with across depth, the influence of climate on decomposition declines, lignin and N concentrations increase and the degradation rate approaches zero. Particle size strongly declines, resulting in a dominance of bacteria over fungi from litter to humus (bar diagram above the model curve). White bars, bacterial biomass; black bars, fungal biomass for litter, fragmented litter and humus, respectively. The *y*-axis shows biomass C in g C/g dry soil/year (after Berg *et al.* 2001). Physico-chemical changes in detritus over time, down the soil profile, result in local, horizon-specific food webs (right), often with different species compositions (Fig. 6.3). Species with high trophic position and an extensive vertical resource use may couple local food webs. The role of soil biota often shifts from the top layer to deeper organic horizons, from immobilization of nutrients in litter to mobilization of nutrients in fragmented litter and humus. L, litter; F, fragmented litter; H, humus.

in many community studies and food models, while variability in community composition over time and space is often ignored (Berg and Bengtsson 2007). To improve our understanding of how structural changes in community composition affect ecosystem processes, and vice versa, we should adopt a more spatio-temporal approach in our studies.

PART IV

APPLICATIONS

CHAPTER 7

Applications of community ecology approaches in terrestrial ecosystems: local problems, remote causes

Wim H. van der Putten

7.1 Introduction

7.1.1 Issues in applied community ecology

There are many examples of the applied value of community interactions in terrestrial ecosystems. Enhancing biological control of pests and plagues, for example, has become a major arena for testing bottom-up versus top-down control, direct versus indirect defence of plants, and constitutive versus induced defence (Karban and Baldwin 1997). This field of applied community ecology has developed important new concepts, such as the concept of multitrophic interactions (Tscharntke and Hawkins 2002). It shows that applied and fundamental research can go hand in hand fruitfully. Another field is that of biodiversity and ecosystem functioning, which started from the general concern about consequences of human-induced loss of biodiversity for the functioning of ecosystems. This led researchers to question how biodiversity loss may influence ecosystem processes, their stability and resilience (Loreau *et al.* 2001). These ecosystem processes and properties have important consequences for the sustainability of human society. However, when compared with biological control, the biodiversity debate is younger and the concepts are now being deepened. Studies that went into more complex community interactions were exciting and have made considerable progress (Cardinale *et al.* 2006). A third well-known applied issue in community ecology is that of the consequences of increased habitat fragmentation for the dispersal and survival of populations and species. Interesting progress in that field of research has been made, for example how species from different trophic levels are affected differently by habitat fragmentation. Habitat fragmentation could affect trophic control of species and, therefore, lead to either outbreaks or extinctions of species (Tscharntke *et al.* 2005).

The above-mentioned examples have received much attention in the past and there are many overviews and reviews written on these interfaces of fundamental and applied community ecology. Here, I want to focus on a current issue: community interactions across system boundaries. My main point is that applied questions concerning the management of species, communities or ecosystems at a given place and time may depend on processes that take place in adjacent subsystems, or in remote ecosystems, or that have taken place in the past. The soil will play an important role in this chapter, because soil has a strong 'memory' and legacy effects can cause historical contingency (see Chapter 4). Many soil organisms are relatively poor dispersers, but they are assumed to exhibit considerable redundancy in ecosystem processes. Nevertheless, redundancy does not apply to all functions and the response rate of soil communities to changes may be slower than that observed above ground.

My message for end-users or stakeholders is that your current problem of interest may very well originate from historical events, or that the cause can be found at your neighbours, or your neighbours' neighbour's yard. My message for scientists is that a local focus or a snapshot in time will not be sufficient to fully understand community interactions that take place here and now. I will argue that applications of community ecology approaches in terrestrial ecosystems may require a spatio-temporal approach in order to predict, anticipate or solve ecological problems as a result of natural, or human-induced, changes in the environment.

7.1.2 Top-down and bottom-up go hand in hand

The long-standing question of what controls the abundance of species was strongly fuelled by the debate between Hairston *et al.* (1960) and Ehrlich and Raven (1964). They disagreed about whether species are controlled by top-down (predator controlled) or bottom-up (resource controlled) forces. Nowadays, however, we realize that top-down forces may become bottom-up (Moore *et al.* 2003) and that evolution takes place in a multitrophic selection arena where selection pressures change from time to time (van der Putten *et al.* 2001). Communities are dynamic, and alternating species abundances may be controlled by a mix of resource and predator controls which, each in their turn, can exert selection pressure on local individuals. Clearly, this is not the end, but only the start of a dynamical perspective of communities and community interactions.

Unlike many simple experimental settings, community composition is not constant; but rather species move in and out of communities all the time. This is observed in lake food webs (Jonsson *et al.* 2005), and in diversity experiments where species-diverse plant communities created by sowing seed were much more stable in composition than non-sown species-diverse plant communities (Bezemer and van der Putten 2007). These examples illustrate that local community interactions can be changed by incoming or outgoing species and that immigration and emigration are quite common in natural communities. That awareness has major implications for the interpretation of well-controlled plant biodiversity experiments, where the intentional sowing of plants promotes spatial and temporal stability when compared with naturally colonized communities (Bezemer and van der Putten 2007).

Temporal instability may involve immigration of species that introduce novel properties. Immigration of species with different traits is a natural process in succession sequences, when grasses and forbs are replaced by shrubs and trees, producing litter that differs in rates of decomposition. While such processes are normal in successional gradients, artificial introduction of species from different continents may change the functioning of entire ecosystems (Vitousek *et al.* 1987). If such invaders also reach disproportionate abundance, for whatever reasons, incoming species may cause switches in community structure and ecosystem processes. Top-down and bottom-up processes play an important role in causing invasiveness and in the consequences of invasions. Invasions are among the major application issues worldwide and understanding why species become invasive, how these invasions can be prevented or how they should be managed is an important challenge for community ecology.

I will first work out how local community interactions may be altered by species that move within or across ecosystems. I will argue that local community interactions are influenced by processes and interactions between adjacent or distant systems. I use the case of plant interactions with above- and belowground organisms and discuss how these organisms may influence each other, through the shared plant resource. Then, I explain how aboveground–belowground interactions can be influenced by legacy effects in soil, and how these delayed influences may affect succession. I will focus on ecosystem restoration on post-agricultural land, which is one of the main application issues in the industrialized world. Subsequently, I will focus on alien exotic plant species, how they may influence local community interactions and how altering above and belowground bottom-up and top-down processes may contribute to invasiveness. Finally, I will speculate about the consequences of (human induced) global warming beyond the climate envelope approach, and I will conclude with some suggestions for users, stakeholders and scientists.

7.2 Community interactions across system boundaries

7.2.1 Linkages between adjacent or distant ecosystems

Adjacent ecosystems may influence each other in a variety of ways. For example, upstream deforestation may lead to downstream nutrient enrichment and sedimentation (Ineson *et al.* 2004). Adjacent ecosystems also may be mutually influenced by animal species that have a wide foraging range (van de Koppel *et al.* 2005). For example, seabirds affect plant productivity, detritus and, consequently, beetles in coastal terrestrial communities (Sanchez-Pinero and Polis 2000). Introduced exotic species may influence these community interactions. For example, introduced rats in New Zealand feed on seabirds, reducing the transport of nutrients from sea to land and leading to cascading effects on belowground organisms and associated ecosystem processes (Fukami *et al.* 2006).

Land use changes can have strong and large-scale consequences far from the actual site of change. The current increasing demand for biofuels to replace fossil fuels in the industrialized world influences production systems in second and third countries. As a consequence, the increasing demand of biofuels in Europe or the USA will influence food production in other parts of the world. This competition for land between food and biofuel production influences food prices and changes the traditional competition between land needed for food production and for biodiversity conservation. Therefore, the desire of human society to counter its impact on climate may lead to food limitation in the third world.

Local subsidies may also disrupt remote community interactions in nature. The increased use of nitrogen fertilizers in the mid-west of the USA is correlated with the increased abundance of mid-continent snow geese. Most likely, agricultural food subsidy to the geese during their overwintering in the mid-west indirectly destroys the breeding grounds in the Canadian Arctic and sub-Arctic; large populations of the snow goose cause overgrazing of shoreline vegetation, transforming the vegetation into bare sediment. Even when the snow goose population is eventually reduced, recovery of the breeding ground vegetation will take decades (Abraham *et al.* 2005).

Climate change may have strong local and non-local effects on phenology. Local effects are, for example, the interruption of prey abundance for hole-breeding birds, such as the great tit (*Parus major*). This forest bird feeds on winter moth, which occurs on birch and oak trees. When they are feeding their nestlings, *P. major* first use winter moth caterpillars from birch and then switch to caterpillars from oak trees. However, climate warming has advanced spring by 10 days over past decades and, while birch trees respond to warming, oaks do not. As a consequence, the birds face a gap in prey availability, limiting their ability to raise their chicks (Visser and Holleman 2001). Some migratory birds, such as the pied flycatcher, overwinter in West Africa. As the effects of climate warming on arrival of spring in North-Western Europe are not apparent in West Africa, the Pied flycatchers still arrive at the same dates, but at a later stage of phenology. This reduces preparation time for nest building and egg laying and constrains the ability of the birds to adjust chick feeding to prey availability (Both and Visser 2001).

7.2.2 Linkages between subsystems: aboveground–belowground interactions

Aboveground and belowground compartments of terrestrial ecosystems traditionally have been considered separately, but now their interlinkages are receiving increased interest (Wardle *et al.* 2004). Community composition and community processes in each linked subsystem depend strongly on the adjacent communities and often there is mutual interference. These dependencies can be caused by organisms that spend part of their life cycle in the soil and the other part above ground, such as many root-feeding insects, which mate and disperse above ground. Also, some carnivorous invertebrates are supposed to use belowground prey subsidies to survive the absence of aboveground preys, for example in between two agricultural crops (Bell *et al.* 2008). Plants play a special role in aboveground–belowground interactions, as they spend most of their life above as well as below the soil surface. Plants influence aboveground and belowground communities by modifying the abiotic environment, providing structure and resources, and by antagonistic interactions involving direct and indirect defences (Wardle *et al.* 2004).

Traditionally, plant community interactions have been explained primarily by variation in the abiotic environment. Interspecific competition and aboveground biotic interactions with vertebrate and invertebrate herbivores, pathogens, pollinators and other mutualists more recently have been recognized as influencing plant community structure. The awareness that plant community interactions are influenced by interactions between plants and soil organisms is even more recent, whereas interest in interactions between plants and below- and aboveground multitrophic communities is less than a decade old (van der Putten *et al.* 2001).

Plants interact with soil organisms through exudates, dead organic material (litter) and through living tissues. Soil organisms can reduce plant growth (via nutrient immobilization, herbivory, pathogenesis) or enhance plant growth (via symbiosis, nutrient availability, enhancement through mineralization) and these processes, and the organisms responsible for them, interact (Wardle *et al.* 2004). Besides primary metabolites, plants also contain secondary metabolites, which are involved in plant defence against natural enemies. In fact, a full understanding of the ecology of plants cannot be attained without including their interactions with both aboveground and belowground organisms.

Abiotic drivers, such as climate and soil type, will have strong predictive power about the composition and dynamics of plant communities on large spatial and temporal scales. However, at smaller spatial and temporal scales, plant community processes will be much more strongly influenced by aboveground–belowground community interactions. For example, aboveground vertebrate herbivores enhance root exudation, which fuels soil decomposer systems resulting in enhanced mineralization and increased plant nutrient availability (Bardgett *et al.* 1998). Therefore, vertebrate grazers may enhance primary production to their own benefit. Aboveground–belowground interactions also influence plant defence. For example, indirect defence of plant shoots against caterpillars by parasitoid wasps is reduced when cabbage root flies (*Delia radicum*) are present in the soil. The root flies reduce the amount of volatiles which normally are released by the shoots and attract parasitoids. The presence of root flies enhances the production of repelling compounds, which distract parasitoid wasps (Soler *et al.* 2007). Aboveground interactions can also be influenced by changing the entire soil community composition. For example, in multispecies grassland mesocosms, soil nematodes enhance the survival and quality of aboveground mites, resulting in enhanced aphid control by their natural enemies, aphid parasitoids. The result is that in the presence of nematodes, but not in the presence of soil microorganisms, both bottom-up and top-down control of aboveground aphids is intensified (Bezemer *et al.* 2005). These examples show that soil communities can play a key role in the management of production ecosystems; the question now is how to translate these findings into applications.

7.2.3 Consequences for application: find the remote cause of local effects

The existence of linkages across ecosystem or subsystem boundaries has major consequences for management of agricultural ecosystems, as well as for understanding and predicting changes in natural ecosystems. The presence and diversity of species, the abundance of populations, the interactions and organization of communities and the functioning of ecosystems all may depend on events that take place in adjacent populations, communities or ecosystems. Although conservation tends to be focused on red list species or species with iconic value, conservation requires a more integrated consideration of community interactions. As many applied problems have remote causes, local solutions will not be effective in the longer term. This applies to many global human-induced changes, such as land use change, climate change, as well as to biological invasions. Solutions for local problems require insight in processes that exceed local spatial and short-term temporal scales. For example, counteracting and mitigating impacts of exotic species requires insight into how these species are controlled in their native range and how these species have evolved during their history of introduction. On the other hand, local changes, varying from changed agriculture or industrial discharge to trade or climate conventions, can have wide and remote implications.

7.3 Community interactions and land use change

7.3.1 Land use change, predictability and major drivers of secondary succession

Land use changes are often driven by external factors, for example overproduction at the world market, or opening up of trade barriers, which lead to land abandonment when prices drop, or to (intensified) cultivation when prices rise. Land use changes often go along with altered physical–chemical inputs, such as soil tillage, fertilization and hydrology measures. These sudden land use changes can have enormous impact on species, community interactions and ecosystem processes. Whereas there is extensive experience with how to bring land into cultivation, ecosystem restoration is of a much more recalcitrant nature. Restoration is a long-term process that often starts with land abandonment and restoring former hydrology and nutrient cycles. Simple steering parameters that enable switching from one state to the other (Suding *et al.* 2004) probably do not exist. One question is whether ecosystems can be restored at all, because they may have become unalterably influenced by their history. The soil plays a crucial role as the main source of historical legacy effects that may constrain restoration efforts (Fig. 7.1). Soil legacy effects cause a major difference between secondary succession, which concerns the transformation from one ecosystem into another (Holtkamp *et al.* 2008), and primary succession, where ecosystems develop from bare soil (Neutel *et al.* 2007).

One of the main explanations for many examples of unsuccessful ecological restoration is the depletion of the soil seed bank and dispersal limitation of plant species from reference sites (Bakker and Berendse 1999). If plant propagules are unlimited, land conversion requires a switch from competition for light to competition for soil resources (Tilman 1982). According to the standard view, highest plant diversity is obtained at intermediate soil fertility (Al Mufti *et al.* 1977). This view, albeit developed by surveying across ecosystems, has provided the major rationale for restoration in industrialized countries, where restoration takes place on relatively nutrient-enriched soils (Marrs 1993). Mowing on wet soils or grazing on relatively dry productive soils may help to reduce nutrient availability (Olff and Ritchie 1998). Selective grazing of the dominant forbs and grasses, which have the highest quality, provides indirect advantage for the rarer, slow-growing and poorly competitive plant species. Introducing aboveground herbivores at mild stocking rates is a practice that is often applied in nature

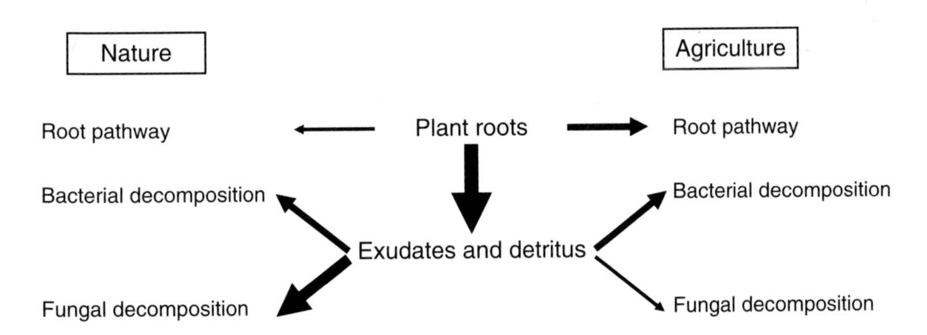

Figure 7.1 Energy transformation in three belowground pathways: the root–root feeder pathway, the detritus–bacteria–bacterivores pathway and the detritus–fungi–fungivores pathway in agricultural soil and a natural soil. Conversion of land use from agricultural to post-agricultural involves reorganization of the energy channels in the soil food web according to arrow thickness: the thicker the arrow, the larger the energy flow.

conservation, although the value of this practice is subject to debate among ecologists.

Besides reducing and concentrating nutrients, herbivores also influence vegetation structure and succession. Aboveground herbivory does not necessarily retard succession in all stages. Aboveground herbivores may slow down succession in some stages, but, initially, they may accelerate succession (Davidson 1993). Aboveground herbivores interact, thereby facilitating or inhibiting each other's effects. In an exclosure experiment, small herbivores (rabbits, voles) did not have consistent effects along a productivity gradient, whereas cattle increased plant diversity at high-production sites, but reduced plant diversity at low-production sites (Bakker *et al.* 2006). Large vertebrate herbivores, especially cattle, are easy to manage and their effects are relatively predictable. Invertebrate herbivores can be controlled far less well and interactions between vertebrates and invertebrates have been rarely explored (Tscharntke 1997). In a study focusing on sap-feeding insects, Schmitz *et al.* (2006) showed that the sap feeders could be controlled by both bottom-up and top-down forces, which change the rate of succession due to abrupt shifts in trophic control.

7.3.2 Secondary succession from an aboveground–belowground perspective

When secondary succession begins in post-agricultural fields, the soil already contains a well-developed food web that includes decomposers, bioturbators, symbionts, herbivores and pathogens. Secondary succession, therefore, requires the transformation of an arable soil food web into a soil food web characteristic of a more natural system that receives less soil disturbance, no mineral nutrients and other fertilizers, no biocide spraying and no monocultures of crop species. As a result, the soil food web will transform from a bacteria-dominated to a more fungi-dominated soil food web, although this process is not necessarily linear over time (Van der Wal *et al.* 2006). The role of mutualistic symbionts (especially of mycorrhizal fungi) and of bioturbators (earthworms) increases when succession proceeds. The initial changes in community composition can be partly due to the absence of regular mechanical soil disturbance (Van der Wal *et al.* 2006). The dynamics of soil pathogens and root herbivores will change from whole field oscillation patterns coinciding with crop presence to finer spatial and temporal patch dynamics as related to the spatio-temporal dynamics in vegetation composition. Therefore, the soil community will acquire (as well as generate) more spatial complexity, whereas the food web will develop from a fast to a slower cycling of nutrients and energy (Fig. 7.1).

Plants interact constantly with other organisms in soil- and aboveground communities and these feedback effects can drive succession. Feedback interactions with soil decomposer organisms proceed indirectly, through root exudation, decaying roots and litter. These interactions most likely take place at a slower rate than feedback interactions with symbionts and root pathogens, which have a more direct and intimate association with plant roots. Feedbacks between plants and their soil community can be assessed by growth experiments and the net effects range from negative to positive. These effects can develop over several months. Negative feedback leads to coexistence, whereas positive feedback results in dominance of plant species (Bever 2003). The sign of plant–soil feedback varies along successional sequences. In a 35-year-old chronosequence, early successional plant species experienced negative soil feedback, mid-successional species had neutral feedback and later successional plant species had positive feedback from the soil community (Kardol *et al.* 2006). This suggests that early succession soil communities enhance and later succession soil communities slow the rate of secondary succession. Moreover, plant–soil feedbacks of early successional plant species make them less competitive and also cause a legacy effect to mid-successional plant species. Such feedbacks can result from soil microorganisms (Kardol *et al.* 2007) as well as from soil fauna (De Deyn *et al.* 2003).

In post-production grassland soils, the soil fauna selectively reduces the abundance of early successional plant species, as well as that of the dominant plants (De Deyn *et al.* 2003). While soil fauna enhanced the rate of succession, it increased evenness in plant community composition (De Deyn *et al.* 2003). Nutrient addition increases dominance of fast-growing plant species, but the soil community counteracts the effects of nutrient addition (De Deyn *et al.* 2004). These inoculation experiments in mesocosms support exclusion experiments in the field by Brown and Gange (1992) and Schädler *et al.* (2004), who used selective biocides to eliminate belowground and aboveground insects. Early secondary succession slowed down when using soil insecticides (Schädler *et al.* 2004) and accelerated when using foliar insecticides (Brown and Gange 1992). Depending on the type of ecosystem, succession may slow down or speed up due to below- and aboveground activities respectively (Davidson 1993; Kardol *et al.* 2006), but these results show that ecosystem restoration is profoundly influenced by above- and belowground biotic interactions and not by changes in the abiotic environment alone.

Aboveground–belowground community interactions enhance temporal variation in natural communities (Bardgett *et al.* 2005). For example, in an extensively grazed pasture, cattle facilitated ant activity, resulting in ant mounds with relatively pathogen-free soil, which was rapidly colonized by pathogen-sensitive plant species (Blomqvist *et al.* 2000). The result was a spatio-temporal mosaic of ant mounds and plant populations. Similarly, mycorrhizal fungi in the plant roots can counteract aboveground insect damage on plants (Gange *et al.* 2003) and the individual effects of soil nematodes, root-feeding wireworms and aboveground grasshoppers resulted in non-additive effects on plant community composition (Van Ruijven *et al.* 2005). Enhanced exposure to aboveground and belowground invertebrates did not necessarily increase plant community diversity. We know now that aboveground and belowground biodiversity may, at least to some extent, interact and also that secondary succession needs to be perceived from an aboveground–belowground community perspective. However, this area of research still needs more examples for convincing generalizations to be made.

7.3.3 Consequences for restoration and conservation

Biodiversity restoration and conservation clearly require an aboveground–belowground approach. Secondary succession is influenced by abiotic conditions, as well as aboveground and belowground community interactions. Probably, these abiotic and biotic interactions act at different spatial and temporal scales and there will be hierarchies in controlling effects. For example, climate determines the type of biome, soil type determines which vegetation types can occur, and aboveground vertebrate herbivores determine vegetation structure and patch structure in the landscape. At the smallest spatial scale, aboveground invertebrate herbivores, plant pathogens and plant–soil organism interactions ultimately determine the composition and dynamics of the plant community by influencing plant competition, performance and abundance (Fig. 7.2). Conservation and restoration

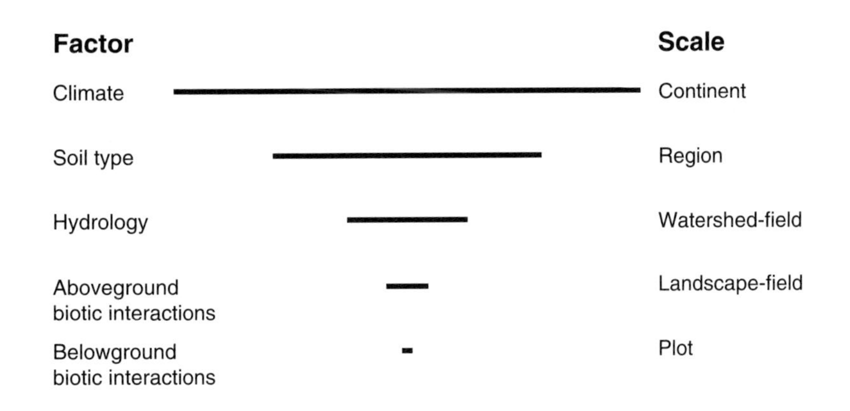

Figure 7.2 Hierarchies in influences of plant community structure and composition. Climate influences plant communities at continental scales, whereas belowground interactions influence plant communities at the smallest (plot, < 1 m^2) scale.

ecologists are challenged to integrate all these abiotic and biotic influences in their management decisions. However, there is relatively little known about how to influence the composition and development of soil communities (Kardol *et al.* 2008). Community ecologists face the enormous challenge of integrating all of these various disciplines to improve current concepts and to develop novel ones, and to facilitate environmental planning and conservation decision-making.

7.4 Biological invasions

7.4.1 Community-related hypotheses that explain biological invasions

Biological invasions of exotic species are causing major problems worldwide, because of their disproportional abundance, negative effects on local biodiversity and alterations of ecosystem processes (Williamson 1996). There are many overviews and reviews that discuss biological invasions in depth. Here, I will focus on biological invasions in relation to aboveground and belowground community interactions. Community ecology theory would predict that the disproportional abundance of exotic plants is caused by altered bottom-up and top-down interactions in the novel environment. Unoccupied niches can provide exotic species with abundant resources, although some studies argue that many invasive species occupy the same niches as the native species (Scheffer and Van Nes 2006). Alternatively, invasive exotic species may have novel traits, such as the capacity to fix nitrogen (Vitousek *et al.* 1987) or novel chemical properties that do not have a coevolutionary history with other organisms in the new range (Callaway and Ridenour 2004).

Prominent hypotheses proposed to explain biological invasions from a community perspective are biotic resistance (BR) and enemy release (ER) (Keane and Crawley 2002). BR was proposed by Elton (1958), who concluded that species-rich communities are invaded less by exotic species than species-poor communities. There is evidence for and against biotic resistance (Stohlgren *et al.* 1999). Certainly, much more work is needed to understand when biotic resistance can prevent invasions. For example, when species-rich plant communities prevent the establishment of invaders, how do these plant communities initially get and maintain their high species richness? Moreover, species-rich plant communities not only provide more potential competitors of exotic plant species. They also may harbour more natural enemies, which increases the chance that there are local enemies suitable to attack invading exotic plants. Therefore, the issue of biotic resistance, which is crucial to predict the long-term development of invasions, needs more multidimensional and multitrophic approaches than have been taken thus far.

The concept of enemy release is supported by examples of successful control of invasive exotic weeds by introduced biological control agents. Other studies that support the ER hypothesis have compared introduced and native plant species in their amount of natural enemy species. For example, in a review including more than 300 introduced plant species, Mitchell and Power (2003) showed that non-native species had fewer aboveground pathogen and virus species than comparable native plant species. The exotic plants also had fewer pathogens in their new than in their native range. Exotic plant species that were actively dispersed by humans, for example crops and ornamental plants, had more pathogen and virus species than other exotic species. It is not clear why this is; possible usage enhances the chance of exposure to potential enemies, or used exotic species did not pass the selection processes as strongly as non-used exotic species. Numbers of enemy species of course are not indicative of enemy effects, but these results at least show that exotic plant species are less exposed to aboveground pathogen and virus species than natives (Mitchell and Power 2003). There are similar studies on exposure of exotic plants to aboveground insects. They show that exotic invaders have less specialist feeders, but that they still can have generalists (Jobin *et al.* 1996; Memmott *et al.* 2000).

All of above-mentioned examples of enemy release and biotic resistance concern release from aboveground enemies. Klironomos (2002) showed that five exotic plants in an old field in Canada exert neutral soil feedback, suggesting that these exotic plants may have become released from natural soil-borne enemies. As soil feedback indicates a net effect, it could be that the neutral effect was due to less pathogenic, or to more symbiotic, activity, as symbionts may overrule the effects of pathogens. However, in the same study, native dominant plants also had neutral or even positive feedback (Klironomos 2002), so that the exotic plants could have shown the same soil feedback as in their native range, when they were dominants. Enemy release includes two components. The first is that exotic species escape from their native enemies and the second is that the exotic species have less enemy exposure in their new range. Therefore, such studies need a comparative approach, including both the native and non-native ranges (Hierro *et al.* 2005).

Further evidence on ER from soil pathogens stems from *Prunus serotina* (black cherry), which is invasive in Europe and native in the USA. In Europe, soil feedback was neutral, whereas, in the USA, the soil feedback was negative (Reinhart *et al.* 2003). In that example, there is still a possibility that symbionts provided an overwhelmingly positive effect, so that the neutral soil feedback effect could have been due to highly effective symbionts. In a study on Kalahari savanna grasses, however, fungi were isolated from a native grass (which expressed negative soil feedback) and from an exotic invader (which had neutral soil feedback). The soil fungi from the native species were pathogenic to their own host, but not to the exotic plant, whereas soil fungi from the exotic plant were not pathogenic to the native and non-native plant species (van der Putten *et al.* 2007). The exotic species was not examined in its native habitat. Therefore, these results all point to ER from soil pathogens; however, the ultimate test, specifically including pathogen species instead of treating soil feedback as a 'black box', is still lacking.

7.4.2 Mount Everest or tip of the iceberg?

In spite of the enormous research effort focused on biological invasions, most studies still are correlative (Levine *et al.* 2003). Moreover, BR and ER have many more dimensions than have been explored thus far. Therefore, it seems as if we have hit the tip of the iceberg, rather than getting a panoramic view from the world's highest peak. Here, I will point out some new viewpoints and conclude that biological invasions provide an enormous challenge to community ecology both from a fundamental and from an applied perspective.

A substantial amount of exotic plants contain chemicals that are novel then native species of the invaded range (Cappuccino and Arnason 2006). These novel chemicals may influence plant–plant interactions by allelopathic compounds (Callaway and Ridenour 2004), or they may reduce the inoculation potential of root symbionts as arbuscular mycorrhizal fungi in invaded habitats (Stinson *et al.* 2006). These views contrast with that of the evolution of increased competitive ability (EICA) (Blossey and Nötzold 1995), which assumes that defences impose trade-offs with growth when plants are exposed less to natural enemies. However, the novel weapons hypothesis has, thus far, not been widely tested for a range of plant species, whereas EICA is contradicted in a number of studies (Wolfe *et al.* 2004). Tests of these hypotheses usually lack the inclusion of negative controls, which would be exotic species that do not become invaders. These are the unseen majority of non-native species, according to the 10s rule of Williamson (1996). The problem is, however, species that do not become established will not be of use in ecological studies. Therefore, it would be possible to include mild invaders in the studies and examine their means of control. Nevertheless, it would still be possible that these species have not yet reached their stage of invasiveness, or that they are already over the top of their invasiveness. Including such 'false positives and negatives' would enhance the objectiveness of ecological studies on causes of invasiveness.

In the debate on EICA and novel weapons, most studies implicitly assume that plant defences are mainly direct. However, indirect defence, through recruiting the enemies of your enemies, may play an important role in plant abundance. Loss of the 'third trophic level' has rarely, if ever, been considered in invasion studies. In contrast, it has been argued that exotic plants may have easy access to symbionts, such as pollinators or mycorrhizal fungi (Richardson *et al.* 2000). This hypothesis has not been rigorously tested and the evidence for the

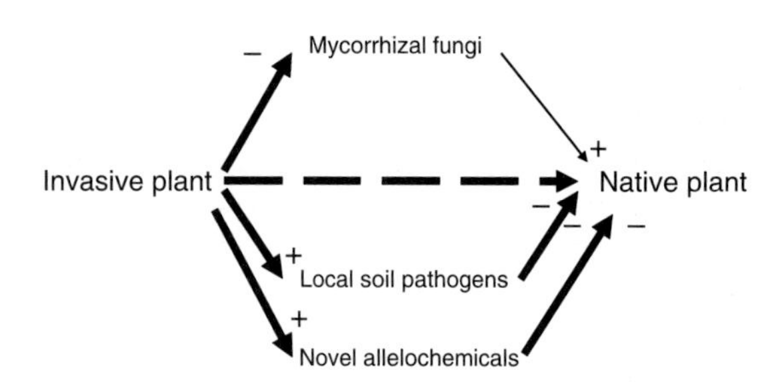

Figure 7.3 Three pathways of how invasive plants can indirectly influence native plants: by reducing, or suppressing, local symbionts, for example arbuscular mycorrhizal fungi; by enhancing local pathogens or parasites; or by excreting allelochemicals that are new to the local plants or enemies. The consequences of these three effects are that the invasive plant has an indirect advantage over native plants, as indicated by the dotted arrow. Arrow thickness indicates the strength of the effect.

positive effect of arbuscular mycorrhizal fungi is quite contrasting, whereas rigorous inoculation studies are lacking (van der Putten *et al.* 2007).

Some exotic plants may accumulate local viruses and pathogens, thereby having an indirect negative effect on native plant species (Fig. 7.3). For example, exotic plants in Californian grasslands accumulate local viruses without suffering from them. These viruses spread into the surrounding native plant communities and cause strong negative effects (Malmstrom *et al.* 2005). Similarly, the invasive tropical shrub *Chromolaena* in India accumulates soil pathogens that have strong negative effects on surrounding local plants (Mangla *et al.* 2008). Models show that such indirect effects could render exotic plants invasive, even when these plants suffer to some extent from mild biotic resistance (Eppinga *et al.* 2006). Although these invasive species make use of local enemies, their effects on native communities could be quite similar to those cases where novel weapons have been proposed to cause invasiveness.

7.4.3 Conclusions and consequences for management

The role of community ecologists in preventing, combating or mitigating effects of biological invasions provides an enormous task for future research. As soon as invasive species become abundant, it is already often too late for their complete eradication. Introducing biocontrol agents may be helpful for some invasive species, but the risk that biocontrol agents may switch to feeding on native plants cannot always be excluded. Moreover, the long-term effectiveness of biocontrol agents could be limited. In a comparison of biocontrol studies, Burdon and Marshall (1981) concluded that biocontrol by aboveground insects and pathogens was least effective against introduced annual plant species and most effective against introduced perennial and clonal plants. They concluded that an annual lifestyle probably enables rapid evolution of defences against pathogens or herbivores. Therefore, managers should consider alternatives, such as exploring how control agents already present in the invaded range might be used, or cultivated, to switch to and control exotic species. In doing so, the full array of enemies, including viruses, pathogens, herbivorous invertebrates, both below and above ground, as well as large herbivores could be considered. There is often little known about what controls exotic species in their native range. However, at the same time it is unclear whether these, or different, controls are required to reduce the abundance of the exotic species in their new range.

7.5 Discussion, conclusions and perspectives

The promise of ecology as a science lies in developing sufficient knowledge to allow us to understand and predict how individuals, species, communities and ecosystems will respond to myriad environmental changes. Applied community ecology should be at the forefront of developing such predictive power and testing the value of the predictions. Are there general laws and how well can we predict? How well do relationships developed in temperate zones, where most ecological studies are carried out, apply to tropical or boreal regions? These and many other applied issues are open for discussion. I will focus on a subset of these issues, show opportunities for end-users and stakeholders, and point out some areas for future work in ecology.

One main message of community ecology to its applied end-users and stakeholders is that many local problems can have remote causes. Whether this concerns local overgrazing, aboveground pests or biological invasions, the problems quite often originate at a different place from where the problems occur. Community ecology could provide help in solving these applied problems by analysis of the complex interactions in which many species are involved. In some cases, such as ecosystem restoration, original key interactions need to be determined and the key players need to be brought into contact. Restoration ecology still makes very little use of community ecological insights, except the use of large herbivores. Major omissions concern the involvement of invertebrates, of soil communities and of interactions of plant communities with above- and belowground organisms.

Biological invasions are a global problem. Increasing globalization of the economy, international tourism and enhanced emigration–immigration ensure that more and more non-native species will move to novel areas where they could become invasive. Predictive systems are necessary to forecast whether certain species have the potential to become invasive in their new habitat and, when so, to identify these species for import limitation. The main problem is twofold. First, only a minor fraction of all introduced species become invasive, so that the predictions need to detect $\sim$0.1% of all species as possible invaders. Usually, this is a range where statistics are uncertain. Second, the current rapid climate warming may change local conditions to become favourable for new introduced exotic species. This, together with range shifts due to climate warming, may elevate the future incidence of biological invasions. Especially when range shifts release plant species from belowground (van Grunsven *et al.* 2007) or belowground and aboveground enemy attack (Engelkes *et al.* 2008). The same has been shown for higher trophic level interactions, such as between insects and their enemies (Menendez *et al.* 2008). These future changes will require considerable attention from community ecologists.

The most striking examples of remote causes for local problems are definitely those where species move over large distances, such as the geese that migrate from mid-west USA to the Arctic zone. To solve these and to prevent other problems, it is crucial to thoroughly analyse all consequences so as to not create additional problems while attempting to solve an initial one. Community ecology should be at the forefront in these analyses, given the complexities involved. Interestingly, these applied questions could also be used as learning opportunities, because the shortcomings of the predictions will reveal the limits of our knowledge. Therefore, community ecologists should take an active role in analysing causes of environmental and biotic change and in forecasting consequences of new policy and management strategies. The challenge is huge, and includes biobased economy, biofuels, sustainable agriculture, biodiversity restoration and conservation, climate warming, and the consequences of genetically modified organisms.

CHAPTER 8

Sea changes: structure and functioning of emerging marine communities

J. Emmett Duffy

8.1 Introduction

The earth is in the midst of a global-scale, uncontrolled experiment involving human alteration of both the abiotic resources that support ecosystems and the trophic linkages that control their structure and function. Inputs of inorganic nutrients have increased substantially worldwide, causing major bottom-up shifts in ecosystems (Cloern 2001). Abundances of large vertebrates have been substantially depleted both on land (Dirzo and Raven 2003; Cardillo *et al.* 2005) and at sea (Pauly *et al.* 1998; Steele and Schumacher 2000; Jackson *et al.* 2001; Myers and Worm 2003; Hutchings and Reynolds 2004), resulting in systematically altered food web structure ('trophic skew'; Duffy 2003; Byrnes *et al.* 2007), and exotic invasions are altering community composition locally and homogenizing communities globally (Sax *et al.* 2005). Finally, rising $CO2$ inputs from fossil fuel combustion are causing climate warming and acidification of the surface oceans (Orr *et al.* 2005); these and associated changes in precipitation and circulation are shifting species ranges, seasonal cycles and interactions (e.g. Stachowicz *et al.* 2002b; Voigt *et al.* 2003; Schiel *et al.* 2004; Winder and Schindler 2004; Hays *et al.* 2005; Perry *et al.* 2005; Parmesan 2006). Together, these processes are producing novel or 'emergent ecosystems' (Hobbs *et al.* 2006) assembled in altered habitats from species that may have had little or no evolutionary history of interaction. The accelerating pace and synergism of these changes create an urgent need for rigorous applied community ecology – we need to understand how complex interactions within real ecological communities play out on the large space and timescales relevant to sustainable management, and how they mediate stressor impacts on ecosystems and their provision of services to humanity.

Much of our knowledge of the organization and dynamics of communities comes from mathematical theory and controlled experiments. These approaches have advanced ecology tremendously over the last 50 years or so. But there is also a widely recognized trade-off between the clarity and elegance of theory and experiments and the complexity of the real world (e.g. Carpenter 1996; Oksanen 2001). For example, the classical experiments of community ecology are generally limited to small spatial and temporal scales, and are logistically prohibitive for the large mobile animals of special conservation interest due to their status as strong interactors (Soule *et al.* 2005). Thus, successfully applying principles from academic community ecology to real-world management and conservation remains a daunting challenge. The ultimate test of ecological theory is its ability to explain, and to forecast, community responses to environmental change in the real world (e.g. Clark *et al.* 2001).

In this spirit, I focus here on one major human impact, fishing, as a case study to evaluate the success of fundamental principles of community ecology in understanding impacts of environmental change in marine ecosystems. I ask whether impacts of human predation are consistent with expectations arising from theoretical and experimental community ecology, particularly regarding trophic cascades, the influence of biodiversity on ecosystem stability and the nature of regime shifts among alternate semi-stable states. I explore what the impacts of human predation on marine communities can tell us about how and whether basic principles in community ecology extrapolate to complex ecosystems at large spatial and temporal scales, and thus assess their value, if any, for conservation and management.

8.1.1 Fishing as a global experiment in community manipulation

Of the several human impacts on marine systems, the strongest and most pervasive is the continuous removal of large quantities of animal biomass through fishing. In general, fishery management aims to reduce a fish stock to ~50% of its unfished biomass in order to maximize productivity. In practice, many stocks are fished well beyond this target (Hilborn *et al.* 2003a; FAO 2007). Modern fisheries, including both landings and by-catch, currently consume 24–35% of global marine primary production in the continental shelf and major upwelling areas (Pauly and Christensen 1995). Thus, any attempt to understand modern marine communities must reckon with the fact that humans are now the dominant predator throughout the world ocean. The many *direct* impacts of human exploitation on marine fish populations and communities are well documented (Jennings and Kaiser 1998; Pauly *et al.* 1998; Jackson *et al.* 2001; Myers and Worm 2003). But this intense predation is expected to have extensive *indirect* effects on marine communities as well. Indeed, effects of fishing provide a uniquely useful case for testing how community models scale up to real ecosystems, for several reasons. First, fishing impacts have followed similar patterns in many regions (Jennings and Kaiser 1998; Pauly *et al.* 1998; Jackson *et al.* 2001; Hilborn *et al.* 2003a; Myers and Worm 2003), providing a degree of replication and potential generality. Second, for pelagic fishes specifically, harvesting provides a relatively 'clean' test of community manipulation in that removal of individuals or species from the system has relatively low impact on the habitat (although 'ghost' nets can continue to ensnare fishes indiscriminately long after they are lost or abandoned); this is in contrast to most human impacts on land (Wilcove *et al.* 1998) and in benthic habitats (Watling and Norse 1998), where habitat destruction or modification confounds species removals with other impacts (Srivastava and Vellend 2005). Third, the major commercial value of fisheries means that there is a large body of detailed data on which, when and how many fish have been removed from the oceans (FAO 2007). Finally, the growing database on ecological changes within marine protected areas (MPAs) offers important large-scale experimental controls against which to evaluate the effects of fishing (Halpern and Warner 2002; Micheli *et al.* 2004).

8.1.2 Physical forcing and the uniqueness of marine ecosystems

The ecology of marine communities and their responses to perturbations are strongly influenced by the unique nature of the marine environment (Steele 1985, 1991). The fundamental physical difference between terrestrial and (pelagic) aquatic systems is the greater density of the liquid medium of water (Strathmann 1990), which has three important consequences for understanding the ecological structure and dynamics of terrestrial compared with pelagic ecosystems. First, in water, buoyancy allows primary producers (and other organisms) to float and obviates the need for large, expensive, metabolically inert structural tissues required to compete for light on land. Thus, the dominant marine pelagic autotrophs are microscopic, fast-growing and highly nutritious (floating *Sargassum* accumulations being a conspicuous exception in the Atlantic gyre). Consequently, compared with land, marine ecosystems show higher grazing rates and production: biomass ratios, a much larger faction of primary production grazed and more efficient conversion of production to herbivore biomass (Steele 1991; Cebrian 1999; Shurin *et al.* 2006). The higher growth rates, nutritional content and vulnerability of marine autotrophs to grazing in turn have important consequences for the structure and functioning of marine communities, including stronger top-down control generally and trophic cascades in particular (Shurin *et al.* 2002), and an altered distribution of biomass among trophic levels. For example, pristine coral reefs support (or did historically) large populations of apex predators (Friedlander and DeMartini 2002) but often little visible plant life, in stark contrast to the rain forests found above the tide line. In short, top-down control and trophic transfer are more efficient in the sea, and marine biomass pyramids tend to be (or were, primevally) less bottom-heavy than those on land (Odum 1971; Del Giorgio and Gasol 1995). These strong trophic interactions should enhance the ability to detect predicted responses to food web alteration in marine systems relative to terrestrial ones.

The second major consequence of water's density and buoyancy of biomaterials is the greater importance in the sea of advection of materials (inorganic nutrients, detritus) and organisms. In certain terrestrial systems, migrating birds, mammals and even insects can transport large quantities of materials over long distances (Polis *et al.* 1997). Nevertheless, constantly moving currents make marine ecosystems more open on average than terrestrial systems. Although pelagic marine communities and populations are more highly structured than might be expected from the superficially featureless appearance of their habitat, many large predators nevertheless can swim between ocean basins, and larvae of many species can drift for hundreds of kilometres before settling. These features mean that between-habitat subsidies, source-sink dynamics and gene flow tend to be considerably higher, on average, in marine communities than on land or freshwater. They also suggest that simple models of community structure and dynamics that implicitly assume closed systems may be less likely to apply in the sea, where metacommunity approaches will probably prove fruitful (see Leibold *et al.* 2004; Chapter 5).

The third important consequence of water's density, stemming from both the microscopic size of most primary producers and system openness, is the much closer and more rapid coupling between physical drivers and biological processes in marine (pelagic) systems than those on land (Steele 1985). For example, nutrient loading can produce responses of primary producers within days or even hours in the sea. Thus, many marine communities, especially pelagic communities, tend both to be more sensitive to disturbances and to rebound more rapidly after disturbance than terrestrial ones. Together with strong trophic interactions, this sensitivity should enhance the ability to detect predicted processes and patterns of community regulation on large spatial and temporal scales.

8.2 The changing shape of marine food webs

8.2.1 Conceptual background

Like other optimal foragers, humans generally target large and abundant prey preferentially, all else being equal. Fishing thus represents not only a strong, but also a selective press perturbation on marine communities, which has been sustained for decades and even millennia in some areas (Wing and Wing 2001; Barrett *et al.* 2004; Lotze *et al.* 2005). The responses of marine communities to this strong top-down influence depend on both 'vertical' components (food chain length, omnivory) and 'horizontal' components of biodiversity (species or functional group richness and composition within trophic levels), and their interactions (Duffy 2002; Duffy *et al.* 2007; Fig. 8.1). These in turn are mediated by organismal traits. Specifically, focusing on the key related traits of body size (Woodward *et al.* 2005), feeding traits that determine trophic level (Pauly *et al.* 1998) and life history (Jennings *et al.* 1998) reveals several apparently consistent patterns in the changing structure and functioning of marine ecosystems, and clarifies the mechanisms involved.

An important consequence of the complex pelagic life histories of many marine animals is that they pass through a large range of body size, and multiple trophic levels, during their lifetimes. Most newborn marine animals receive no parental care – in stark contrast to the birds and mammals that dominate upper trophic levels on land – and are therefore highly vulnerable to predation, starvation and abiotic stress. Importantly, the larvae of apex predators frequently serve as prey of fishes that the apex predators hunt as adults. Thus, compared

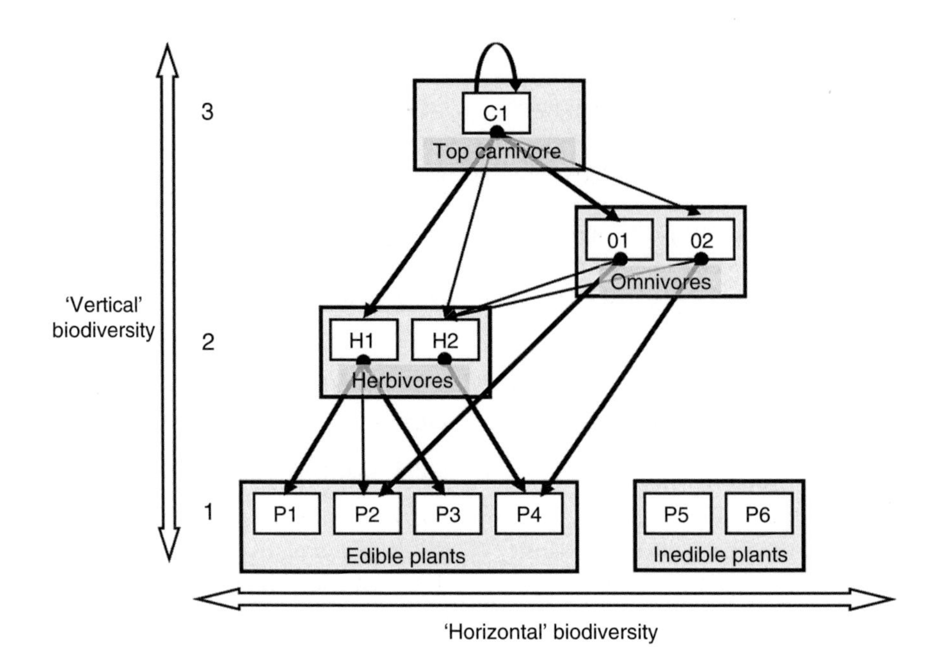

Figure 8.1 Components of horizontal and vertical diversity in a schematic food web. Vertical diversity includes average food chain length and degree of feeding from one (e.g. herbivores) compared with more than one (omnivores, and cannibalistic top carnivore) trophic level. Horizontal diversity includes number of functional groups and degree of feeding specialization (e.g. species H2) compared with generalism (e.g. H1, O1 and O2). Reproduced with permission from Duffy *et al.* (2007).

with terrestrial food webs, pelagic marine food webs tend to be more strongly structured by body size than by species, to contain more cannibalistic and omnivorous species (Dunne *et al.* 2004) and to contain more 'loops'. This shifting ontogenetic niche identity is well illustrated by stable nitrogen isotope data for North Sea fishes, which show that the trophic level of a species, averaged across ontogenetic stages, was unrelated to its maximum size, but was strongly related to body size at the individual level (Jennings *et al.* 2001; Fig. 8.2). Despite such indeterminacy, however, both experiments (Menge 1995) and observational data (Williams and Martinez 2004; Thompson *et al.* 2007) support the reality of discrete trophic levels, at least at lower positions in food webs. Moreover, maximum trophic level in a system is clearly related to the presence of species that can attain large size.

Cohen *et al.* (2003) showed for the pelagic food web of Tuesday Lake, USA, that there are consistent relationships among body size, numerical abundance and trophic level of species, with primary producers being both much smaller and more abundant than predators. These relationships are presumably also characteristic of pelagic marine systems, which are similarly based on microscopic algal producers. In the pelagic system, species richness also shows a pyramidal distribution with few apex predators and many primary producer species (Cohen *et al.* 2003; Petchey *et al.* 2004). Predators, as distinguished from parasites and pathogens, must generally be larger than their prey (Brose *et al.* 2006), and simple allometric theory dictates that the few larger, less abundant species high in the food web also are slower growing than basal species.

These considerations have several general implications for responses of marine communities to human influence (Duffy 2002, 2003; Petchey *et al.* 2004). First, the smaller populations and slow population growth rates of top predators should raise their risk of extinction due to demographic and

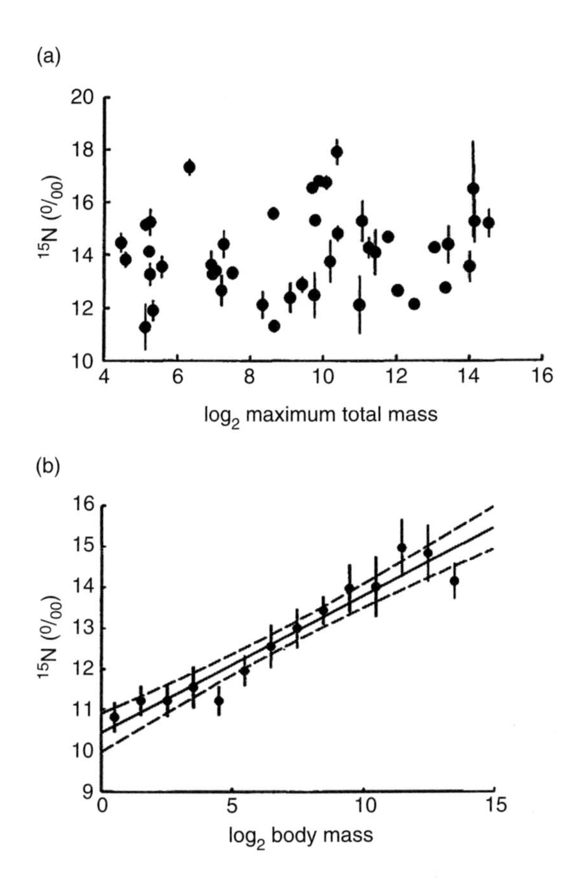

Figure 8.2 Among North Sea fishes, trophic level (as indexed by δ^{15}N signature) is unrelated to maximum body mass of the species (a) but closely related to individual body mass (b). Thus, trophic level is a property of individuals, not species, and increases through ontogeny. Reproduced with permission from Jennings *et al.* (2001).

environmental stochasticity, and may also lower their resilience to demographic perturbation. Second, harvesting disproportionately targets large animals near the top of marine food chains (Botsford *et al.* 1997; Pauly *et al.* 1998; Jackson *et al.* 2001). Because most ecosystems support few species of apex predators, this combination of demographic vulnerability and targeted human pressure means that marine (and other) ecosystems under human influence are inherently vulnerable to loss of an entire functional group or trophic levels at the top of the food web (Jackson *et al.* 2001; Duffy 2002; Dobson *et al.* 2006). The predicted pattern that results is 'trophic skew', a vertical compaction and blunting of the trophic pyramid due to proportionally greater losses of higher level species (Duffy 2003).

8.2.2 Empirical evidence for trophic skew in the ocean

Several lines of evidence illustrate that human impacts cause predictable changes in the functional structure of marine communities. The importance of life history traits in mediating responses is illustrated by data from the North Sea. Sustained size-selective fishing during the late 20th century shifted the demersal fish community toward increased aggregate growth rate and decreased average age and length at maturity as smaller, faster maturing species gained in relative abundance, while larger, slower growing species declined (Jennings *et al.* 1999a). A similar pattern was found on fished coral reefs (Jennings *et al.* 1999b).

The decline of apex predators in the oceans has taken on iconic status since Pauly *et al.* (1998) presented evidence for 'fishing down the food web', i.e. a worldwide decline in the mean trophic level of fishery landings, which they suggested resulted from sequential depletion of large-bodied predators. Although this pattern results in part from addition of new fisheries at lower trophic levels ('fishing through the food web'; Essington *et al.* 2006), there is abundant evidence that extinction and depletion of marine animals are consistently biased toward loss of large animals at high trophic levels (Pauly *et al.* 1998; Jackson *et al.* 2001; Dulvy *et al.* 2003; Byrnes *et al.* 2007), resulting in broad-scale declines in both abundance and body size of marine predators (Baum *et al.* 2003; Myers and Worm 2003; Hsieh *et al.* 2006).

But marine ecosystems are also increasingly affected by exotic invasions, which could in principle counteract the loss or depletion of species (Sax and Gaines 2003). Byrnes *et al.* (2007) explored this possibility by synthesizing data from global tallies of marine extinctions (Dulvy *et al.* 2003) and marine invasions in four well-documented areas, and assigning each species a trophic level. They found that the gains and losses of biodiversity did not compensate one another functionally but instead led to consistent directional change in the shape of

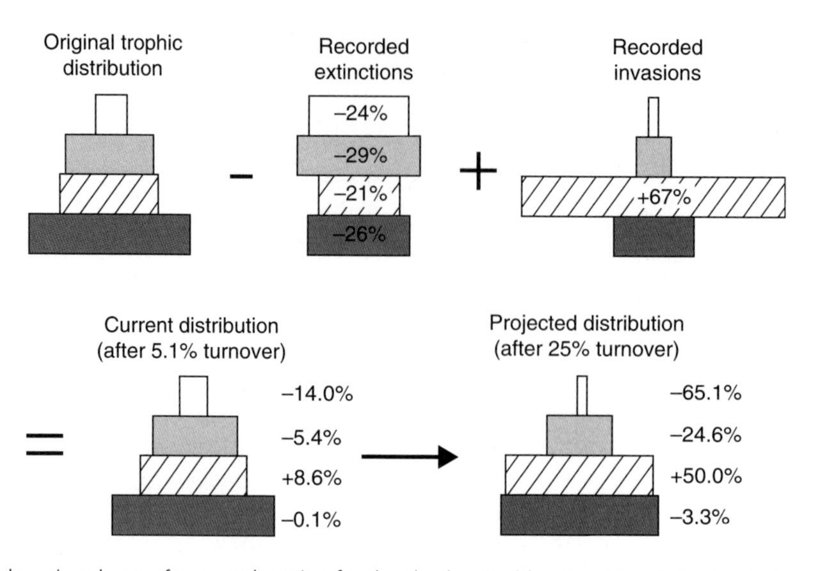

Figure 8.3 The changing shape of a coastal marine food web, the Wadden Sea, The Netherlands, based on Byrnes *et al*. (2007). Bars represent successive trophic levels: primary producers (dark grey); herbivores, deposit feeders, detritivores and zooplankton (hatched); omnivorous consumers (light grey); and carnivores and parasites (open).

food webs (Fig. 8.3). On a global basis, 70% of documented extinctions were of predators, while most invasions were at lower trophic levels, truncating the typical trophic pyramid into a vertically compressed food web dominated by filter-feeders and scavengers. Thus, invasion by exotic species did not counteract the trophic skew caused by predator depletion but in fact exacerbated it.

One important exception to the general pattern of declining species richness with trophic height, with important implications for community dynamics, involves so-called 'wasp-waist' ecosystems, in which one or a few species dominate intermediate trophic levels, such that their particular biological characteristics control trophic transfer from primary production to upper levels of the food web (Cury *et al*. 2000; Hunt and McKinnell 2006). The classic example involves the herbivorous sardines and anchovies that dominate upwelling ecosystems throughout the world. Typically, wasp-waist has been used to refer to pelagic ecosystems. But these pelagic systems also bear some functional similarities to benthic systems in which the herbivore trophic level is dominated functionally by a single species of sea urchin. In both cases, factors influencing abundance of these key intermediate species have a major influence on overall ecosystem structure and functioning, and can shift the system between alternate semi-stable states (Sala *et al*. 1998; Bakun 2006).

In summary, many marine food webs historically had a characteristic 'shape', with diversity and abundance generally declining, and body size increasing, with height in the food chain. Human impacts change the shape of marine food webs predictably, tending to reduce average food chain length and skew communities toward dominance by small-bodied, fast-maturing omnivores, detritivores and suspension-feeders. Trophic skew thus appears characteristic of human-influenced marine systems, and altered top-down control should be a central consequence in the sea.

8.3 Trophic cascades in the sea

8.3.1 Conceptual background

What are the broader consequences of the systematic shortening of marine food chains? Hairston, Smith and Slobodkin (HSS; Hairston *et al*. 1960) initiated one of the longest running controversies in ecology with their assertion that trophic-level abundance should be controlled alternately by bottom-up and top-down processes as one descends through successive levels of the food chain. The chain of indirect effects emanating from top predators has since been called a trophic cascade (Paine 1980; Carpenter *et al.* 1985). Despite the obvious greater complexity of real communities and the several factors that would appear to work against trophic cascades (Strong 1992; Polis 1999), many controlled experiments in marine and other systems have supported the general HSS hypothesis, showing that loss of predators indeed often releases prey from top-down control, leading in turn to strong reductions in the prey's resources (Menge 1995; Shurin *et al.* 2002). Similar trophic cascades have been documented experimentally in a wide variety of ecosystems (Pace *et al.* 1999; Borer *et al.* 2005). Moreover, controlled experiments show that changes in abundance of apex predators often have cascading impacts throughout ecosystems (Pace *et al.* 1999; Shurin *et al.* 2002), and that top-down influences penetrate farther, on average, through food chains than do bottom-up influences of nutrient loading (Borer *et al.* 2006). Given the characteristic shortening of marine food chains under human influence, these generalizations imply that a consistent consequence of human impacts will involve cascading indirect effects of reduced predation pressure.

8.3.2 Evidence for trophic cascades in open marine systems

Marine ecosystems are open, with propagules and apex predators moving over large distances, and are subject to climate forcing and other influences that could attenuate trophic cascades (Jennings and Kaiser 1998). Perhaps surprisingly, the expectations from simple theory of alternating bottom-up and top-down control at adjacent trophic levels are nevertheless supported by accumulating evidence that fishing can drive trophic cascades. This evidence includes time-series data from kelp beds (Estes *et al.* 1998; Davenport and Anderson 2007), coral reefs (Dulvy *et al.* 2004), open ocean plankton (Shiomoto *et al.* 1997), the demersal communities (Worm and Myers 2003) and pelagic communities (Frank *et al.* 2005, 2007) of continental shelves, and

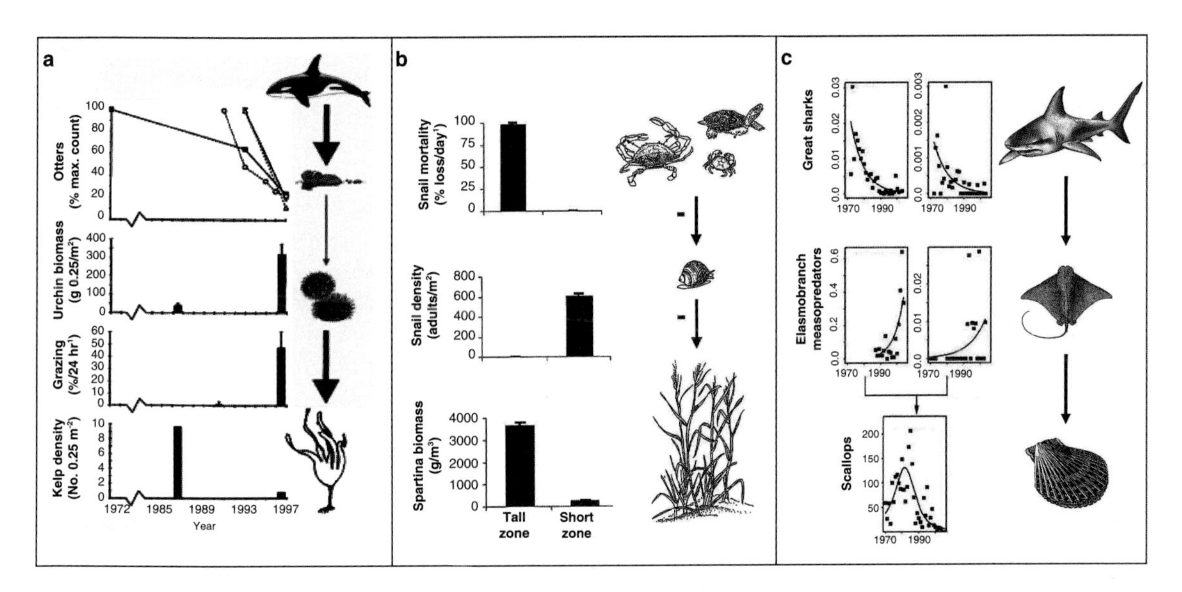

Figure 8.4 Evidence for trophic cascades in disparate coastal marine ecosystems. (a) Killer whales to sea otters to sea urchins to kelps in northeast Pacific kelp beds. Reproduced with permission from Estes *et al.* (1998). (b) Shell-crushing crabs and terrapins to snails to cordgrass in west Atlantic salt marshes. Reproduced with permission from Silliman and Bertness (2002). (c) Great sharks to cownose rays to bay scallops in west Atlantic estuaries. Reproduced with permission from Myers *et al.* (2007).

temperate estuaries (Deason and Smayda 1982; Myers *et al.* 2007). These examples involve a wide range of organisms and environments (Fig. 8.4), and suggest that systematic depletion of predators from the oceans is producing far-reaching indirect impacts on the structure and functioning of many marine ecosystems. I divide these examples into rocky bottoms, continental shelves and pelagic systems.

8.3.2.1 Rocky bottoms

The most famous and dramatic marine trophic cascade involves the four-level food chain from predatory orcas (killer whales) to sea otters to grazing sea urchins to dominant kelps in the northeast Pacific Ocean (Fig. 8.4a). After fur traders exterminated sea otters on several islands in the 18th century, their sea urchin prey exploded and in turn eliminated kelp forests, whereas kelp beds remained vigorous on islands too remote for otter harvesting (Estes and Palmisano 1974; Estes and Duggins 1995). Loss of kelps on the otter-free islands in turn led to pervasive ecosystem-level changes, including local extinction of several marine species associated with the kelp habitat, possibly reduced abundances of coastal raptors that depend on fishes, and increased coastal storm damage stemming from the loss of buffering by kelp forests (Mork 1996). In recent decades, killer whales (orcas) began attacking sea otters, evidently as the pelagic food chains supporting them withered, and the effects of killer whale predation cascaded down through sea otters and sea urchins to reduce kelp again, demonstrating a four-level trophic cascade (Estes *et al.* 1998). Similarly strong cascades involving vertebrate predators, sea urchins and macroalgae have since been documented on rocky bottoms throughout the world. In kelp beds of the western North Atlantic, archaeological data and time series suggest that overfishing of cod and other groundfish released grazing sea urchins from predatory control, and cascaded down to decimate kelps (Steneck *et al.* 2004). In the warmer Mediterranean Sea, experiments and comparisons of marine protected areas with nearby fished areas also showed that harvesting of predatory fishes allowed urchins to proliferate and overgraze macroalgae, converting large areas to 'barrens' of structure-free coralline algal pavements (Sala *et al.* 1998; Guidetti 2006). Finally, on rocky reefs of Tasmania, predation by spiny lobsters and fishes on urchins cascades to macroalgae (Shears and Babcock 2003; Pederson and Johnson 2006). In all these systems, loss of top predators shifts a structurally complex, diverse community dominated by macroalgae to a depauperate 'urchin barren' dominated by crustose coralline algae and maintained by intense grazing.

But cascades are not limited to urchin-dominated communities. An intriguingly similar example involves small herbivorous crustaceans as the intermediate link. There have been several reports of perennial seaweeds such as rockweeds (*Fucus*) and giant kelp (*Macrocystis*) being decimated by anomalous outbreaks of grazing amphipod and isopod crustaceans (e.g. Kangas *et al.* 1982; Haahtela 1984; Tegner and Dayon 1987). For example, after an El Niño warm-water event destroyed kelp communities in California during the early 1980s, recovering kelp beds were left without their normal assemblage of fishes, and populations of the kelp-curler amphipod (*Peramphithoe humeralis*) exploded, devastating the kelps again, probably as a result of relaxed predator control (Tegner and Dayon 1987). Mesocosm experiments suggested that particular species of grazing amphipods might mediate these shifts in dominance by large brown algae (Duffy and Hay 2000). Recent field experiments in California support this hypothesis (Davenport and Anderson 2007), showing that invertebrate-feeding fishes reduced amphipod abundances and grazing impact, and that these effects in turn cascaded to increase kelp blade growth by 100–300%, with a trend toward also reducing kelp mortality. A survey of unmanipulated kelp reefs similarly showed that amphipod abundance was negatively related to that of fishes. Intriguingly, many of these communities resemble the 'wasp-waist' architecture found in some pelagic communities, in which the intermediate trophic level is dominated by one or a few strongly interacting species. This hints that low diversity is an important mediator of these strong trophic dynamics, as suggested by verbal theory (Strong 1992; Duffy 2002), a point to which I return below.

8.3.2.2 Continental shelves

Increasingly, trophic cascades are being detected even in large open coastal and oceanic ecosystems. Time-series data from offshore fisheries reveal, for example, that the collapse of Atlantic cod (*Gadus morhua*) stocks in recent decades as a result of fishing have been accompanied by increases in both their benthic crustacean prey (Worm and Myers 2003) and in overlying pelagic communities (Frank *et al.* 2005). Worm and Myers (2003) analysed time series from nine cod stocks throughout the North Atlantic and searched for correlations of boreal shrimp and cod abundance with one another and with water temperature. They found that cod and shrimp abundance showed opposite trends through time at most sites, consistent with a trophic cascade from humans through cod to benthic shrimp. They were also able to reject an alternative bottom-up hypothesis that these trends stemmed from changing climate: although water temperature affected cod, it had no detectable influence in shrimp abundance.

Clearly such marine trophic cascades are of more than academic interest. Cod, shrimp, and lobster all support major commercial fisheries, for example. Another sobering case involves the decline of sharks, which have been especially hard hit by fishing (often as by-catch) throughout the world oceans (Stevens *et al.* 2000; Baum *et al.* 2003). Along the eastern seaboard of North America, time-series data reveal that precipitous declines of large sharks since 1970 were accompanied by increases of many 'mesopredator' rays and small sharks, most of which are eaten almost exclusively by large sharks. These patterns suggest that the mesopredators have been released from predation by larger enemies (Myers *et al.* 2007), as has been suggested for the increase of small, mammalian predators on land as well (Crooks and Soule 1999; Johnson *et al.* 2007). Notable among these marine mesopredators is the cownose ray (*Rhinoptera brasiliensis*), which feeds largely on bivalve molluscs, and which increased by an order of magnitude over the last three decades. Both survey data and experiments show that the growing population of cownose rays has inflicted heavy mortality on bay scallops in North Carolina (Fig. 8.4c), resulting in a collapse of the century-old fishery for this species. Reports from Chesapeake Bay similarly suggest that cownose rays are now causing severe damage to seagrass beds as they forage for the infaunal bivalves living there, some of which are commercially important. Similar patterns of observed increases in mesopredatory rays and decreases in bivalve populations have also been observed in the northeast Atlantic and in Japanese waters (Myers *et al.* 2007). One interesting question arising from such dramatic changes in community structure is whether and how predators are sustained after their prey are driven to such low levels. Evidence consistent with a trophic cascade has also been found in salt marshes, where areas inaccessible to marine predators, notably blue crabs, have much higher densities of rasping snails and marsh grasses achieve accordingly lower biomasses; there is concern that declining populations of heavily fished blue crabs may result in deterioration of salt marshes as a result of this cascade (Silliman and Bertness 2002).

8.3.2.3 Pelagic systems

Trophic cascades were first documented in freshwater pelagic (Carpenter *et al.* 1985) and benthic (Power 1990) systems, and are common in ponds and lakes (Carpenter and Kitchell 1993; Brett and Goldman 1996). While they appear less common in the more functionally diverse marine pelagic communities (Micheli *et al.* 1999; Shurin *et al.* 2002), there is growing evidence for trophic cascades in both estuaries and the open ocean. In estuaries, ctenophores (comb jellies) are voracious predators of other zooplankton, and often reach very high densities in summer, where in some systems they can crop 20% of standing crustacean zooplankton stock per day. A 6 year field study showed generally synchronous but opposite fluctuations of predatory ctenophores, their copepod prey and phytoplankton, particularly the dominant diatom species (Deason and Smayda 1982). Integrating abundances of each group by month or season revealed clear evidence of a trophic cascade as ctenophore 'blooms' were followed by decimation of herbivorous zooplankton and subsequent phytoplankton blooms (Deason and Smayda 1982). Field observations suggest that such trophic cascades also occur in a variety of other marine pelagic systems, particularly those dominated by gelatinous zooplankton (Verity and Smetacek 1996). The striking example of the Black Sea illustrates both the complexity of processes driving cascading ecosystem shifts and the potentially high stakes for human society. Analysis of four decades of time-series data revealed patterns consistent with a cascade through five levels, from predatory fishes, to planktivorous fishes, to zooplankton, to phytoplankton, to water-column nutrient stocks (Daskalov 2002; Daskalov *et al.* 2007). A major shift from a clear-water phase supporting abundant large fish to a turbid phase began in the early 1970s after industrial fishing depleted apex predators (Daskalov 2002), a situation strongly reminiscent of phase shifts in north temperate lakes (Scheffer and Jeppesen 2007). Because these changes also coincided with eutrophication and the invasion of an exotic predatory ctenophore, mass balance models were developed to evaluate the relative importance of bottom-up and top-down mediation of these changes. The model simulations produced cascading changes in biomass of lower trophic levels quite similar to the observed pattern when predators were removed, whereas simulated eutrophication produced biomass increases across all levels, in contrast to observed patterns (Daskalov 2002). In this system, then, it appears that restoration of predatory fishes should be at least as effective in restoring water quality as reducing nutrient loading. Negative correlations across trophic levels, from predatory fishes through zooplankton to phytoplankton and even down to water-column nutrient stocks, in time series from the North Pacific (Shiomoto *et al.* 1997) and coastal north Atlantic Oceans (Frank *et al.* 2005), suggest that cascading trophic interactions can occur even in open pelagic ecosystems.

The question remains whether these patterns are general. Micheli (1999) conducted a meta-analysis of marine pelagic systems, using data from both mesocosm experiments and time series from unmanipulated systems to ask whether nutrient loading and predation penetrated through the food chain. Although both zooplanktivorous fishes and nutrient loading significantly affect the adjacent trophic level (zooplankton and phytoplankton respectively), these effects attenuated rapidly through the food chain. Trophic cascades thus seemed to be the exception rather than the rule in marine pelagic ecosystems. But the story may be more complex. Detailed analysis of experiments found that removal of predators in the marine pelagic frequently did cascade to affect phytoplankton biomass, but that the sign of predator influence on phytoplankton depends on food chain length, which in turn depends on cell size and thus taxonomic composition of the dominant algae (Stibor *et al.* 2004). When data from three- and four-link experimental food chains were averaged, the strong influence of predators on phytoplankton was masked. It remains uncertain, however, whether these pelagic cascades are also common in unmanipulated, open marine systems where food chains with three (classical) and four (microbial loop) links operate in parallel.

8.4 Biodiversity and stability of marine ecosystems

8.4.1 Conceptual background

Early ecologists, from Darwin (Hector and Hooper 2002) to MacArthur (1955) and Elton (1958), believed that diverse communities were more stable and better able to resist disturbances than depauperate ones. These ideas have received renewed attention as concern about declining biodiversity has grown (McCann 2000). There are several mechanisms by which biodiversity might increase stability of community- or ecosystem-level properties. The most general is simple statistical averaging (also called the 'portfolio effect'): as long as temporal fluctuations of co-occurring species are not perfectly correlated, variance of their aggregate abundance in response to stochastic environmental variance will be lower than the average variability of component species (Tilman *et al.* 1997; Doak *et al.* 1998). Biodiversity may also stabilize ecosystem properties against perturbations by enhancing the system's ability to absorb a stress without changing (resistance) or the rapidity with which it returns to its original state after perturbation (resilience), through at least two biological mechanisms. First, niche differentiation (functional diversity) among species increases the probability that at least some species will thrive as environmental conditions change. Second, and in contrast, functional redundancy can enhance stability against perturbations that cause extinction by reducing the probability that extinction removes a functionally unique species. The last two phenomena are sometimes combined as the insurance hypothesis (Yachi and Loreau 1999).

In the context of applied marine ecology, a question of particular interest is how diversity influences resistance to human predation, i.e. fishing. Are more diverse marine food webs more stable in response to such anthropogenic perturbations? The theory most relevant to the case of fishing effects in marine systems involves the role of prey richness in buffering the community from predator impacts. Leibold (1989, 1996) argued that a resource base with more species is more likely to contain at least one species that is resistant to consumption and can dominate in the presence of a consumer, such that a more diverse prey community will maintain higher aggregate biomass under predation. Similarly, if different prey species are resistant to different predators, or gear types in the case of fishing, then a diverse prey community will maintain higher biomass under fishing pressure. This general argument has been extended to suggest that trophic cascades also should be most prevalent in low-diversity systems, with one or a few important species at each trophic level (Strong 1992; Duffy 2002).

8.4.2 Evidence linking diversity and stability in marine systems

Growing empirical evidence suggests that changing horizontal (i.e. within-trophic level) diversity can have several important consequences for marine food web interactions and ecosystem processes (Emmerson and Huxham 2002; Duffy and Stachowicz 2006; Stachowicz *et al.* 2007). Experimental research has supported a stabilizing effect of diversity on community biomass in some competitive plant assemblages, aquatic microbial food webs and soil microfaunal communities (reviewed by McCann 2000; Cottingham *et al.* 2001; Loreau *et al.* 2002). There is also some experimental evidence that marine species diversity can enhance trophic level resistance to top-down control (Hillebrand and Cardinale 2004; Duffy *et al.* 2005) and to invasion by other native (France and Duffy 2006) and non-native species (Stachowicz *et al.* 1999, 2002a).

8.4.2.1 Comparisons through time

Observational evidence from fishery science corroborates predictions (Tilman *et al.* 1997; Doak *et al.* 1998) that diversity can enhance stability of aggregate biomass or production (i.e. overall catch), both in the general sense of reducing long-term fluctuations and in the specific sense of providing resistance to perturbations of fishing and environmental forcing. For example, time series of fish biomass from the North Sea show that aggregate biomass is less variable than that of individual fish species (Fig. 8.5; Jennings and Kaiser 1998), supporting the suggestion that diversity enhances general stability. Diversity can also provide resistance to specific perturbations. Among the most intriguing cases is the link between stock diversity and productivity of Alaskan salmon under decadal-scale climate variation (Hilborn *et al.* 2003b). Because salmon return to the streams or lakes of their birth to spawn, populations are genetically highly structured into distinct genetic populations or 'runs' that differ substantially in life history, phenology and ecology. Hilborn *et al.* (2003b) used historical catch records for Bristol Bay sockeye salmon (*Onchorhyncus nerka*) dating back to the 1890s to show that the relative contributions of different populations to total salmon catch differed greatly through time as individual populations responded differently to long-term variation in climate forcing by the El Niño Southern Oscillation (ENSO) and the Pacific Decadal Oscillation (PDO). The population-specific variability in response to changing environmental conditions resulted in an aggregate salmon catch that was more stable through time than was that of any individual population (Hilborn *et al.* 2003b). This link between stock diversity and stability in response to human predation is consistent with the proposed importance of niche differentiation as a mechanism by which biodiversity can stabilize biomass and production (Loreau *et al.* 2002).

Similarly, there is considerable evidence that functional diversity of herbivores is important to maintaining coral dominance over algae on tropical reefs. In the Caribbean, overharvesting of herbivorous

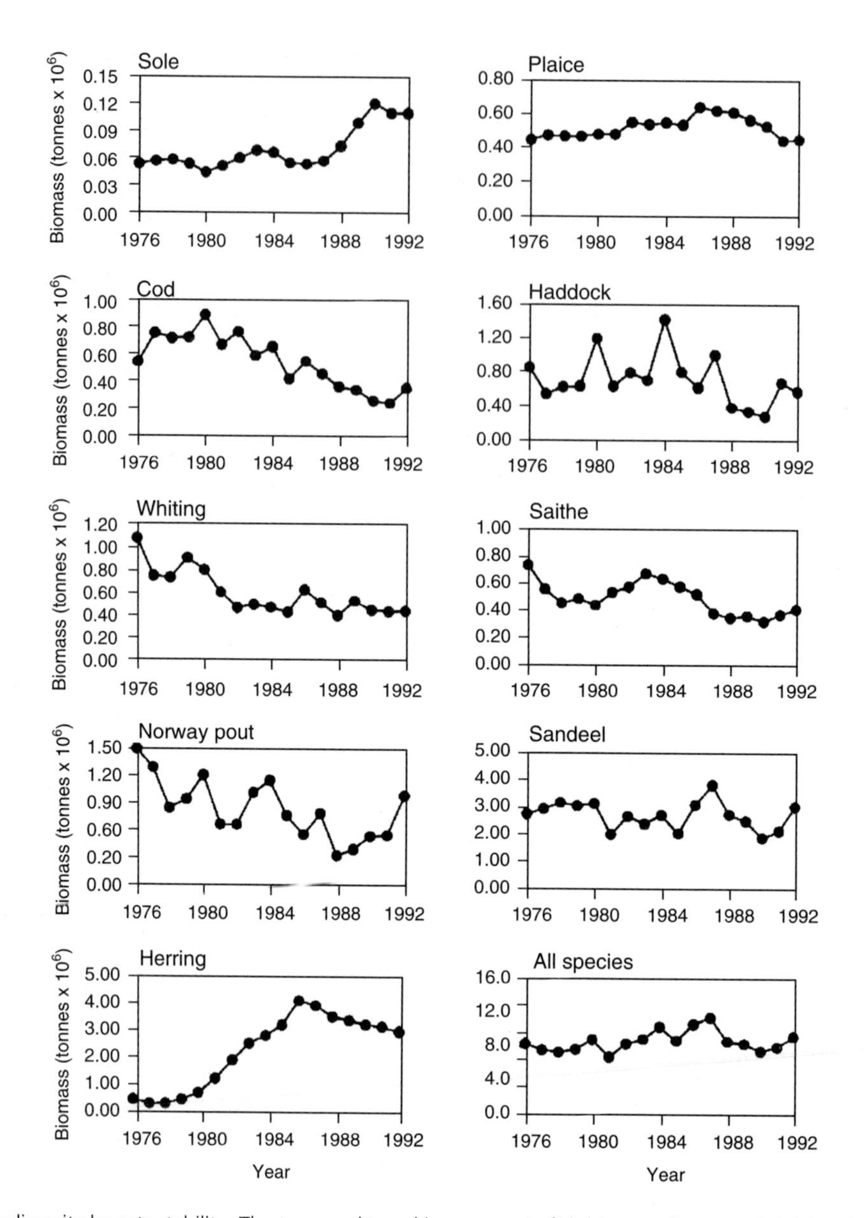

Figure 8.5 Biodiversity begets stability. The temporal trend in aggregate fish biomass (bottom right) is substantially less variable than the time series for any individual species. Reproduced with permission from Jennings and Kaiser (1998).

fishes had little effect on algal biomass initially, perhaps because sea urchin grazing compensated for the reduced fish grazing (Hay 1984; Hughes 1994; Jackson *et al.* 2001). But when sea urchins suffered mass mortality from disease, algal biomass exploded (Carpenter 1990). Macroalgal blooms proliferated only after both fishes and sea urchins were reduced. These observed patterns are consistent with experiments and comparisons among protected and fished areas on Kenyan reefs, which showed that sea urchins were more abundant on overfished reefs than in protected areas, but that experimental reduction of urchins allowed large macroalgae to dominate only on fished reefs, where herbivorous fishes were scarce (McClanahan *et al.* 1996; see also Hay and Taylor 1985). That is, the presence of two effective herbivore groups (fishes and urchins) buffers the system against algal overgrowth if one group is lost.

While these examples involve community (or, more accurately, guild) stability at the expense of population fluctuations, experiments show that interactions within diverse natural assemblages may also foster stability at the population level by imposing density dependence in demographic rates (Hixon and Carr 1997; Carr *et al.* 2002). Field experiments showed that per capita mortality rates of recruiting coral-reef fishes in the Bahamas were strongly density dependent in the presence of multiple predators (Hixon and Carr 1997) or predators and competitors together (Carr *et al.* 2002), whereas mortality was independent of density when competitors alone, or only one type of predator, was present. These experiments provide an intriguing contrast with the pattern found in a meta-analysis of predator–prey experiments, in which the presence of predators tended to destabilize temporal dynamics of their prey (Halpern *et al.* 2005). It would be quite interesting to know whether the destabilizing effects of predators in the meta-analysis might be an artefact of experiments including only one or a few species of predators.

8.4.2.2 Comparisons across space

Several prominent spatial patterns are consistent with the hypothesis that diversity enhances stability in marine systems. Across geographic regions, the least stable ecosystems are those low in fish diversity, with a few strong predator–prey links and little capacity for prey switching (Jennings and Kaiser 1998). For example, the dramatic cascades emanating from sea otters to kelp in the relatively low-diversity community of Alaska were not observed in the more diverse kelp beds of southern California (Dayton *et al.* 1998), even though both regions lost sea otters by the early 19th century. A possible explanation is that, in southern California, spiny lobsters and sheephead, which also feed on urchins, compensated for the reduced impacts of sea otter predation; indeed, when lobsters and sheephead were heavily exploited in the 1950s, kelps did decline in southern California (Dayton *et al.* 1998). Evidence consistent with such an explanation again comes from a coral reef in Kenya, where experimental reductions of sea urchins allowed algae to proliferate to about twice the level on fished reefs as on unfished reefs (McClanahan *et al.* 1996).

A stabilizing role of biodiversity in exploited marine communities is also suggested by regional comparisons of trophic dynamics in northwest Atlantic continental shelf ecosystems (Frank *et al.* 2006). These authors analysed time-series data for several trophic levels, from phytoplankton to harvested fishes, at nine heavily fished sites to assess the strength and direction of trophic control. Correlations between adjacent trophic levels varied among sites, being predominantly negative at higher latitudes, indicating top-down control, but positive at lower latitudes, indicating bottom-up control. Interestingly, the strength and sign of trophic control varied systematically with species richness, with stronger top-down control in the more depauperate northern sites and bottom-up control at the more diverse southern sites. This weakening of top-down control with prey species richness is consistent with results of experiments (Steiner 2001; Hillebrand and Cardinale 2004; Duffy *et al.* 2005). However, the link to diversity in the northwest Atlantic data is confounded by strong correlations of species richness with latitude and temperature, which are also expected to influence the strength of top-down control through effects on demographic rates (Frank *et al.* 2006). Disentangling these influences is not possible at present.

Finally, Worm *et al.* (2006) conducted a comprehensive analysis of the links between marine biodiversity and response to fishing (Fig. 8.6). They analysed relationships between species richness and fishery production for the world's 64 Large Marine Ecosystems (www.fishbase.org). Regions with naturally low fish diversity supported lower average fishery productivity, and had more frequent 'collapses' (strong reductions in fishery yield) and lower resilience (degree of recovery after overfishing) than naturally species-rich systems. Worm *et al.* (2006) suggested that the greater resilience of more diverse ecosystems may be explained by the greater ability of fishers to switch among target species in diverse ecosystems; when abundance of a species declines to a low level, it is

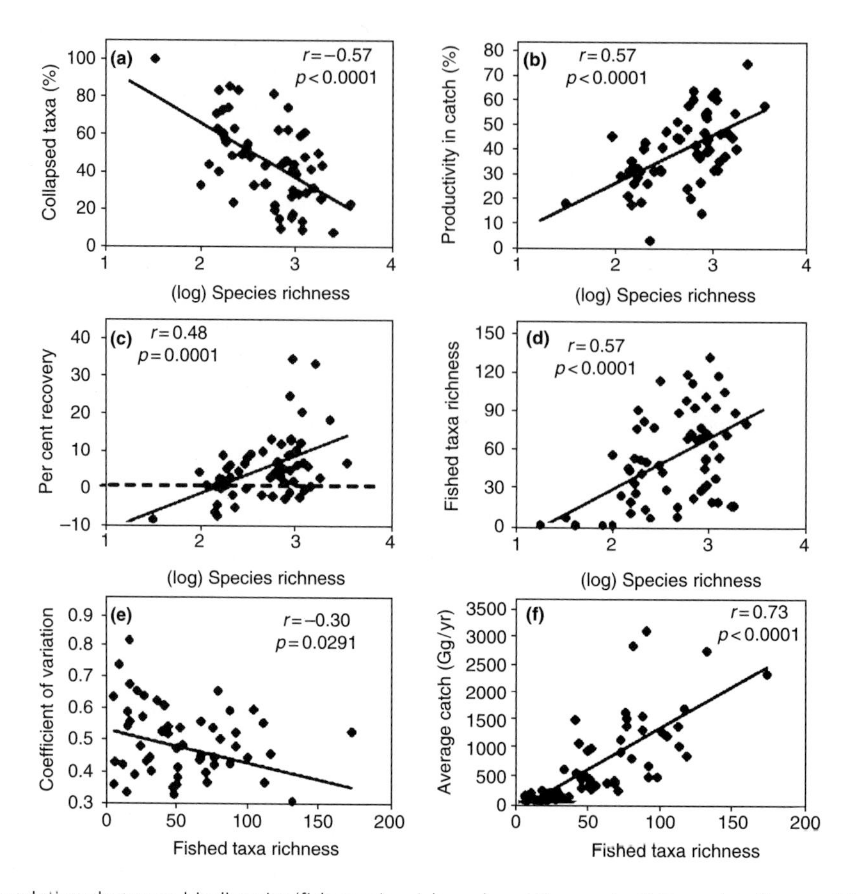

Figure 8.6 Correlations between biodiversity (fish species richness) and the productivity and resilience of fishery catches across 64 large marine ecosystems. Reproduced with permission from Worm *et al.* (2006).

more profitable for fishers to target other species, which provides the overfished species with a refuge and allows them to recover. This explanation is consistent with the negative relationship between fished taxa richness and interannual variation in catch (Worm *et al.* 2006, see also Fig. 8.5).

8.4.2.3 Mechanisms

What is the mechanism for these putative effects of biodiversity on ecosystem resilience? One likely candidate is functional compensation (or functional 'redundancy') among species within a guild or functional group, such that decline of one species is compensated by increase in another species with similar functional characteristics, e.g. through relaxed competition. Evidence potentially consistent with this mechanism comes from the tropical Atlantic, where longline fisheries for billfish show a pattern of sequential depletion of species (Myers and Worm 2003), with decline of blue marlin in the 1960s accompanied by a rise in catch of sailfish, which then declined in turn as swordfish catches increased through the late 1970s and 1980s. The result was that total billfish catch remained relatively stable through time despite boom and bust patterns in the catch of individual species. Similar patterns have been observed in demersal ecosystems, using both catch data (Myers and Worm 2003) and fisheries-independent data (Shackell and Frank 2007), in which declines of targeted fish species were accompanied by compensatory increases in other groups. These patterns of species turnover are very similar to that predicted by theory when a consumer imposes mortality on prey species differing in competitive ability and edibility (Leibold 1996).

8.5 Interaction strengths and dynamic stability in marine food webs

8.5.1 Conceptual background

Predicting the cascading indirect effects of predator depletion in food webs depends both on the topological structure of webs, which the previous section showed are changing systematically, and on the distribution of interaction strengths. One approach to the challenge of scaling up ecological processes to real ecosystems uses empirical data on topology and interaction strengths to explore dynamic responses to perturbations through simulation and modelling (de Ruiter *et al.* 2005). A principal challenge is the paucity of data on interaction strengths to parameterize models realistically (Wootton and Emmerson 2005). Most existing food webs are 'connectance webs' or 'energy webs', based on qualitative trophic links between species or on patterns of energy flow, respectively. Paine (1980; see also Raffaelli and Hall 1996) showed that links in the food web that are important energetically are often not the same links that are functionally (dynamically) important, i.e. that have strong impacts on the structure and organization of the community and, by inference, on ecosystem processes. It is the latter, functional food web structure that is of primary interest in understanding a system's stability and response to perturbations (Berlow *et al.* 2004; Wootton and Emmerson 2005).

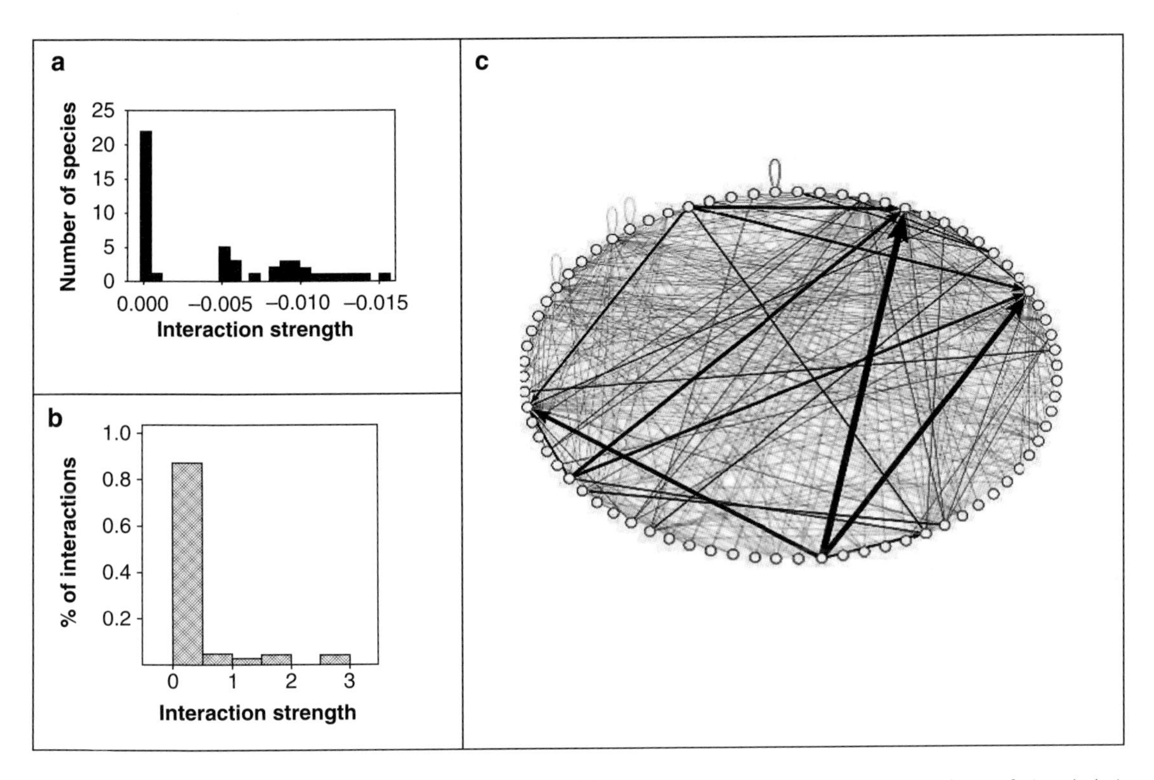

Figure 8.7 Skewed distributions of interaction strengths in (a) 45 herbivore species feeding on sporelings of giant kelp in California (reproduced with permission from Sala and Graham 2002), (b) three bird species feeding on 23 species of intertidal animals in Washington, USA (reproduced with permission from Wootton 1997) and (c) fishes feeding on diverse prey in a large Caribbean coral reef food web (reproduced with permission from Bascompte *et al.* 2005). Note that, in all food webs, only a small percentage of interactions have strong effects (thickness of arrows is proportional to interaction strength in (c)).

The small but growing number of communities in which interaction strength has been estimated suggests that strong skew in interaction strength may be a general feature of communities, with many links between species having negligible effects on the dynamics of either party, and a few links having very strong impacts (reviewed by Wootton and Emmerson 2005, Fig. 8.7). This has important implications for community stability, as skew in interaction strength has been shown to confer stability on trophic networks in theory (McCann *et al.* 1998; Emmerson and Yearsley 2004) and in simulation studies of empirical food webs (Emmerson and Raffaelli 2004).

8.5.2 Empirical evidence

Simulation-based approaches to exploring food web stability have generally measured stability as resistance to small perturbations imposed on species randomly. A critical question for applied ecology is how interaction strength of a species covaries with its vulnerability to real perturbations such as fishing, and how those perturbations ripple through the web. Soberingly, analysis of a diverse Caribbean reef food web found that, while combinations of strong interactions capable of generating trophic cascades were quite rare, human impacts fell disproportionately on species involved in such interaction combinations, primarily because large- bodied species are both strong interactors and disproportionately targets of human impacts (Bascompte *et al.* 2005). Thus, in contrast to the stabilizing effects of many weak interactions predicted by theory in diverse communities such as reefs, the targeted harvesting of large predators can have important cascading impacts because fishing imposes strong, persistent and non-random perturbations.

There is reason to expect that the results of Bascompte *et al.* (2005) may be common in that strong interactors are typically the larger species in a community (although see Sala and Graham 2002 for an exception), and at least for direct harvesting such as fishing, large species are also those targeted preferentially. Thus, it is likely that there will often be positive covariance between a species' interaction strength and its vulnerability to fishing. Although these conclusions are preliminary, this suggests that the preponderance of weak interactions in many food webs is unlikely to protect them from the specific sorts of impacts imposed by fishing.

8.6 Alternate stable states and regime shifts in marine ecosystems

8.6.1 Conceptual background

An ecological phenomenon of growing concern in understanding marine ecosystem dynamics and their implications for society is the phenomenon of regime shifts. Regime shifts can be defined as relatively rapid transitions between distinct and relatively long-lasting, semi-stable states of a system (Knowlton 2004; Steele 2004). The potential existence of alternate stable states has been a subject of keen interest and controversy in ecology for decades (Lewontin 1969; Sutherland 1974). In recent years, regime shifts have gained new prominence in the context of conservation and management (Scheffer *et al.* 2001). The term has been used in two ways. The first definition is phenomenological, and refers to multi-year periods of relative stability in time series of observational data that are separated by abrupt shifts to intervals that fluctuate around a different mean. The second, stricter definition involves the dynamics of the system and refers to the existence of two or more semi-stable states, or alternate attractors, in an ecosystem. Regime shifts in the phenomenological sense have been described in a number of marine ecosystems, and appear often to track decadal-scale climate variation (e.g. Overland *et al.* 2006).

Recent syntheses (Collie *et al.* 2004) have defined three types of regime shifts, which are actually points along a continuum (Fig. 8.8a): (1) a smooth regime shift, defined by a quasi-linear relationship between a forcing variable and a response variable, (2) an abrupt regime shift, in which the relationship is non-linear and (3) a discontinuous regime shift, in which the relationship is not only non-linear but the trajectory of the response variable differs when the forcing variable is declining compared with when it is increasing. The latter situation results in two possible states of the response variable at a given value of the forcing variable and is also

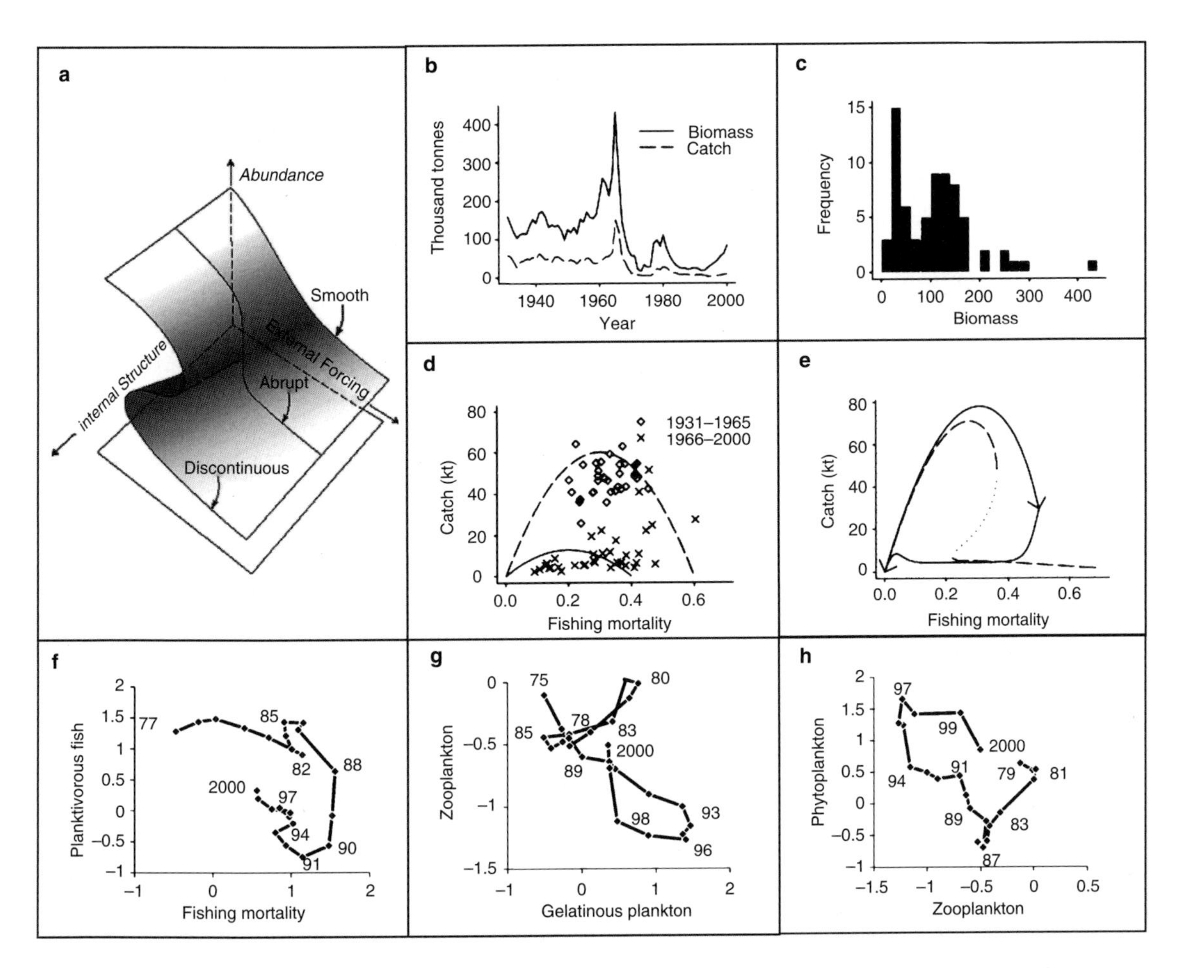

Figure 8.8 (a) A conceptual model of the three forms of regime shift, and empirical evidence consistent with discontinuous regime shifts in (b–e) Georges Bank haddock and (f–h) the Black Sea pelagic ecosystem. In Georges Bank haddock, the regime shift is illustrated by (b) a discrete shift from higher to lower abundance after ~1965, (c) a bimodal distribution of biomass across the time series, (d) a different functional relationship between fishing mortality and catch in the two time periods and (e) hysteresis in this functional relationship in simulations of a model fit to empirical data. Reproduced with permission from Collie *et al.* (2004). In the Black Sea, the regime shift is similarly supported by (f–h) changing functional relationships through time between predators and prey. Reproduced with permission from Daskalov *et al.* (2007).

referred to as hysteresis (Fig. 8.8a). Scheffer and Carpenter (2003) described a series of criteria for identifying regime shifts, none of which alone diagnoses a regime shift but which, together, constitute strong evidence. In field data these include abrupt shifts in time series, a bi- or multimodal frequency distribution of states in a time series, and dual (or multiple) relationships between ecosystem state and a forcing variable. Experimental evidence includes dependence of final state on initial state (e.g. order of colonization during succession), shift towards a distinctly different stable state after a pulse perturbation and hysteresis, i.e. change of the ecosystem along different pathways when the forcing variable is increased compared with when it is decreased. Two important questions for applied ecology are whether rapid shifts between relatively long-lasting states of an ecosystem can be forced by gradual changes in conditions, and whether these shifts are reversible.

8.6.2 Empirical evidence for regime shifts in marine ecosystems

At the population level, Collie *et al.*(2004) applied Scheffer and Carpenter's (2003) criteria to test for regime shifts in Georges Bank haddock, based on age-structured abundance time series for the period 1931–2000. Although some of these criteria are difficult or impossible to evaluate in open marine ecosystems (e.g. whether the system goes to different states after a perturbation or under different starting conditions), others can be addressed with available time-series data (Fig. 8.8b–e). First (Fig. 8.8b), they demonstrated a discrete step in the average stock biomass, which dropped abruptly after a spike in catch due to influx of foreign fishing fleets in the early 1960s, and remained low for most of the remaining century. Second (Fig. 8.8c), the distribution of biomass values was bimodal. Third (Fig. 8.8d), catch showed a different functional relationship to fishing mortality before and after the shift in the 1960s. These three observations support the existence of two distinct regimes during the time series. Moreover, simulations of a population model fit to empirical data showed that catch followed different trajectories when fishing mortality was increased versus decreased, i.e. the system exhibited hysteresis (Fig. 8.8e), a key piece of evidence for a discontinuous regime shift between alternate semi-stable states (Collie *et al.* 2004).

An ecosystem-wide regime shift, evidently forced in part by overfishing, has also been documented in the Black Sea (Daskalov *et al.* 2007). Here strong fishing pressure on top predators caused their decline and eventual collapse in the 1970s, which was accompanied by cascading changes in lower trophic levels, leading to phytoplankton blooms and nutrient depletion; a subsequent change in the focus of fisheries to smaller planktivorous fishes such as sprat and anchovy ('fishing down the food web'; Pauly *et al.* 1998) then led to a subsequent collapse of these planktivores and corresponding increases in the jellyfish that compete with them (Daskalov 2002). Plotting the time trajectories of the various trophic groups shows that, for several interactions, the relationships between consumer and prey abundances differed in early compared with later years, suggestive of the hysteresis characteristic of a discontinuous regime shift (Fig. 8.8f–h). Models confirm that overexploitation can trigger such shifts between alternate states (Daskalov 2002; Collie *et al.* 2004).

Are such regime shifts common in marine ecosystems and, if so, how do they relate to forcing mechanisms? Feng *et al.* (2006) explored this question using dynamic simulation of mass-balance models (Ecopath with Ecosim; Christensen and Walters 2004) applied to 24 marine ecosystems. The models imposed a simulated perturbation of 10 years of intensified fishing, then relaxed fishing to the initial level and followed the system's long-term (70 years) trajectory, asking whether each system returned to its original state or equilibrated to an 'alternative attractor'. Six scenarios considered fishing on top predators compared with intermediate levels (wasp-waist system) under bottom-up, mixed or top-down control. The simulations showed that, under top-down or mixed control, 11–28% of the ecosystems showed alternate attractors, i.e. shifted into a new regime that persisted after fishing pressure was relaxed, whereas none of the ecosystems showed alternate attractors under bottom-up control (Feng *et al.* 2006). These model results, together with a few well-documented empirical examples such as the Black Sea, suggest that intense fishing pressure can produce shifts to new ecosystem states that are difficult to reverse, supporting suspicions that such regime shifts may be at least partly responsible for the failure of many heavily fished species to rebound even decades after fishing moratoria were enacted (Hutchings and Reynolds 2004). If such regime shifts are indeed common responses to top-down perturbations, they have serious implications for management of natural ecosystems.

8.6.2.1 Mechanisms

Several mechanisms potentially can produce regime shifts between alternate attractors or semi-stable states (Collie *et al.* 2004; Folke *et al.* 2004; Knowlton 2004). At the population level, the most general mechanism involves the Allee effect, i.e. depensation or positive density dependence at low population sizes (Knowlton 1992), although this process requires some additional factor to prevent extinction at low population density. Alternative stable states of a population can also result from size dependence of vital rates such as recruitment, growth and fecundity (Botsford 1981). For simple predator–prey systems, the range in types of regime shifts can be generated from the same model simply by changing parameters, particularly the ratio between prey carrying capacity and predator half-saturation constant (K/D), the relationships between maximal predation rate and prey growth rate, and the minimum timescale for shifts (Collie *et al.* 2004). Alternate attractors may also arise from idiosyncrasies of species behaviour. For example, pelagic ecosystems often show rapid shifts between decadal-scale states dominated by different planktivorous fish species, such as sardine and anchovy. At high population density such fishes are usually found in 'pure', monospecific schools, but, when reduced to low density by fishing or other processes, their strong schooling inclination causes them to join schools of other species, which may place them into conditions that are poor for feeding and reproduction, driving their population further toward decline (Cury *et al.* 2000).

Regime shifts in predator–prey interactions may also be mediated by the ontogenetic shifts in trophic level (Fig. 8.2) common in both benthic (Barkai and McQuaid 1988) and pelagic (Swain and Sinclair 2000; Bakun 2006) ecosystems. In many such systems, the dominant species in one of the alternative states feeds on early life history stages of the alternative dominant, generating an unstable feedback loop that prevents the alternative dominant from gaining abundance. A potential example of this phenomenon involves the collapse of cod in the Baltic Sea, which was accompanied by a shift to dominance by the cod's 'wasp-waist' prey, planktivorous herring and sprat (Bakun 2006). Cod have failed to rebound from their initial collapse throughout much of the north Atlantic, despite severe fishing restrictions, probably in part because abundant herring and sprat feed heavily on cod eggs and larvae. Thus, for pelagic marine food webs, the ontogenetic size structuring of trophic interactions may be a key factor in mediating commonly reported regime shifts between alternate stable states (Bakun 2006).

Regime shifts may also result from effects of organisms on the environment. In benthic systems in particular, ecosystem engineers or other species may modify the environment such that it becomes less hospitable to species characteristics of the alternative regime. Certain taxa of infaunal invertebrates, for example, are both more tolerant of mobile sediment resuspension and more active in resuspending it; these may prevent establishment of species that would otherwise dominate in stable sedimentary environments (Peterson 1984; van Nes *et al.* 2007). Similarly, in lakes and probably also in estuaries, clear-water phases are maintained in part by growth of benthic macrophytes, which bind sediment, preventing its resuspension; when macrophytes are lost for whatever reason, the mobility of both sediments and sediment-bound nutrients foster turbidity in the water column, which resists re-establishment of benthic macrophytes.

Finally, a link between changing biodiversity and regime shifts, though not rigorously studied, is suggested by several lines of evidence. First, rapid transitions in ecosystem state appear to be better documented in relatively low-diversity systems, including temperate lakes, the Black Sea (Daskalov *et al.* 2007) and the North Atlantic (Frank *et al.* 2007). In particular, wasp-waist ecosystems appear especially prone to rapid, pronounced 'regime shifts' between alternate semi-stable community states, and this vulnerability has been attributed in part to the low diversity and low resilience of this intermediate trophic level. Second, experiments show that invasion of marine communities by exotic species, which can trigger irreversible shifts in ecosystem structure and function, is generally more frequent in communities of low diversity (Stachowicz *et al.* 1999, 2002a).

8.7 Emerging questions in emerging marine ecosystems

The ocean of the 21st century is changing at rates and in directions never before seen in human history. The causes involve both abiotic changes – including eutrophication, habitat alteration, and, increasingly, climate warming and acidification – and direct alteration of community structure via intense predation by humans and invasion of non-indigenous species. The studies reviewed here suggest that these emerging marine ecosystems show several recurrent patterns that raise the following intriguing questions for future research.

The decline in large animals through overharvesting constitutes an important loss of functional diversity, which is eroding complex food webs into topologically simpler, and probably more strongly linked, food chains. This simplification may lead to fundamental change in community and ecosystem dynamics with several potential consequences. Are these simpler food chains indeed less stable generally and, specifically, more vulnerable to dramatic regime shifts than naturally diverse communities (Folke *et al.* 2004)?

There is some evidence that intense exploitation may change not only the strength of interactions but also the mode of control, from bottom-up in lightly exploited, naturally diverse systems to top-down in heavily exploited systems, as in the northwest Atlantic (Frank *et al.* 2006). Indeed, this review suggests that trophic cascades may be more common in marine systems than concluded previously (Jennings and Kaiser 1998). The different conclusions may stem in part from newer, fine-scale data (e.g. Frank *et al.* 2006); but it is also conceivable that marine ecosystems have changed even in the last decade toward states more vulnerable to perturbation. Does heavy exploitation generally shift marine communities towards stronger top-down control?

Accelerating invasions of non-indigenous species are changing marine communities worldwide (Ruiz *et al.* 2000). Do these emerging communities interact in different ways as a result of the lack of shared evolutionary history among species? For example, do marine consumer–prey interactions involving non-indigenous species differ systematically from those involving only native species, as they do on land where non-indigenous consumers promote 'invasional meltdown' (Parker *et al.* 2006)?

Finally, it is becoming increasingly clear that evolutionary change often occurs on similar timescales to ecological interactions among species, and can be critical to understanding the dynamics of those interactions (Thompson 1998; Hairston *et al.* 2005). This is especially true of human predation on marine fishes, which generally targets larger, more economically valuable individuals, and accordingly has produced declines in average body size in many exploited marine fish species over recent decades (Hsieh *et al.* 2006). If length and age at maturity are at least partially heritable, the resultant size- and age-selective mortality means that rapidly maturing genotypes will be favoured under fishing mortality (Law 2000) and this truncation in size structure will produce not only ecological ramifications through the ecosystem but also evolutionary change. In particular, length and age at sexual maturity are key life history traits affecting fitness. Controlled experiments in both laboratory (Conover and Munch 2002) and field (Reznick and Ghalambor 2005) confirm that size-selective mortality can produce substantial genetically based changes in age and size at maturity within a few generations. Data from commercially exploited fishes also indicate that age and size at maturity have substantial heritabilities, and many stocks indeed have shown predicted declines in age and size at maturity over recent decades (Hutchings and Baum 2005). A critical question for future research is how much of the change in life histories of wild fish stocks results from evolution versus other factors such as release from competition, and whether this evolution reinforces the hysteresis between exploited and unexploited states of marine ecosystems, as suggested for cod-dominated ecosystems (Olsen *et al.* 2004; de Roos *et al.* 2006).

Acknowledgments

I am grateful to James Douglass, Peter Morin and Herman Verhoef for comments that improved the manuscript, and to the National Science Foundation for support (OCE-0623874).

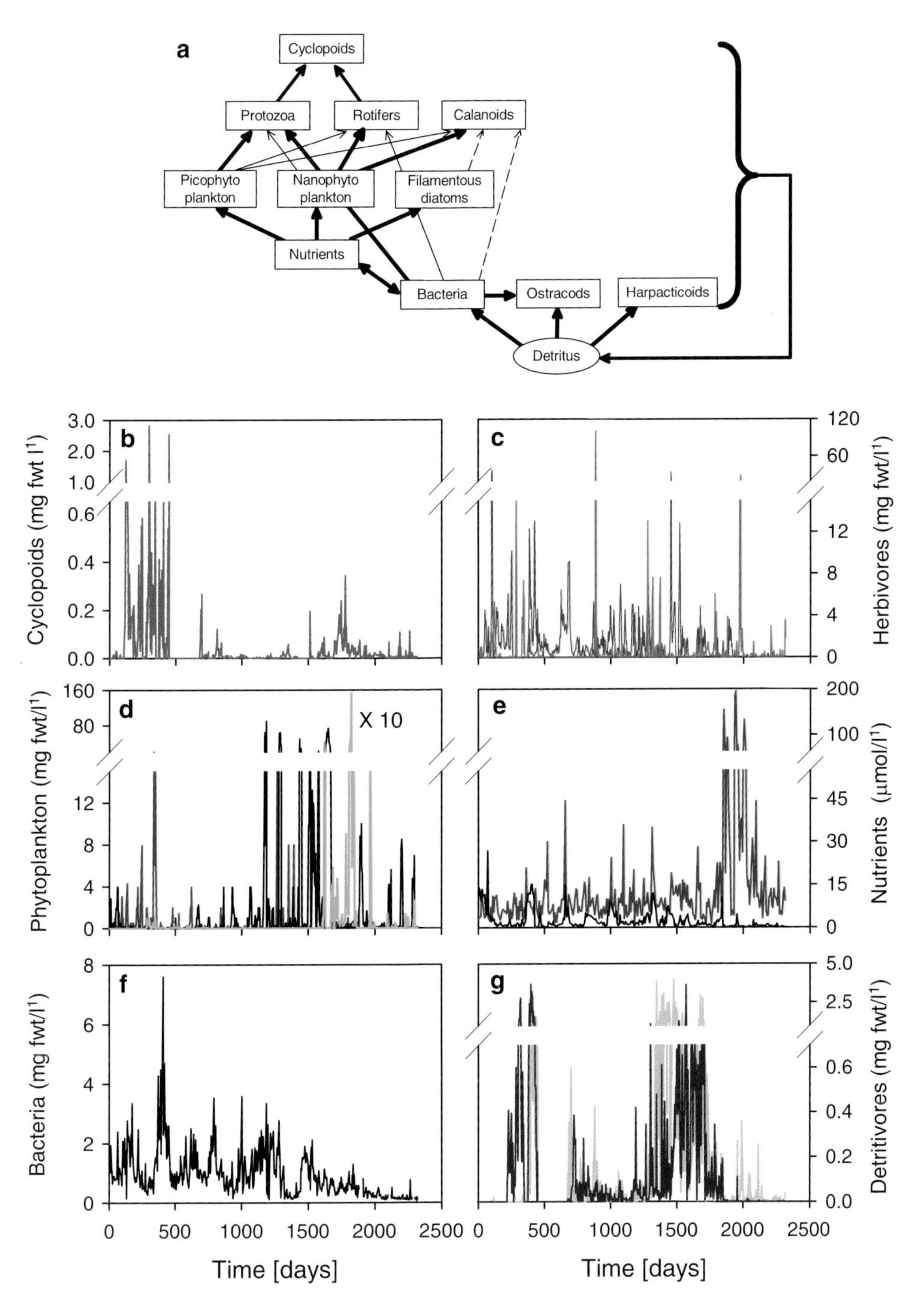

Plate 1 Description of a plankton community in a mesocosm experiment. (a) Food web structure of the mesocosm experiment. The thickness of the arrows gives a first indication of the food preferences of the species, as derived from general knowledge of their biology. (b–g) Time series of the functional groups in the food web (measured as freshweight biomass). (b) Cyclopoid copepods; (c) calanoid copepods (red), rotifers (blue) and protozoa (dark green); (d) picophytoplankton (black), nanophytoplankton (red) and filamentous diatoms (green); note that the diatom biomass should be magnified by 10; (e) dissolved inorganic nitrogen (red) and soluble reactive phosphorus (black); (f) heterotrophic bacteria; (g) harpacticoid copepods (violet) and ostracods (light blue). (Reproduced with permission from Beninca *et al.*, 2008). See page 31.

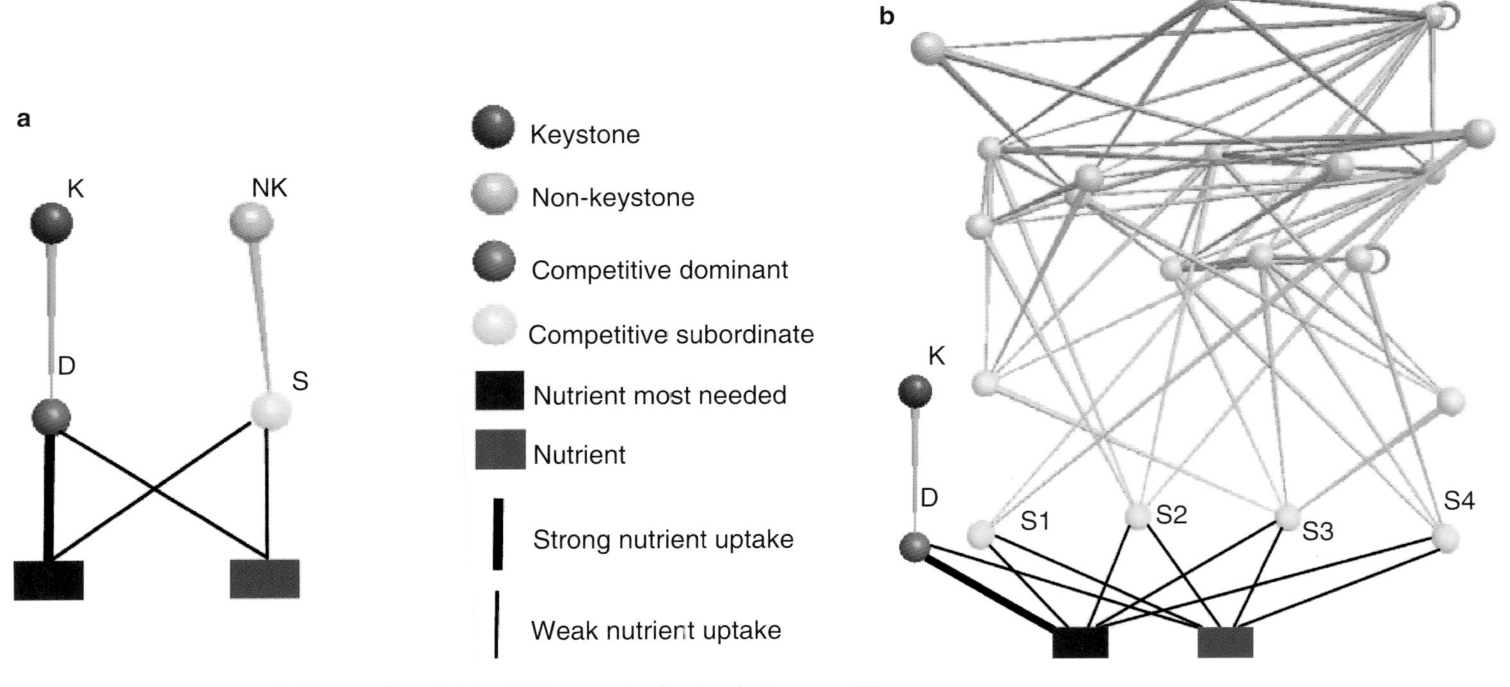

Plate 2 Keystone consumer in (a) a small module and (b) a complex food web. See page 39.

	Island biogeography	**Metapopulation dynamics**
Important characteristics	Diversity External mainland species pool Islands sinks Stochastic extinctions	Single – few species Internal pool of colonizers Patches equal (can be relaxed) Stochastic extinctions.
View of the world		
Basic equation solutions	**Diversity** $S^* = IM/(E+I)$ **For single species** Equilibrium proportion occupied islands $P^* = m/(e+m)$	**For single species** $P^* = 1 - e/m$ For competitors: $e = e_0 + e_{ij}$
Importance for management	Historically large but in practice useless because islands depend completely on the mainland	Large Diversity and species persistence dependent on isolation and regional commonness

Plate 3 Contrasting views on the world in the basic theories of island biogeography and metapopulation dynamics. S, species number; M, mainland pool species number; E and I, number of extinct and immigrating species per unit time (in island biogeography model); e and m, extinction and migration rates (in metapopulation model); P*, equilibrium proportion of occupied islands/patches. See page 116.

Plate 4 *Butea monosperma*, an important legume from a deciduous forest ecosystem of India, is a keystone mutualist forming mutualistic associations with some animal species for pollinator services. *Butea monosperma* also forms belowground mutualisms with rhizobia and arbuscular mycorrhizal fungi. Inset, a rose ringed parakeet (*Psittacula longicauda*) foraging on the flowers of *Butea monosperma*; the parakeet is a nectar robber and does not provide any reciprocal benefit to the plant. See page 182.

CHAPTER 9

Applied (meta)community ecology: diversity and ecosystem services at the intersection of local and regional processes

Janne Bengtsson

9.1 Introduction

The most pressing ecological questions today concern the management of natural resources. At the heart of natural resources lies biodiversity. The common biodiversity in ecosystems provides goods and services for society, such as food (e.g. crops, grazing animals, fish), materials for buildings and other human artefacts; bioenergy; and processes such as plant or animal production, biological control and pollination that result in those goods (Daily 1997). Ecosystems are also utilized to treat the waste products from human activities (e.g. Folke *et al.* 1997). Behind most of these ecosystem services are *interactions* between species or functional groups, which are – no matter whether we are conscious of them or not – modified by human activities. The ultimate test for the validity of community ecology, as outlined in the previous chapters, is whether it can provide useful knowledge for managing ecological interactions and biological resources in a sustainable way.

Humans are now considered to be the dominating ecological and evolutionary force on Earth (Vitousek *et al.* 1997; Palumbi 2001). Despite this, in the past, community ecology had an awkward relationship with human-dominated landscapes. On one hand, applied questions have been crucial for theory and empirical studies in both population and ecosystems ecology. Well-known examples range from biological control, regulation of wildlife and fisheries to nutrient cycling and crop production. On the other hand, there is a long tradition among ecologists to separate ecosystems into natural versus disturbed ones, in which the latter were disturbed mainly by human activities (Worster 1994). At times, it was even questioned whether such human-dominated ecosystems had any 'real ecology', and, if so, if it was the same ecology as that for undisturbed ecosystems. Hence, studies of communities had to be made in pristine systems where the unnatural humans played a negligible role. Traces of such notions can still be found within conservation biology, especially in the USA, when emphasis is put on 'wilderness' and 'naturalness'. However, the emergence of recent major issues in ecology, such as biodiversity conservation, ecosystem services and climate change, show that this is an unrealistic approach.

In this chapter, I argue that community ecology ought to understand ecological processes that are crucial for providing goods and services to society. Applied questions can both provide large-scale tests of general ecological theories and lead to crucial insights into basic ecological theory. Using the theoretical framework of spatial population and community dynamics, now more trendily referred to as metacommunity dynamics (Chapter 5), I start with a short background in classical patch and niche theory. The juxtaposition of local versus regional dynamic processes in metacommunity theory is then used to suggest guidelines for management of biodiversity and ecosystem services in human-dominated landscapes. As the literature on these subjects is increasing rapidly, I do not attempt to cover the recent literature. Instead, I use some examples to illustrate how applied and basic questions in ecology can gain from being studied in tandem in the same managed system. My examples mainly derive from studies on biodiversity and ecosystem services in terrestrial agricultural landscapes. I use the terms *ecosystem functioning* and *ecosystem processes* to refer to any process carried out in ecosystems by organisms maximizing fitness, for example biomass production or decomposition, while *ecosystem services* are the subset of ecosystem processes that benefit humans (or society). *Diversity* is used synonymously with species richness, unless stated otherwise.

Basic in biodiversity and ecosystem functioning research is the role of diversity. It is important to remember that diversity in itself does not carry out any process. Ecosystem processes are carried out by species (populations and individuals); diversity is an indicator of species being present to do the work in ecosystems (Bengtsson 1998). Management of ecosystem services in heterogeneous landscapes ultimately relies on understanding communities of interacting populations.

9.2 A theoretical background

9.2.1 A simplified historical narrative

Island biogeography and metapopulation theory emerged during the 1960s almost simultaneously and actually within the same intellectual environment (Worster 1994). Many ecologists seemed to regard them as essentially the same, but from several aspects and especially for conservation these theories offer different views of the world. A major conceptual difference is that island biogeography (MacArthur and Wilson 1967) saw the world as mainlands sending off migrants to islands, an extreme source–sink situation with respect to colonization–extinction dynamics (but see Schoener 1976). On the other hand, in metapopulation theory (Levins 1969) patches are colonized from other

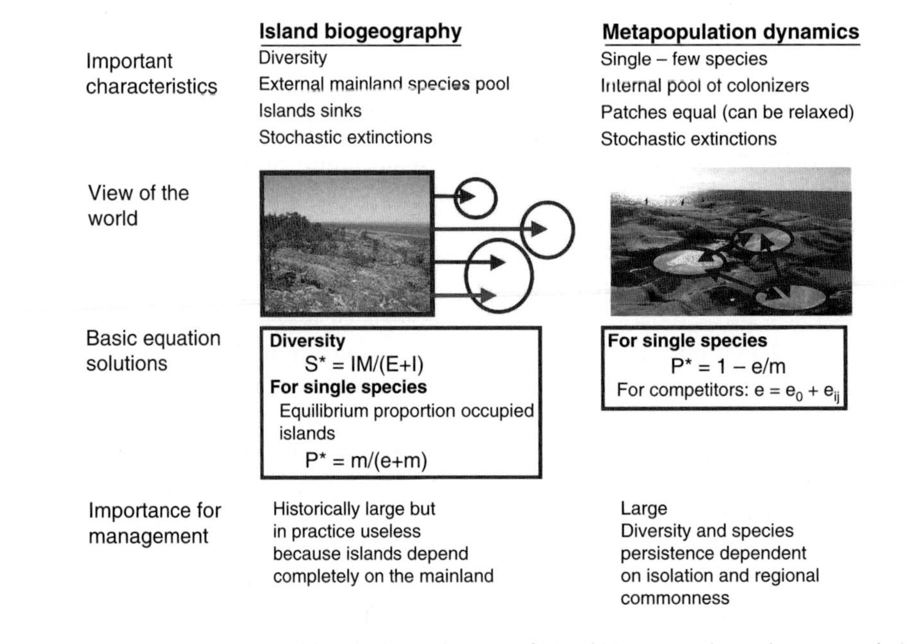

Figure 9.1 Contrasting views on the world in the basic theories of island biogeography and metapopulation dynamics. S, species number; M, mainland pool species number; E and I, number of extinct and immigrating species per unit time (in island biogeography model); e and m, extinction and migration rates (in metapopulation model); P*, equilibrium proportion of occupied islands/patches. See plate 3.

patches in the region, not from a mainland (Fig. 9.1). Island biogeography was important in early conservation biology, despite the conceptual problem that islands (patches) require a mainland to sustain their diversity, which makes it practically useless to apply in reserve design. Metapopulation models in which the number of colonizers depends on how many patches are inhabited in the region, and how isolated those patches are, provide a more relevant model of the world, especially for conservation of species in fragmented landscapes. The general acceptance of the metapopulation concept for single species was crucial for the later development of metacommunity theory.

Although both island biogeography and metapopulation theory could be formulated to incorporate species interactions – in both cases by allowing interspecific competition to increase extinction rates (e.g. Hanski and Ranta 1983) – their basic formulations and later interpretations, especially in conservation biology, modelled the world as essentially consisting of single species with independent dynamics. This limited the usefulness of metapopulation theory for interacting communities. Only after other areas of ecology, such as food webs (Polis *et al.* 1997), had provided a spatial perspective on species interactions and community structure, the time became ripe for an emerging metacommunity theory that linked landscape ecology, food webs, metapopulation dynamics and ecosystem functioning.

In the 1960s, niche theory for local communities was also developed by MacArthur and Levins (see Chapter 5). For example, the theory of limiting similarity was used to explain the limits to local diversity (see Chase and Leibold 2002). Classical niche theory, despite its falling popularity, forms an important background for community ecology, and is important for the discussion of why diversity might affect ecosystem functioning. The complementarity hypothesis (Loreau *et al.* 2001), which predicts increases in various aspects of ecosystem functioning as diversity increases, is based on niche differences in resource utilization between species. However, not until the last decade had niche theory developed to be integrated into a metacommunity framework. Important ideas in this process were as follows: (1) The recognition that a species' environmental *requirements* (resources and environmental conditions allowing survival) and a species' *impact* on the environment (withdrawing resources but also incorporating ecological engineering and other aspects of niche construction; Odling-Smee *et al.* 2003) are *both* major components of the niche (Chase and Leibold 2002). (2) The idea that, when environments vary over time or in space, stability of ecosystem functioning is ensured by species with similar effects on ecosystems while at the same time showing a diversity of responses to environmental variation (response diversity, Elmqvist *et al.* 2003; insurance hypothesis, Loreau *et al.* 2003). (3) The neutral community theory (Bell 2001; Hubbell 2001) was important by emphasizing the role of dispersal limitation for community composition, despite dismissing niches as being important for species coexistence (building on earlier notions in especially plant ecology).

Niche theory, as well as ecology in general, regarded the local habitat to be of prime importance. This was especially so when dealing with conservation of single species or diversity. Local habitat patches were the objects to be managed because they were threatened by human activities. Habitats of high value for conservation were set aside as reserves or land-owners given payments to manage them in an economically less rational way, often by traditional practices. Metapopulation and landscape ecology theory allowed researchers and managers to see effects of landscape and patch configuration on species, but the incorporation of this insight into management was slow. However, the rise of the corridor concept – despite the controversies surrounding it – opened up for a recognition of the importance of landscape context, in this case connectivity, for managing diversity in individual patches.

In addition, insect ecologists had long recognized the importance of larger-scale processes and dispersal in population dynamics. In fact, already in the 1950s, metapopulation ideas were developed by Andrewartha and Birch (1954). Migration between crop fields and other habitats, both among pests and among natural predators, was identified early as important for conservation biological control (Barbosa 1998).

From these sources emerged the modern view of local habitat patches interacting via dispersal in a landscape consisting of habitat patches of varying quality, and a matrix of varying but lower quality between patches. Crucial for this view being accepted was the accumulation of observations that patterns in local diversity and local interactions could not be fully explained by local processes and local environmental conditions (e.g. Ricklefs and Schluter 1993; Polis *et al.* 1997; Holyoak *et al.* 2005). This led a new generation of ecologists to design studies in which the effects of local conditions and regional dynamics or landscape composition could be separated more rigorously. Many such studies tested very applied questions. Some examples are the importance of the amount of natural habitats in the surrounding landscape for parasitoids on pest insects in crops (Thies and Tscharntke 1999), and the effects of different agricultural practices and landscape heterogeneity on insect diversity (Weibull *et al.* 2000). In both these as well as many other cases, the effects of landscape were significant and sometimes larger than effects of local management.

9.2.2 Implications of metacommunity theory

Allowing theoretical predictions to guide management is risky. Many applied questions in ecology require both knowledge of ecological theory and intimate understanding of the natural history of the system under study. Furthermore, for many applied questions the system that needs to be understood includes not only the ecosystem, but also many aspects of humans as part of linked social–ecological systems, including sociology, economy, politics and the conflicting interests of those advocating different ideological views of the world (see, for example, Lawton 2007). Nonetheless, it is valid to examine what the emerging metacommunity theories might predict about the management of diversity and ecosystem services in human-dominated landscapes (see, for example, Bengtsson *et al.* 2003 in the context of reserves). An analogy from evolutionary biology is useful: Predictions on evolutionary changes in pests or pathogens in response to human actions are of interest, although we know that decisions in society do not fully take these into account.

Leibold *et al.* (2004; see also Chapter 5) offered four simplified perspectives of how metacommunity theory might view the world, i.e. *neutral, patch dynamics, species sorting* and *dispersal-driven* metacommunities (also termed *mass effects*, but in this author's view this term is ambiguous and should be replaced to emphasize the main role of dispersal for the dynamics of this type of communities). Each of these offer interesting insights into how diversity, species composition and ecosystem services are affected by local (sorting) and regional (dispersal) processes, but, as noted in Chase and Bengtsson (Chapter 5), the four perspectives are not exclusive. The basic processes and management implications of the four metacommunity perspectives are summarized in Table 9.1.

The metacommunity view of the world is mainly one of discrete patches and a matrix between patches that is non-habitable environment. The quality of the matrix in this view mainly affects dispersal (migration rates). However, as pointed out by Vandermeer and Perfecto (2007) the matrix is often habitat for many species in agricultural and forest systems. The effect of a matrix that is habitable or even similar in quality to patches is most easily included in the *dispersal-driven* (*mass effects*) perspective. This view of the matrix highlights the points of Oksanen (1990) and Holt (2002) that the relative productivity of patches or of the matrix compared with patches is of great importance when examining the effects of spatial heterogeneity. Spill-over from more productive habitats or patches can greatly affect species interactions. It may complicate management, but is important for the delivery of ecosystem services such as biological control and pollination (see below).

Some general insights for ecosystem management can be deduced from these perspectives (e.g. Mouillot 2007), although for particular systems and questions predictions have to be much more specific. In the following, unless otherwise stated, communities are assumed to be at one trophic level, structured by competition for space or other resources, and there is a trade-off between competitive ability and traits associated with dispersal.

The *species sorting* perspective implies that management of local conditions and local habitat heterogeneity is most important to maintain diversity in a patch (Table 9.1). Because *species sorting* allows the most efficient competitors to dominate patches,

Table 9.1 Implications of different perspectives on metacommunity dynamics (Leibold *et al.* 2004) for management of biodiversity and ecosystem services (inspired by Mouillot 2007)

Perspective	Ecological processes	Management implications
Species sorting	Local processes determine diversity. Coexistence by niche separation, local heterogeneity creates niches leading to higher diversity. Disturbances decrease local diversity.Low dispersal into patches, but some dispersal needed for local communities to track environmental changes. Sorting of most efficient competitors results in high ecosystem functioning	Proper local management most important. Local diversity: maintain local conditions by management. Regional diversity: maintain diversity of conditions on regional scale; manage patches differently when appropriate. Maintain some connections between patches to allow environmental tracking. Local management to maintain diversity increases resource use and local ecosystem services
Patch dynamics	Metapopulation dynamics, patches fairly equal. Extinctions can be stochastic, caused by competitors or predators, disturbance-driven or deterministic. Colonization often distance dependent, dispersal limitation may be important. Regional coexistence and diversity depends on trade-offs, e.g. between dispersal and competitive ability. Stochastic extinctions lead to lower ecosystem functioning	Local and regional diversity and species occurrence maintained by higher connectivity between patches, and maintaining disturbance regimes. Local diversity maintained by maintaining the regional species pool. Ecosystem services depend on dispersal of functionally dominant species or sets of species with similar effects on functioning Low dispersal rates and low regional occurrence result in lower average ecosystem services with a high variability among patches
Dispersal-driven metacommunities (mass effects) (1) between patches, or (2) from matrix to patches	Island biogeography roots. Source-sink dynamics, patches of varying quality. Dispersal from outside local patches, i.e. other source patches or matrix, has major effects on local dynamics and composition (diversity). Local diversity may increase with increasing dispersal of new and less competitive species into patches, but then decrease as immigration of good dispersers dominate dynamics, which also leads to homogenization of regional diversity	Maintain dispersal but do not homogenize region. Important to identify and manage source patches. Promote diversity of conditions regionally to maintain sources for different species. Ecosystem services depend on dispersal from sources, but high regional dispersal may prevent local sorting of efficient species. High-quality matrix important if dispersal from matrix enhances ecosystem functioning
Neutral	Acts on longer time scales. Individuals of different species are similar in fitness, competitive ability, etc., but are not necessarily equal. Balance between ecological drift and speciation maintains diversity	Short-term management implications less clear. Maintain regional dynamics and habitat diversity. Do not change fitness relations drastically (e.g. many forestry practices predominantly decrease fitness of old-growth species, thus decreasing diversity)

The perspectives are patch-based, and the uninhabitable matrix affects only migration rates. However, in many managed systems the matrix is often habitat for many species and is most easily included in the *dispersal-driven* (*mass effects*) metacommunity perspective.
Note: Perspectives are not exclusive (see text).

ecosystem functioning increases with local diversity, and thus by proper local management.

In the *patch dynamics* perspective, which is based on classical metapopulation theory, local communities are assembled from the regional species pool and then subjected to local sorting (Bengtsson *et al.* 2003). At least moderate dispersal between patches in the region is required for the maintenance of local and regional diversity, and to ensure reliability of ecosystem services (Table 9.1).

When dynamics in local patches are largely determined by dispersal from source patches or matrix habitats (*dispersal-driven metacommunities*), local diversity may first increase and then decrease as immigration of good dispersers increases (Mouquet and Loreau 2003). Regional diversity is maintained by patches with different conditions having source communities with different composition. Dispersal may prevent adaptation of the local community to local conditions, which may decrease ecosystem functioning (Table 9.1). However, when local conditions vary over time, dispersal may still be important to maintain ecosystem services (Norberg *et al.* 2001; Loreau *et al.* 2003). If local diversity and ecosystem functioning also depend on the matrix, the matrix has to be managed in an appropriate way.

Management implications of the *neutral* perspective are less obvious (Table 9.1). This is mainly because it concerns the long-term evolutionary dynamics of large-scale systems, while its short-term applied ecological consequences are unclear.

Some simple rules of thumb for management can be suggested by combining these perspectives (Box 9.1a). First, it is not enough to manage habitats patch-wise. Instead whole landscapes must be managed as networks or mosaics (Bengtsson *et al.* 2003; Lindenmayer *et al.* 2008). Second, this regional perspective means that a diversity of conditions and management strategies should be maintained in a region, and that a certain degree of connections between habitat patches is needed to manage diversity and ecosystem services properly (Box

Box 9.1. Metacommunity theory, landscape management and land use intensification

a Rules of thumb for landscape management following from one or several metacommunity perspectives

- Maintain local conditions by management (*species sorting*)
- Manage not only single patches but whole landscapes (*patch dynamics, dispersal-driven, neutral*)
- Maintain diversity of local conditions in region (*species sorting, dispersal-driven, neutral*)
- Maintain connections between patches without homogenizing the landscape (*species sorting, patch dynamics, dispersal-driven, neutral*)
- Maintain disturbance regimes close to natural (*patch dynamics, neutral*)

b Summary of predictions from metacommunity theory on the effects of land use intensification

1. Landscape heterogeneity and higher connectivity in the landscape result in higher diversity and a larger regional species pool. This increases the possibility for local ecosystem services to be maintained.
2. The intensity of dispersal influences the effect of landscape context on metacommunity dynamics.
3. Homogenization of landscapes and land use intensification create a weedy world.
4. Intensification of land use will affect trophic structure and strength of local interactions. The effect will vary depending on relative patch productivities and the importance of the matrix.
5. Homogeneous landscapes have lower diversity and biomass of many functional groups and thus less efficient ecosystem services.

9.1a). Third, natural or close to natural disturbance regimes and landscape mosaics should be maintained, as most species are likely to be well adapted to these over evolutionary and ecological time (Bengtsson *et al.* 2003; Lindenmayer *et al.* 2008).

9.2.3 Metacommunities in human-dominated landscapes: effects of habitat loss and fragmentation

From an applied perspective, a crucial question concerns the effects of land use changes under the different metacommunity perspectives. The intensification of agriculture and forestry during the last century has resulted in an unprecedented loss of natural habitats and increased distances between the remaining ones (fragmentation). In addition, management intensification has resulted in an increase in the area covered by monocultures of certain crops and tree species in large patches. At the same time, intensification has resulted in an increased disturbance frequency and intensity, to a larger extent in agriculture than in forestry. Note that a similar drastic increase in disturbance frequency also has taken place on most fishing grounds. The high frequency of disturbance is expected to favour certain species only. Traits such as rapid exploitation and high dispersal ability may make these specific species more likely to persist in such landscapes (Tremlova and Münzbergova 2007).

What are the expected consequences of these processes on different types of metacommunities? To discuss this question, we first need an idea of what kinds of metacommunity dynamics were dominant in the landscape *before* the onset of land use changes. Then a short scenario related to the effects of fragmentation and habitat loss will be discussed.

Assume a fairly homogeneous landscape in which communities can mainly be described as close to the *neutral* or *dispersal-driven* metacommunity types, because dispersal distances between patches are negligible. Although species may be dispersal limited, the short distances between suitable patches allows persistence of many species and a high local and regional diversity. If this landscape is broken up into smaller fragments by human activities, the situation will first approach the patch dynamics perspective. Locally and regionally, sorting by dispersal ability will take place, because species that are poor dispersers but, presumably, good competitors will not be able to persist (Nee and May 1992). As intensification of land use proceeds, species with poor dispersal will approach remnant metapopulations in which extinctions are not balanced by colonizations (Eriksson 1996). Long-lived species such as plants may persist for a long time, but finally go extinct. The end result will be scattered patches with similar community composition, dominated by the best dispersers in the region or those common in the intensively managed matrix. That is, the metacommunity will again appear to be dispersal driven, but with low local and regional diversity and impaired ecosystem services.

Alternatively, assume an originally heterogeneous and mosaic landscape, in which niche differences and environmental variation result in a high diversity of species. This metacommunity will be closer to a *species sorting* perspective, and to variable degrees *patch dynamics*. When such landscapes are fragmented and dispersal is made more difficult, the result will first be sorted remnant metacommunities in patches of varying quality and composition. Further land use intensification may have different consequences. In remnant patches of sufficient quality and size, species adapted to local conditions and with high competitive ability can remain for a long time, and the species sorting perspective prevails. Alternatively, if local extinction rates are higher, there will be a shift from competitive dominants to species sorted by dispersal, and the dispersal-dominated perspective will be more applicable.

A third scenario concerns disturbance-driven mosaic landscapes, where disturbances, e.g. fire or wind-throws, are major causes of local extinctions (e.g. Bengtsson *et al.* 2003). At moderate to high disturbance frequencies the role of species sorting will be reduced, and communities will be dominated by fast resource trackers with high dispersal ability. Depending on disturbance frequency such landscapes will originally appear to be dominated by *dispersal* or *patch dynamics*. However, some species may also persist in these landscapes by escaping in time rather than space, e.g. by having dormant stages waiting for the next disturbance. Human activities altering disturbances and increasing fragmentation will increase the role of dispersal for species

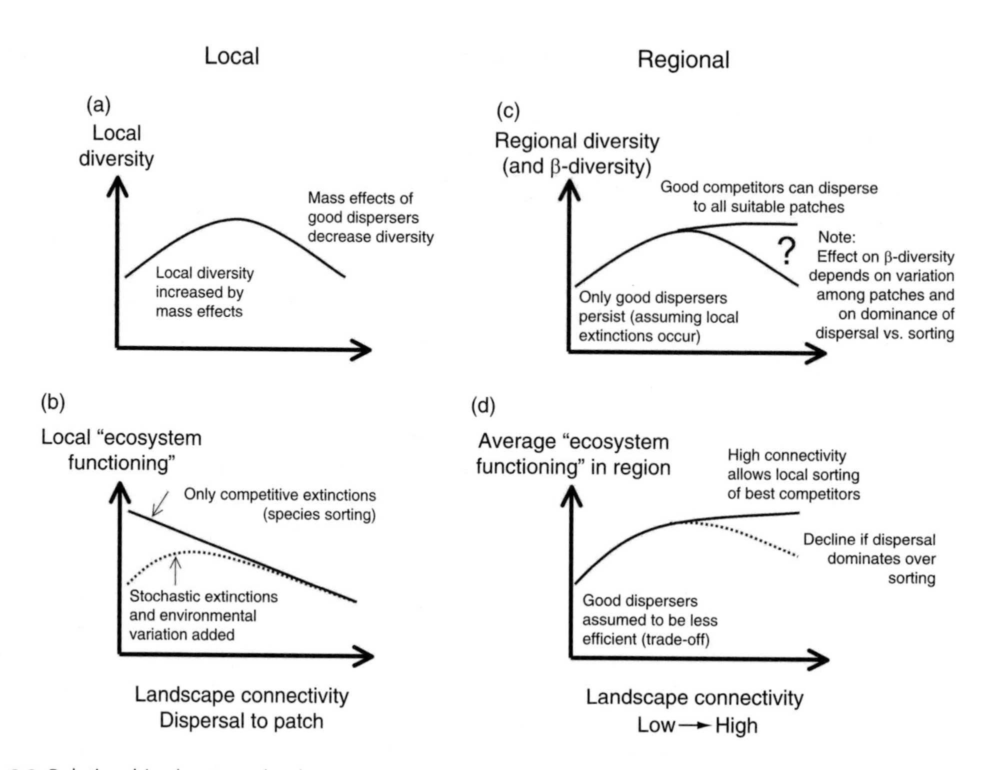

Figure 9.2 Relationships between landscape connectivity and (a) local diversity, (b) local ecosystem functioning, (c) regional diversity and (d) regional average ecosystem functioning. Suggested relationships are based on metacommunity theory but may not always hold (see text).

persistence, for many species by first approaching a patch dynamics perspective, but as land use intensification proceeds the end result will be low-diversity communities of *r*-selected good dispersers.

Taken together, human alterations of habitats and disturbance regimes will likely result in substantial decreases in habitat and landscape quality for many species. However, for some species, the landscape created by humans is of exceptionally high quality. These species are admittedly a minority but often abundant and obnoxious ones, with good dispersal ability and either generalistic requirements or being specialists on monocultures.

Some general expectations from metacommunity theory on the effects of land use intensification are summarized in Box 9.1b and Fig. 9.2.

First, in human-dominated landscapes, increased heterogeneity and connectivity is likely to lead to higher diversity and a larger regional species pool (Benton *et al.* 2003). However, high dispersal and high connectivity may homogenize landscapes and decrease diversity. This is because species in the region then will behave as one population (patchy populations, *sensu* Harrison 1991), decreasing both local and, if dispersal dominates over sorting, regional diversity. A larger regional species pool increases the possibility for local ecosystem services to be maintained. Niche theory suggests that increased diversity increases utilization of available resources, at least at low levels of diversity. Also, a diverse species pool increases the probability that at least some species survive more variable environmental conditions (Elmqvist *et al.* 2003; Loreau *et al.* 2003).

Second, dispersal ability and intensity influence the effect of landscape context (connectivity) on metacommunity dynamics. It is predicted that low dispersal increases the importance of local sorting in patches, while at intermediate dispersal landscape factors are more important. At high dispersal, local conditions in patches may again seem more important, because then sorting can be expressed in all patches.

Third, homogenization of landscapes and increased disturbances associated with land use intensification 'select' species with particular traits such as high dispersal ability, generalistic resource and habitat requirements, and ability to rapidly exploit abundant resources. In essence, homogeneous landscapes are creating *a weedy world* (this term was originally coined by Carl Folke; note the similarity with classical *r*-selected species (MacArthur and Wilson 1967), such as ruderals and rats).

Fourth, because different organisms are likely to differ in their sensitivity to habitat loss and increased isolation, and in their response to intensification, different landscapes will contain different sets of species (metacommunities) and the strength of trophic interactions in patches will vary between landscapes. For example, if predators have lower population sizes or lower dispersal rates than their prey, then increased isolation, smaller habitat patches and increased intensification is predicted to decrease ratios of predators to prey, resulting in less predator control of prey, e.g. pests. However, predators may spill over from more productive matrix habitats to small patches or to intensively cultivated areas, resulting in higher predation. Thus, the mechanisms for different levels of predation need to be understood in a spatial context (Oksanen 1990; Polis *et al.* 1997; van de Koppel *et al.* 2005; Chapter 5)

Finally, as a consequence of the third and fourth points above, homogeneous landscapes are likely to show a lower diversity and biomass of many functional groups, except those deliberately favoured by agriculture or forestry through planting (note that planting is not the case in fisheries, which exacerbates these effects). Changes in diversity and species composition are often associated with changes in size distributions, which in turn may entail less efficient ecosystem services (Bengtsson 1998).

9.3 A selection of empirical studies

9.3.1 Applied questions allow experimental studies on management scales

A major issue in ecology is on which scales phenomena and processes should be studied (e.g. Peterson and Parker 1998). Scale refers to the size or duration (spatial and temporal scale) of phenomena, and it is a general problem that conclusions about a system depend on the scale at which we view it. For example, there have been thousands of laboratory and small plot experiments in applied ecology. Often, such small-scale studies have been used to guide management at larger scales, as in forestry where the majority of experiments have been made on plots smaller than a hectare (100 × 100 m) monitored over much less than a forest generation. Still it is assumed that they inform us about the large-scale and long-term effects of various forestry practices, despite the fact that this assumption often is unlikely to hold. Management advice based on small-scale studies is risky.

However, management of real ecosystems is carried out on large scales, and different management practices can provide information about the dynamics of ecosystems on larger scales. While small-scale experimental model systems can be instructive, any advice on, for example, corridors or land use should be relevant at the scales at which decisions about land use are made. In agricultural landscapes such decisions are made at the scale of single fields to farms or even landscapes with many farms. These decisions are made by farmers, advisors and land use planners based on socioeconomic factors, naturally given landscape and soil features, but also on individual feelings and prevailing ideologies. This creates variation in management that can be used to examine the effects of various land use practices in a pseudo-experimental way.

An example of this is how different farming systems affect biodiversity and ecosystem services. The original question was if organic farming – a system characterized by using no pesticides and no inorganic fertilizers and consequently usually more perennial semi-permanent grasslands with nitrogen-fixing plants – had higher biodiversity than conventional farming systems. Clearly, plot experiments on the scale of a few metres are almost useless for answering such a question. It should be answered at the management scale of fields or preferably farms. However, at this larger scale, landscape structure, such as the amount of natural habitat or the diversity of habitats, is important for species richness of many organisms from plants to insects and birds (Steffan-Dewenter *et al.* 2002; Benton *et al.* 2003). How can we distinguish farming system effects from such landscape effects?

A straightforward way is to account for the effect of landscape by examining farms (or fields) situated in different landscapes, but making sure that organic and conventional farms are evenly distributed over the landscapes. The simplest way is to choose matched pairs of organic/conventional farms along a gradient of landscape heterogeneity, and then analysing the results with an analysis of covariance or similar models. This approach was used by Weibull *et al.* (2000) and has since then been extensively used by a variety of authors in Germany, the UK and Sweden (Roschewitz *et al.* 2005; Fuller *et al.* 2005; Rundlöf and Smith 2006). With this design, it is also possible to examine whether there are interactions between the farming system and landscape structure. For example, organic farming might have a large effect in homogeneous agricultural landscapes but not in small-scale mosaic landscapes.

However, in order to extract the variables that best describe the habitat and landscape factors that organisms are likely to respond to, the biology and ecology of the organisms under study must be understood. Here, we often encounter an imbalance in our knowledge: Most field biologists working on an organism group have an intuitive feeling for which local habitat factors influence the presence of species. On the other hand, understanding regional influences on species distributions require that we also have a good idea about the factors affecting dispersal and movements of organisms in the landscape. Such knowledge is more difficult to obtain and for many organisms quite scanty. Hence, while the local factors included in an analysis often capture what organisms really respond to, regional scale variables are often merely informed guesses, and determined by the availability and resolution of the landscape maps used to measure, for example, proportion of habitats, landscape heterogeneity and edge zones. Thus, almost by default regional influences will contribute less to the explained variation in species distributions and diversity than the more precisely known local habitat factors. In view of this, major advances in metacommunity ecology will require more studies of how organisms disperse and how they utilize the matrix between focal patches.

9.3.2 Biodiversity in human-dominated landscapes: local or landscape management?

The intensification of agriculture in recent decades is a good example of how management practices affect whole landscapes as well as the quality of habitat patches. How these qualities affect organisms and their interactions are exactly the kind of applied question that metacommunity theories should try to explain. In the UK, the major changes in agricultural landscapes took place in the 1970s as it joined the European Union (Chamberlain *et al.* 2000), similar to many other Western European countries.

It is well established that modern agriculture (and forestry, e.g. Bengtsson *et al.* 2003) has resulted in declines in biodiversity (Donald *et al.* 2001; Tscharntke *et al.* 2005; Biesmeijer *et al.* 2006). In agricultural landscapes the reasons are well known: Simplified crop rotations, farm specialization, increased use of pesticides and herbicides, heavy fertilization and loss of marginal and natural habitats, singly or together, have affected various organisms negatively. For organisms depending on traditionally managed habitats such as semi-natural grasslands, their habitats nowadays occur as isolated patches in a sea of crops or forest.

To reverse this trend, intuitively we would suggest that diversity could be restored by increasing the amount of seminatural habitats and landscape heterogeneity, and decreasing the use of pesticides and other intrants. This has led to suggestions that organic farming practices would result in a higher farmland biodiversity, because organic farming is supposed to have, among other things, a higher crop diversity, more complex crop rotations with more semi-permanent fields like leys and pastures, and not using pesticides, herbicides and inorganic fertilizers. Organic farms are also more likely to be mixed farms with both animal and crop production. However, all these features of organic farming may not be realized on single farms, although pesticides, herbicides and inorganic fertilizers always are prohibited. In many landscapes organic farmers

Table 9.2 Organism mobility in relation to contributions of landscape, farming system and local habitat type to (A) species richness (diversity) and (B) variation in species composition on 16 farms in mosaic agricultural landscapes in east central Sweden

Mobility:	Plants→	Carabids→	Butterflies
A. Diversity			
Landscape	(+) $r = 0.47$	(+) $r = 0.58$	(+) $r = 0.52$
Farming system	ns	(+) $P = 0.02$	ns
Habitat	+	+	+
B. Species composition			
Landscape	± 0%	≈ 5%	≈ 10%
Habitat	≈ 10%	≈ 5%	> 10%
Model r^2	15%	15%	20%

Mobility is assumed to increase from plants to carabids to butterflies, based on data on adult movements. Habitat types were pasture, semi-permanent ley and field margin. The table summarizes results in Weibull *et al.* (2003) and Weibull and Östman (2003).

For diversity (A), only habitat was significant in the full model, but effects of landscape (highest correlations with landscape variables shown) and farming system (carabid diversity lower on organic farms) were found in other analyses (see Weibull *et al.* 2003). +, $P < 0.05$; ns, not significant.

In (B) the figures refer to the proportion of the variation in species composition explained by canonical correspondence analysis (CCA) axes related to landscape or habitat (Weibull and Östman 2003) and interpreted by this author. Model r^2 is the total variation explained by the CCA model.

may for economic reasons specialize on a single or a few crops. This is especially a risk as contracts with large retailer chains such as Wal-Mart or Sainsbury become more common.

The expected positive effects of organic farming on biodiversity are one of the reasons it has been included in agri-environmental schemes. An important applied question is then whether organic farming really delivers a higher biodiversity. For more basic questions, the advantage of using organic farming is that it is a *farm-scale* manipulation of landscape structure. Organic farming mainly affects how fields are managed, and, for organisms in natural habitat patches, it may mainly increase the quality of the matrix (Vandermeer and Perfecto 2007). This is likely to have effects on metacommunity dynamics and trophic interactions.

Our studies were conducted in a mosaic agricultural landscape in central Sweden, where fields are interspersed with areas of glacial till and bedrock which are not used for crops but often contain semi-natural pastures or small forests. We did not find any positive effects of organic farming on diversity among butterflies, carabid beetles, plants in field margins or spiders, although landscape heterogeneity at the farm scale influenced these groups positively (Weibull *et al.* 2000, 2003, Öberg *et al.* 2007). In addition, local habitat type was more important for diversity than landscape factors (Table 9.2).

The importance of local habitat and landscape for species composition appeared to vary between the three organism groups (plants, carabids and butterflies) in relation to mobility (Table 9.2). The proportion of explained variation accounted for by landscape measurements increased from almost zero for plants to almost 50% in butterflies, which in general are the most mobile of the three groups. Other studies have also found that responses to the landscape may vary with differences in mobility (Steffan-Dewenter *et al.* 2002; Tscharntke *et al.* 2005 for an overview). This is in accordance with metacommunity theory. The importance of the landscape, either other patches or the matrix, should increase with organism mobility and dispersal rates (second expectation, above; Fig. 9.2a).

Such patterns may also be found for the same group in different landscapes, as suggested by a study of plants in field buffer zones in Finland

(Ma 2006). In the most connected landscape, *dispersal effects* seemed most important and resulted in a lower regional diversity and more similar local communities, associated with the dominance of a number of competitive species which were argued to disperse more easily in this landscape. In the least connected landscape, species such as dandelions (*Taraxacum* spp.) and *Aegopodium podagraria* were more common, supporting the view that fragmentation of habitats may result in local communities being more dominated by weedier species with high dispersal ability. The landscape with intermediate connectivity had the highest regional diversity, resulting in the predicted hump-shaped pattern of regional diversity in relation to connectivity (see Fig. 9.2c).

Our results showing that organic farming had no effect on diversity were at odds with generally held notions, and also with the majority of other studies (analysed by Bengtsson *et al.* 2005). Why was Sweden different? The large variation in effect size among studies in the literature, coupled with closer inspection of individual results, suggested that the effect of farming system might be larger in more intensively managed and more homogeneous landscapes than in small-scaled mosaic landscapes with many semi-natural habitats. This is in essence a test of the hypothesis that at low dispersal the importance of local sorting increases, while at intermediate dispersal the landscape is more important (see above). If this is true, an interaction between landscape and farming system is expected in studies conducted along a sufficiently long gradient of landscape heterogeneity. Exactly this was found independently in southern Sweden for butterflies (Rundlöf and Smith 2006) and bees (Rundlöf 2007), and in central Germany for arable weeds (Roschewitz *et al.* 2005) and bees (Holzschuh *et al.* 2007). Such results are highly informative from a basic metacommunity perspective, but if they are common they also pose problems when implementing ameliorative measures such as agri-environmental schemes: Should organic farming be encouraged only in homogeneous and intensively managed landscapes which have lower regional diversity, because this is where the largest gains in diversity can be achieved?

In fact this may not be the case. The highest biodiversity is usually found in heterogeneous landscapes, but this is also where farming is presently less profitable and farmland risks being abandoned with current policies. Any scheme that results in continued farming in these areas will counter decreases in biodiversity and also have other socio-economic benefits, and organic farming is one such scheme. In the most homogeneous landscapes, the increase in landscape quality that organic farming entails will mainly favour a few generalistic species that are quite common in other parts of the human-dominated landscape, such as villages and urban areas. The fact that the effect appears large is that the most intensively used landscapes are so extremely depauperated that any increase in diversity will appear substantial. For example, in many agricultural areas in The Netherlands, regional bumble bee diversity is in essence only four species (Kleijn and Langevelde 2006), whereas more than 10 species are regularly found in agricultural and urban areas in southern and central Sweden (Rundlöf 2007; Andersson *et al.* 2007; J. Risberg, J. Bengtsson and B. Cederberg, unpublished data).

In addition, there appears to be a positive effect of the amount of organic farming in the landscape. Rundlöf *et al.* (2008) found that for butterflies there was a higher local diversity in landscapes with a high proportion of organic farming within a 1 km radius, irrespective of farming system. This indicates that, along these long gradients from homogeneous to mosaic landscapes, the landscape effect on diversity can be substantial.

An important message for managers emerging from these studies is that landscape effects on diversity or species composition of plants and insects often may be found on the scale of management, in agricultural landscapes usually individual farms. In our studies in Sweden, landscape variables measured at the 25 km^2 scale were not significant, whereas heterogeneity at the farm and multiple-field scales was (approx. 0.5–2 km^2) (Weibull *et al.* 2000). This means that the choices of individual landowners often can have a large influence on diversity, and identifies the farmers as important decision-makers for conservation. However, there may be more immediate gains for farmers managing their land for higher diversity, to which we will now turn.

9.3.3 Local and regional effects on ecosystem services

The species most likely to profit from organic farming and other agri-environmental schemes decreasing agricultural intensity can be termed 'the common biodiversity' (I owe this term to Anki Weibull, and have found it very useful to distinguish these species from the red-listed species that often – and rightly so – are the major concern in conservation biology). These species are not rare, but may be declining. They are common enough to have direct effects on ecosystem services, such as predators controlling pests, insects pollinating fruits and vegetables, or earthworms maintaining soil fertility. The common biodiversity provides farmers, landowners and society with a number of ecosystem goods and services whose values usually are unknown and thus never enter the economic calculations that constrain how ecosystems are managed. Clearly, a useful community ecology should provide a better understanding of 'the ecology of ecosystem services'.

Many ecosystem services provide interesting cases of natural habitats interacting with crops, with the quality of the matrix of agricultural fields between natural habitats playing an important role. Pollinators often rely on semi-natural and less disturbed habitats, but still use flowering crops for nectar and pollen. However, biological control of pests in crops provides an even better example of how metacommunity theories concerning species interactions in spatially structured habitats naturally interact with very applied questions. There is also a direct economic benefit of biological control for farmers.

Two types of biological control are discussed in the literature. One is control of pests by specialized predators. In this case the trick is to maintain both predators and pests so that control continues over time, either naturally or by continuous addition of predators. This has been quite successful in, for example, greenhouses. The second is the control of pests by generalist predators, which do not depend on, but readily feed on, the pest. The latter type often relies on predators being present in the fields when potential pests emerge or colonize. These predators usually depend on habitats surrounding the fields for persistence. Early examples were polyphagous insects such as carabids feeding on cereal aphids (Ekbom *et al.* 1992) and parasitoids on rape seed beetles (Thies and Tscharntke 1999), but

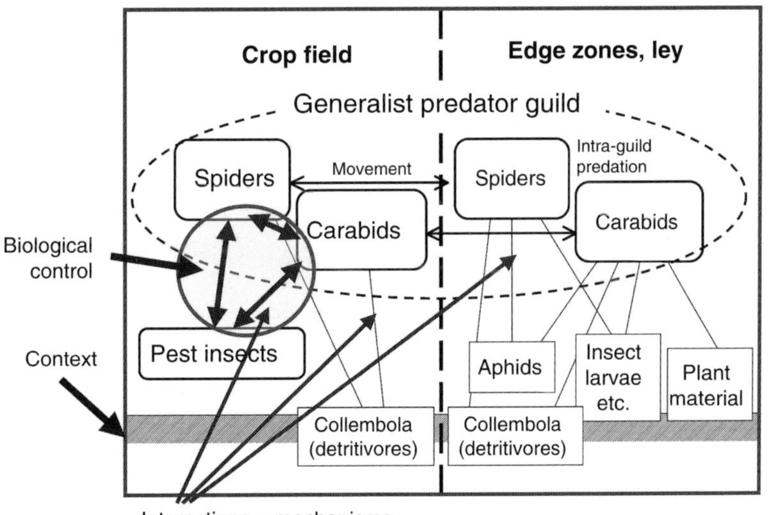

Figure 9.3 The biological control of a pest in a crop, for example aphids, is only a part of the interactions that the predators (and pest) are involved in, and proper management requires that all interactions in the food web are managed. In this case, including the edge habitat results in a basic food web question in a spatial context, incorporating predator–prey dynamics and spatial subsidies.

spiders have also been studied in this context (Schmidt and Tscharntke 2005; Öberg *et al.* 2007).

The basic system is shown in Fig. 9.3. In crop fields, aphids on cereals can be serious pests, but they are preyed upon by predators which require more permanent habitats like field margins for survival and overwintering. The predators also feed on other prey, both in the field and in the margins. Presented in this way, this clearly is a system in which the basic theories for trophic interactions in spatially structured habitats developed by Holt (2002), Oksanen (1990) and Polis *et al.* (1997) are applicable. It is a metacommunity problem where the strength of trophic interactions depends on local and regional (landscape) factors. The strength of biological control will be affected by the productivity of predators in the second habitat (field margins) and their movements, which can depend on distance but also on the abundance of other prey in the crop.

For control of pests such as aphids with multiple generations during a season, it is crucial that the predators are present in the fields when aphids colonize (Ekbom *et al.* 1992). Aphid abundance in one year does not depend on particular fields and their margins; their dynamics are on a much larger scale. Hence aphids can be considered as 'donor controlled' and we cannot expect the predators in a field to regulate aphids, only keep their numbers below damage thresholds.

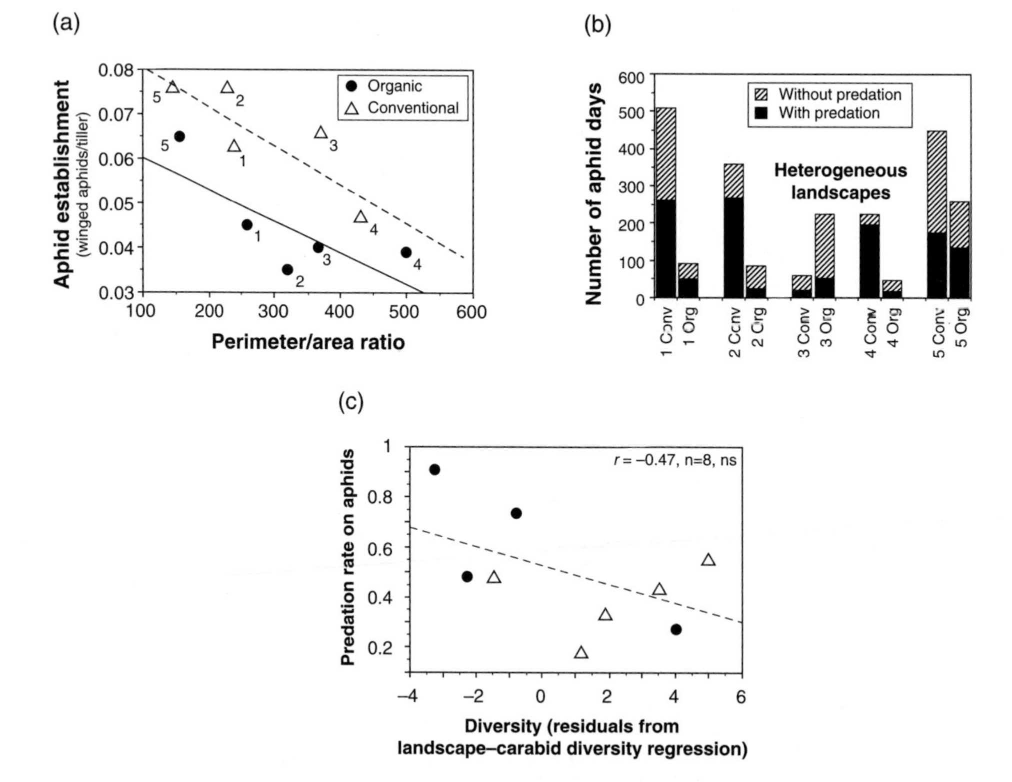

Figure 9.4 Effects of organic farming and landscape heterogeneity on aphid establishment in spring cereal fields, and the total number of aphid days (an indicator of yield loss) on 10 farms in central Sweden in 1999. Landscape heterogeneity was measured as the perimeter-to-area ratio, high values indicating mosaic landscapes. Organic farms are indicated with closed symbols, conventional farms with open symbols. (a) Aphid establishment during the early colonization phase. (b) The number of aphid days in areas within (hatched+black) and outside (black) predator exclosures. Effects of organic farming and landscape were both significant (Östman *et al.* 2001). (c) Predation rates on aphids glued on paper and placed in the field in relation to diversity. Diversity was measured as the residuals from the regression of carabid diversity (*y*-axis) versus landscape heterogeneity (*x*-axis). Extracted from data in Östman *et al.* (2001) and Weibull *et al.* (2003).

We examined the biological control of the bird-cherry oat aphid *Rhopalosiphum padi* by the guild of generalist predators, and could by exclusion experiments roughly estimate the gain in cereal yields resulting from predation on aphids and thus the economic value of predation in one particular year in one particular region (e.g. Östman *et al.* 2001, 2003). The results are summarized in Figs 9.4 and 9.5.

First, predators had a larger impact on aphid populations in heterogeneous landscapes and on organic farms. In both cases, aphid numbers early in the season were lower (Fig. 9.4a). This could be attributed to a higher predation rate on aphids (Östman *et al.* 2001), resulting in lower numbers of aphids during the growing season. Further, exclusion of predators resulted in a much higher abundance of aphids (Fig. 9.4b). Because there is a direct correspondence between the number of aphid days on the crop and yield loss, predation clearly affects the yield a farmer gets (Östman *et al.* 2003). However, there was no relation between carabid diversity and predation rates (Fig. 9.4c). This suggests that either the abundance of particular species rather than predator diversity plays a role for predation, or predator diversity in this area (15–35 spp.) was sufficiently high for the relation between diversity and ecosystem function to saturate towards an asymptote.

Finally, using the price of cereals we could estimate the value of natural predators for farmers in 1999 as approximately €40 or $37 per hectare. We do not know whether this particular year and these landscapes are representative, and a generalization clearly requires more studies. Nonetheless, because organic farmers rely on natural predators for pest control, they represent a substantial value for farmers, indicating that landscape management is an important tool for achieving economically sustainable farms.

However, when attempting to estimate economic value, we encounter different perceptions of economists and ecologists. For an economist, the economic value on the market is the marginal increase in yield values under the present rate of service delivery (Fig. 9.5a). This value usually levels off when the service is available in abundance, implying that ecosystem services are worth less when they function well. For ecologists, the value of a service such as predation is the total effect of predators, which we measured as the effect of excluding all (or most) predators (Fig. 9.5b). This means that economists and community ecologists may perceive the value of an ecological service quite differently.

9.3.4 What have we learned in the context of metacommunity ecology?

First, metacommunity theories will be important tools in analyses of interactions between species living in adjacent habitats, such as crop fields, field margins and semi-natural habitats. These situations are common and crucial to understand for applied questions such as biological control and pollination.

Second, the ecological processes involved in maintaining biodiversity and ecosystem services in agricultural landscapes often act on larger scales than

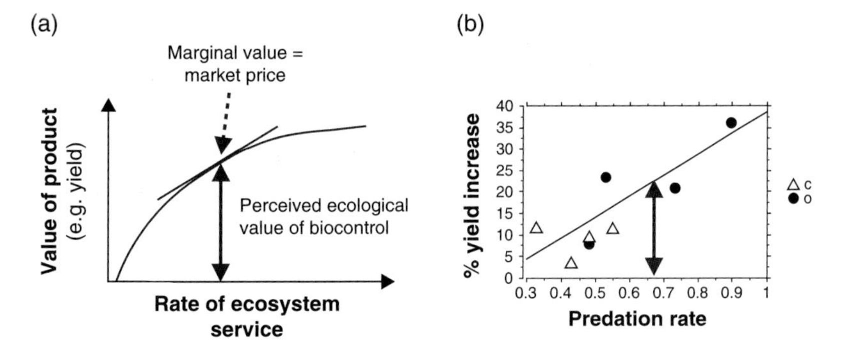

Figure 9.5 Different perceptions of the value of ecosystem services such as biological control in ecology and economy (a). In (b) the perceived ecological value of natural predators for biological control is indicated. Data from Östman *et al.* (2001, 2003).

individual fields, but for many organisms such as insects still on scales that are possible to manage by individual decision-makers such as farmers, i.e. on whole farms or multiple field-matrix mosaics. This may be different from many birds and game animals, and from forestry, which may require management of larger landscapes to include the relevant disturbance regimes (see Bengtsson *et al.* 2003).

Third, these examples illustrate that metacommunity theories can be useful because they address the interplay between local and regional dynamics at scales that are relevant to managers. These theories can thus be used to guide management strategies for conservation and ecosystem services in human-dominated landscapes. At the same time, applied studies can provide tests of metacommunity theory on the roles of landscape and local factors for diversity and ecosystem functioning, and of theory on trophic interactions in spatially heterogeneous ecosystems. Variation in management may provide the treatments needed to more rigorously examine these questions, despite the fact that full control over the treatments is not achieved. However, this is balanced by an increased realism and applicability of the results in management and policy, and the possibility of applying theoretical frameworks on larger scales rather than extrapolating from small-scale model systems whose relevance for management often is questioned.

Acknowledgements

I thank Tomas Pärt, Jens Åström, Erik Öckinger and Henrik Andrén for comments on various parts of the manuscript, my previous and present PhD students and collaborators for their intellectual input, and the organizers of the Community Ecology course in 2005 in The Netherlands, where the two lectures on which this chapter is based were held. I have been funded by the Swedish Research Council and FORMAS for the research reported here.

CHAPTER 10

Community ecology and management of salt marshes

Jan P. Bakker, Dries P.J. Kuijper and Julia Stahl

10.1 Introduction

Salt marshes are ecosystems at the edge of land and sea. They are influenced by tidal movement. It is the interaction of the vegetation and sediment trapped from inundating water that creates a salt marsh. Currently, there are about 176 000 ha of salt marsh around the Baltic and Atlantic coasts of Europe. For the Wadden Sea the area of the salt marshes can be subdivided into ~13 000 ha of salt marshes on the barrier islands and ~26 000 ha of salt marshes along the mainland coast (Bakker *et al.* 2005a). Back-barrier marshes develop at the lee side of the sand dune system of barrier islands in front of the mainland coast, where foreland marshes develop.

Salt marshes are considered to represent one of the few pristine ecosystems in North-West Europe. That may be true for some marshes, others are distinctly influenced by humans (Davy *et al.* 2009). The role of salt marshes along the coast has been transformed from primarily coastal protection tasks to a combination of the former with nature conservation interest. Large areas are nowadays assigned to nature reserves or national parks. These designations initiated critical debates on naturalness and suitable management of marshes and concern especially the need and intensity of livestock grazing (Bakker *et al.* 2003a).

Naturally developed salt marshes feature a self-stimulated development and geomorphological condition and growth that are not affected by humans. They show a natural drainage system with meandering creeks and levees with higher elevation than the adjacent depressions. Erosion protection measures, coastal defence or agricultural purposes play no critical role. They occur in sandy back-barrier conditions on islands such as Mellum, Spiekeroog (Germany), eastern parts of Ameland and Schiermonnikoog (The Netherlands). On the other hand, semi-naturally developed salt marshes either have an extensive wide-stretched natural creek system but are affected by measures to enhance livestock grazing (e.g. back-barrier conditions at the peninsula of Skallingen (Denmark) or feature a salt marsh within sedimentation fields with a man-made drainage system by ditches and are grazed by livestock or left fallow after previous grazing (e.g. artificial marshes along the mainland coast of The Netherlands, Germany and Denmark; Bakker *et al.* 2005a).

Abiotic conditions on salt marshes are related to the inundation period and frequency depending on an elevation gradient running from the upper marsh at the foot of a dune at the back-barrier marshes, or the foot of the seawall along the mainland coast to the intertidal flats. This elevational gradient also influences the rate of sedimentation, which is the main driver of plant succession. The rate of sediment input on salt marshes varies from < 5 mm/year on sandy back-barrier marshes to up to 20 mm/year on marshes in sedimentation fields (Bakker *et al.* 2002). This results in a distinct zonation of plant communities (Bakker *et al.* 2002), invertebrate communities (Andresen *et al.* 1990), avian herbivores (Stahl *et al.* 2002) and mammals (D.P.J. Kuijper unpublished data).

In this chapter we will discuss the naturalness of salt marshes and their plant cover and the interaction of the vegetation with abiotic conditions, such as sediment and nutrient input, and with biotic conditions, such as wild herbivores and livestock. We will particularly address the long-term dynamics of salt-marsh communities. We will demonstrate to what extent the findings of small-scale experiments on individual salt marshes can be generalized to add to our understanding of community ecology of salt marshes, and how this knowledge can be applied for management purposes.

10.2 Natural salt marsh: the back-barrier model including a productivity gradient

Barrier islands in the Wadden Sea feature sandy beaches along the North Sea and silty salt marshes along the Wadden Sea. Sedimentation of fine suspended material (silt or clay) can take place in the shelter of dunes. The geomorphological conditions of the sandy subsoil show a gradual slope from the foot of the dunes towards the intertidal flats. As the period of inundation is longer and the frequency higher at low elevation, the input of sediment is higher at the low marsh than at the higher marsh. Apart from the zonation from low to high marsh, the thickness of the sediment layer changes over time from a young marsh to an older marsh. The back-barrier salt marsh of the Dutch island of Schiermonnikoog shows such a successional pattern. The eastern part of the island gradually extends further eastward. Hence, a chronosequence representing vegetation succession (De Leeuw *et al.* 1993; Olff *et al.* 1997) has established with very young marsh (from 0 years onwards) at the far east and older marshes (up to 150 years) more to the west (Fig. 10.1). Increasing age of the marsh coincides with a thicker layer of sediment resulting from tidal inundation. Thus, the eastern part of Schiermonnikoog features a matrix of two phenomena: zonation and succession. While walking from east to west at high or low elevation levels, succession of the higher and lower marsh can be studied, respectively. With the sediment, organic matter including nitrogen is imported. The nitrogen pool of the top 50 cm of the soil, i.e. the rooting depth of most plant species, is positively related to the thickness of the sediment plus underlying sandy soil. By comparing various back-barrier systems

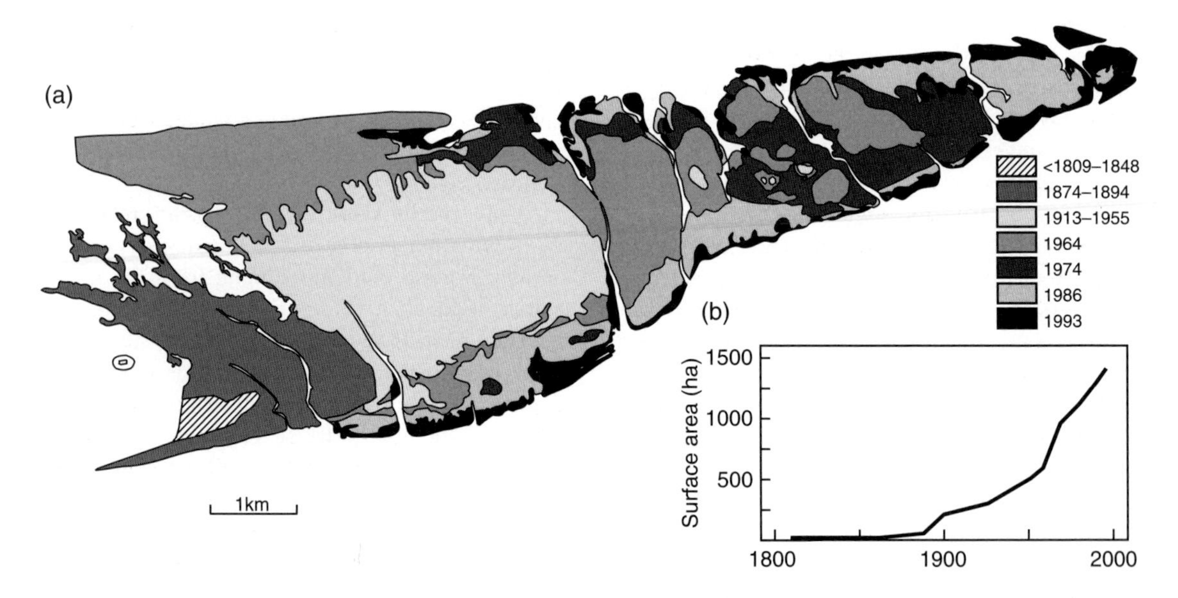

Figure 10.1 (a) The development history of the eastern part of the Dutch Wadden Sea island of Schiermonnikoog. The different shadings represent different age classes on the basis of maps and aerial photographs. (b) Development of the size of the vegetated marsh and dune area on the eastern part of Schiermonnikoog from 1989 onwards. After Van der Wal *et al.* (2000b).

During undisturbed succession at the high marsh in temperate European marshes, the low-statured species *Festuca rubra* eventually will be replaced by the tall-growing grass *Elymus athericus* (Leendertse *et al.* 1997). Both species were affected when herbivores were excluded, indicating local effects of grazing by intermediate-sized herbivores, because the herbivores are not able to prevent the increase of *Elymus athericus* at the high marsh (Kuijper and Bakker 2005). The main reason for this may be that *Elymus athericus* is not preferred by any herbivore (Prop and Deerenberg 1991; Van der Wal *et al.* 2000a; Kuijper *et al.* 2008), and grazing pressure drops dramatically once this species dominates the vegetation (Kuijper *et al.* 2008).

10.5 Large-scale effects of an intermediate herbivore on salt-marsh vegetation

The small-scale exclosure experiments and studies on individual plants on the salt marsh on Schiermonnikoog revealed that plant species replacement is retarded by herbivory. The effects of hare grazing especially were dominant and were most pronounced in young salt marshes (Kuijper and Bakker 2005). Grazing by hares retarded succession by more than 25 years (Van der Wal *et al.* 2000c). This implies succession should proceed fast when hares are not present at the initiation of salt-marsh development. Hence, late successional species should dominate at an earlier stage of development compared with salt marshes that developed in the presence of hares. This idea was tested by comparing the hare-grazed salt marsh on Schiermonnikoog with those of two Wadden Sea islands without hares, namely Rottumerplaat (The Netherlands) and Mellum (Germany).

On all three islands, sites were selected where salt-marsh development had started in the early 1970s. Transects of 1000 m running from the foot of a dune towards the intertidal flats were matched for surface elevation with respect to the level of mean high tide and sediment thickness (Kuijper and Bakker 2003). Early to mid-successional plant species *Puccinellia maritima* and *Plantago maritima*, which are the preferred food plant of geese, occurred at a similar elevation with higher cover on Schiermonnikoog than on Rottumerplaat and Mellum (Fig. 10.4). *Plantago maritima* was rarely found on Rottumerplaat and Mellum. *Festuca rubra*, a preferred food plant for both geese and hares, occurred over a large part of the elevation gradient on Schiermonnikoog, but was found at only a small part of the gradient on Rottumerplaat and Mellum (Kuijper and Bakker 2003). In contrast, the typically late successional species *Atriplex portulacoides* dominated the lower elevations on both Rottumerplaat and Mellum, whereas it had low cover on Schiermonnikoog (Fig. 10.4). *Elymus athericus*, a characteristic late successional species of the high marsh, occurred with higher cover at both low and high elevation on Rottumerplaat and Mellum compared with that on Schiermonnikoog. At the upper part of the elevation gradient on Rottumerplaat and Mellum a monoculture of *Elymus athericus*, covering 100%, was found. In contrast, on Schiermonnikoog, *Elymus* cover did not reach values higher than 70% (Kuijper and Bakker 2003).

It can be concluded that the small-scale exclosure experiments on Schiermonnikoog are not applicable only to understanding the local effects of grazing, but can also be extrapolated to a larger scale. Intermediate-sized herbivores affect the community structure of large-scale salt-marsh systems on the back-barrier Wadden Sea islands.

10.6 Interaction of herbivory and competition

Apart from experiments focusing on the level of the entire vegetation, detailed experiments with individual plant species may reveal which mechanisms play a role in plant species replacement along the productivity gradient. In addition to plant–plant competition, plants have to deal with changing levels of herbivory. The small highly herbivore-preferred *Triglochin maritima* is hardly present at the very young and old marshes, but is very abundant at intermediate-aged marshes. Competition and grazing are closely linked: when grazing pressure is relaxed, competition with neighbouring plants is intensified. Grazing is shown to influence these competitive interactions between plants, acting

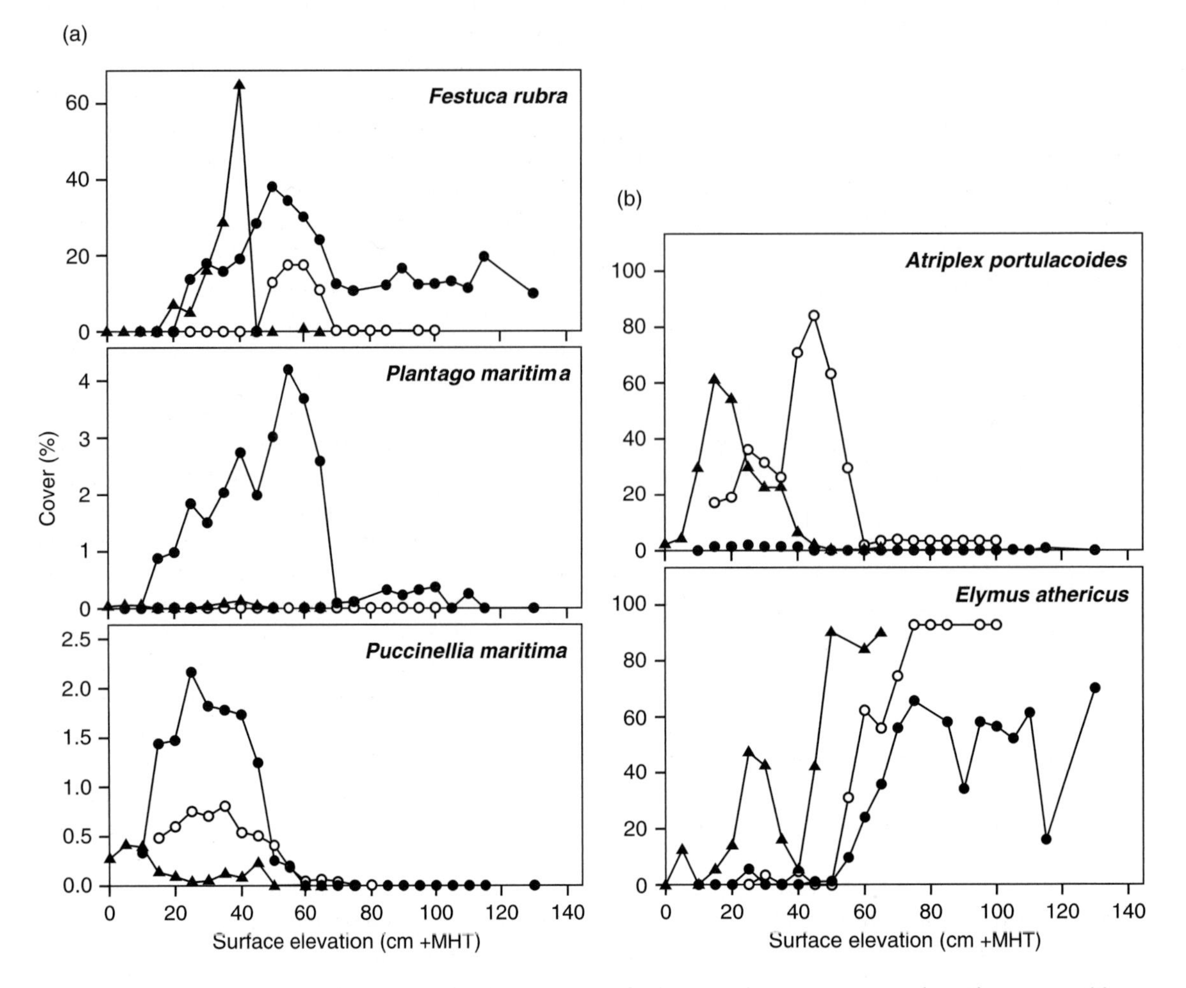

Figure 10.4 Average cover of (a) important food plant species for hares and geese, *Festuca rubra, Plantago maritima* and *Puccinellia maritima*, and of (b) typically late successional, unpalatable species, *Atriplex portulacoides* and *Elymus athericus*, at different marsh surface elevation (cm + mean high tide (MHT)) for Schiermonnikoog (island with hares), Rottumerplaat and Mellum (islands without hares). Closed circles, Schiermonnikoog; closed triangles, Rottumerplaat; open circles, Mellum. After Kuijper and Bakker (2003).

both directly on the target plant and indirectly through its neighbours. The significance of competition and herbivory largely depends on plant stature relative to the neighbouring vegetation. Although establishment of *Triglochin maritima* starts from seed, the high grazing pressure at younger marshes determines its abundance in the sward. However, at productive old marshes this small-statured plant is outcompeted by tall-growing late successional species. The distribution of *T. maritima* is 'sandwiched' between intense grazing in the younger marsh and increasing competition for light in the older marsh (Van der Wal *et al.* 2000a).

Adult plants of *Elymus athericus* are tall and not preferred by any of the herbivores. However, experiments in which grazing and competition were manipulated along the productivity gradient show that herbivory negatively affects the survival of seedlings (being a good food source) in the unproductive sites. At the productive sites, plant competition becomes an overruling factor. When seedlings grow in natural vegetation, the increased competition prevents any increase in biomass, whereas in the absence of competition the plant can grow fast because of high nutrient availability along the productivity gradient. Even though *Elymus athericus* is an unpalatable superior competitor as an adult plant at highly productive sites, in its seedling phase its growth is strongly reduced by herbivory at unproductive stages and competition with neighbouring plants at the productive stages (Kuijper *et al.* 2004).

10.7 Competition and facilitation between herbivores

10.7.1 Short-term competition and facilitation between hares and geese

For a large part of the year hares and geese forage on the same food plants, hence competitive interactions may also occur. Exclusion of brent geese at scales ranging from 30 m^2 to 1 ha at the salt marsh on Schiermonnikoog enhanced the level of utilization by hares in both *Festuca rubra*- and *Puccinellia maritima*-dominated marshes. The more geese were excluded from a site, the stronger the increase of hare grazing pressure. When geese were excluded, the 'original' decrease in *Festuca* consumption by geese was completely matched by increased hare grazing, while for *Puccinellia* only part of the surplus was grazed. Apparently, competition for food between hares and brent geese also occurs and plays a role in the habitat use of hares (Van der Wal *et al.* 1998).

Competitive and facilitative interactions between geese (barnacle and brent geese) (Stahl 2001) and geese and hares were studied on Schiermonnikoog (Stahl *et al.* 2006). Biomass (through temporary exclosures) and quality (by fertilizer application) of grass swards were manipulated and the foraging preferences of the herbivores were recorded. Captive barnacle geese were used to set the stage for a choice experiment with captive brent geese, as the latter species normally exploits the vegetation 'on the heels' of the former. Brent geese preferred to forage on vegetation previously grazed by barnacle geese, probably reacting to enhanced quality of the regrowth, in spite of the higher biomass of the ungrazed swards (Stahl 2001). In another experiment with captive barnacle geese, it was demonstrated that grazing affected the sward characteristics significantly: the proportion of dead biomass in the vegetation was reduced, and the production of additional axillary tillers increased

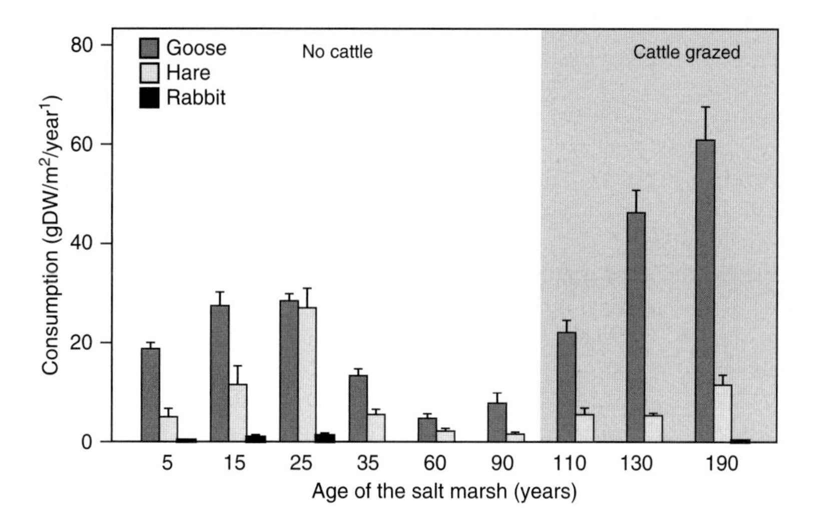

Figure 10.5 Total plant consumption of geese, hares and rabbits in salt marshes of different ages at Schiermonnikoog. Cattle grazing occurs only at the older marshes. Consumption was calculated on the basis of total droppings weight by multiplying the cumulative amount of droppings during 1 year (1999–2000) by the droppings weight per species. Subsequently, consumption was calculated from: total faecal mass/(1 – DE). Digestive efficiency (DE) for hares and geese was obtained from literature (Van der Wal *et al.* 1998). After Kuijper (2004).

(Van der Graaf *et al.* 2005). Both barnacle and brent geese selected plots with plants that have a high nitrogen content. Barnacle geese avoided plots with high biomass. Geese mainly selected plots that have been previously grazed by either geese or hares within the same season. Grazing by both geese and hares leads to an increased quality of the sward. Under these circumstances, herbivores profit from the increased tissue quality as a result of an elevated rate of nutrient intake. However, when the forage resource is used jointly by more than one herbivore species, a shift towards less preferred plots by one species may take place. Hares prefer the combination of high biomass with high plant quality in the absence of geese (Stahl *et al.* 2006). Van der Wal *et al.* (1998) suggested that large flocks of socially foraging geese rapidly deplete preferred salt-marsh sites in spring and evict hares to alternative less favourable foraging sites.

10.7.2 Long-term facilitation between herbivores

The previous section showed that the cover of species that are selected as food plant by both geese and hares, such as *Puccinellia maritima*, *Plantago maritima* and *Festuca rubra*, is higher at hare-grazed islands, whereas the cover of unpreferred plants, such as *Atriplex portulacoides* and *Elymus athericus*, is lower. Hare grazing may thus facilitate food supply for geese (Kuijper and Bakker 2003). This idea was tested experimentally at the salt marsh on Schiermonnikoog. The woody shrub *Atriplex portulacoides* is unpalatable for geese. It can overgrow the preferred food plant *Pucinellia maritima*. When *Atriplex portulacoides* was removed, goose grazing, expressed as the number of droppings found, was higher than in the control plots. In contrast, goose grazing declined when *Atriplex portulacoides* individuals were planted in a *Pucinellia maritima* sward (Van der Wal *et al.* 2000c). Knowing that hares forage on *Atriplex portulacoides* during winter, this experiment clearly demonstrated the effect of grazing facilitation by hares for geese.

Although hares can retard vegetation succession for several decades (Van der Wal *et al.* 2000c; Kuijper and Bakker 2005), they eventually lose control in the higher ranges of the productivity gradient. Large herbivores, such as livestock, are needed to set back the successional clock. Indeed, at the older cattle-grazed salt marsh in the chronosequence on Schiermonnikoog, grazing pressure of hares and geese increases again compared with the ungrazed older marsh (Kuijper 2004; Fig. 10.5). An experiment with exclosures on the cattle-grazed marsh revealed that after 30 years of cessation of cattle grazing no hares grazed inside the exclosures when the cover of tall plants, such as *Elymus athericus*, was $> 30\%$. Thus, clear facilitative effects of cattle on the feeding opportunities of hares were found (Kuijper *et al.* 2008). This finding is in contrast to studies from other areas that reported only competitive interaction between hares and livestock (Hulbert and Andersen 2001; Smith *et al.* 2004). The contrasting conclusions of these studies may be the result of the timescale of the experiments. Facilitative effects between cattle and hares on Schiermonnikoog were observed only when looking at the long-term effects, including the effect of cattle on the competitive replacement of plant species. Only when species replacement did occur in the absence of cattle was an effect on the abundance of hares observed. In contrast, in a short-term experiment on Schiermonnikoog in which cattle were excluded for 5 years, plant biomass increased inside the exclosure, but the period was too short for plant species replacement to occur. In this short-term study no effect on the abundance of hares was detected (Kuijper *et al.* 2008). This suggests that at a short timescale no effect of cattle grazing on hare abundance is apparent, whereas at a longer timescale facilitation occurs (Kuijper *et al.* 2008).

It can be concluded that competition between different species of herbivores occurs only in the short term, i.e. within one spring season. In the long-term, facilitation plays an important role. At the salt marsh on Schiermonnikoog, barnacle geese facilitate for brent geese within one season, hares facilitate for geese for several decades, and ultimately cattle facilitate for hares and geese, when hares have lost control of the vegetation.

10.8 Exclusion of large herbivores: effects on plants

10.8.1 Natural marshes

The effects of large herbivores on salt marshes is restricted to that of livestock. In fact, livestock grazing is the most common land use of North-West European salt marshes (Bakker *et al.* 2005a; Davy *et al.* 2009). Hence, the obvious way to study the effects of livestock grazing is to establish exclosures. In 1973 at the oldest part of the chronosequence on Schiermonnikoog (> 150 years) that was always cattle grazed, two exclosures were established, one at the higher and one at the lower marsh. At the higher marsh *Elymus athericus* was already present in the grazed area. The *Elymus athericus* community established at the expense of the *Juncus maritimus* community within five years after the cessation of grazing. The deposition of driftline material initiated temporary spots with the annual *Atriplex prostrata*, but within two years these were taken over again by the *Elymus athericus* community. This community also spreads at the transition to the low dune, but only gradually, and after 27 years remnants of the *Festuca rubra* community with *Armeria maritima* were still present. It seems that *Elymus athericus* is also spreading in the grazed area, but this is mainly due to the fact that the tall *Juncus maritimus* is not preferred by cattle and protects *Elymus athericus* from grazing, thus acting as a 'natural' exclosure (Bakker *et al.* 2003a).

At the lower marsh *Elymus athericus* was lacking in the grazed area at the start of the experiment. The *Artemisia maritima* community dominated within five years in the relatively higher parts inside the exclosure. It took 12 years before the first clone of *Elymus athericus* found its window of opportunity and became established. After 22 years the *Elymus athericus* community expanded. The initially bare soil at the lowest places became covered by the *Plantago maritima/Limonium vulgare* community after about ten years, after which the *Atriplex portulacoides* community took over after 22 years. The last has locally been replaced by the *Elymus athericus* community, 27 years after the cessation of grazing (Bakker *et al.* 2003a).

Taking into account the aforementioned natural succession without livestock grazing, it is likely that the oldest part of the salt marsh with a thick layer of clay in most sites will eventually be covered by the *Elymus athericus* community at both the high and the low salt marsh. That is exactly what happens after the long-term exclusion of livestock. The cessation of livestock grazing produces two main conclusions. Initially, the vegetation transforms into a 'flower garden' as many existing species have the opportunity to flower during the first few years before tall species become dominant and replace the present plant community with another one. Eventually, most plant communities are replaced by the *Elymus athericus* community at the salt marsh on Schiermonnikoog. Another part of the salt marsh on Schiermonnikoog was abandoned in 1958 for cattle grazing and grazed anew from 1972 onwards. Permanent plots in exclosures revealed that different plant communities converged into the *Elymus athericus* community after various periods of cessation of grazing: the *Juncus maritimus* community, the *Plantago maritima/Limonium vulgare* community and the *Artemisia maritima* community after 30 years and the *Juncus gerardi* community after 35 years. The only exception was the *Festuca rubra/Armeria maritima* community, which was not replaced 35 years after cessation of livestock grazing (Van Wijnen *et al.* 1997). Perhaps the combination of a thin layer of sediment (low nutrient pool) at this high elevation site and evapotranspiration during dry summer periods with subsequent high soil salinity have until now prevented replacement.

The natural marsh of Süderhafen (Germany) developed in the shelter of the former salt-marsh island of Nordstrand after 1925. The site was hardly grazed before 1968, and not at all since 1971. Repeated vegetation mapping in 1968 and 1995 revealed an expansion of the *Elymus athericus* community at the expense of the *Festuca rubra* community, and of the *Atriplex portulacoides* community at the expense of the *Puccinellia maritima* community (Bakker *et al.* 2003a).

Combining permanent plot data from experimentally ungrazed sites on the back-barrier marshes on Schiermonnikoog (The Netherlands), Terschelling

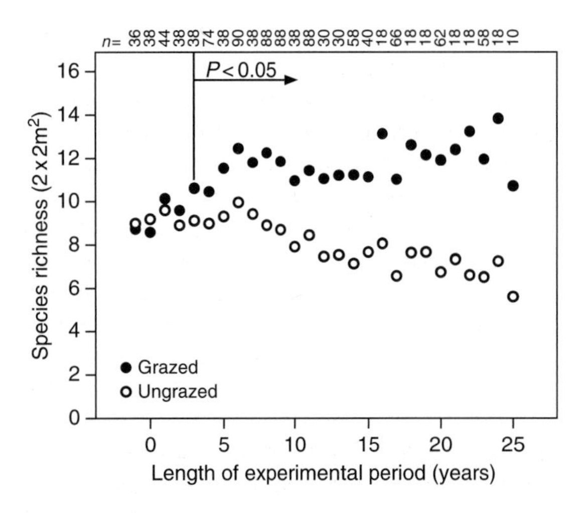

Figure 10.6 The development of plant species richness over time in paired livestock-grazed and ungrazed permanent plots in the Wadden Sea. Sample sizes (*n*) per year since the start of the treatment are indicated in the top of the diagram. After a period of three years the differences between grazing treatments were significant, indicated by the arrow with text $P < 0.05$. Observations for two sites started one year before the treatment was established. After Bos *et al.* (2002).

(The Netherlands) and Skallingen (Denmark) revealed that the convergence to the *Elymus athericus* community after the exclusion of livestock grazing is a general phenomenon (Bos *et al.* 2002).

Not only the diversity of plant communities declined after the cessation of livestock grazing. The species richness within plant communities in paired permanent plots in experimentally ungrazed and control plots also decreased significantly after five years (Fig. 10.6). (Data have been combined from the back-barrier marshes on Schiermonnikoog, Terschelling and Skallingen (Bos *et al.* 2002; Bakker *et al.* 2003a).) These permanent plots also revealed that out of 30 frequently occurring plant species only four had a significantly higher occurrence at ungrazed than at grazed marshes, namely *Artemisia maritima, Atriplex portulacoides, Atriplex prostrata* and *Elymus athericus*. Three species were indifferent, namely *Festuca rubra, Juncus maritimus* and *Lotus corniculatus*. All remaining 23 species had a significantly higher occurrence at grazed than at salt-marsh sites excluded for more than 20 years (Bos *et al.* 2002).

10.8.2 Artificial salt marshes

There are experimentally ungrazed plots in artificial marshes in Dollard Bay, The Netherlands. In these brackish, highly productive marshes the exclusion of cattle resulted in the increase of *Elymus repens* within six years, mainly at the expense of *Puccinellia maritima*. Species richness was higher in grazed than in excluded plots (Esselink *et al.* 2002). When salt marshes are broad enough, a gradient in grazing intensity emerges. Cattle and sheep tend to concentrate near the seawall, where fresh drinking water is available. Hence a reduction in grazing is found at the seaward site of salt marshes resulting in a taller canopy. Indeed, gradients of increasing canopy height towards the marsh edge were reported in the Dollard (Esselink *et al.* 2000), the Ley Bucht (Andresen *et al.* 1990) and Sönke-Nissen-Koog (Germany) (Kiehl *et al.* 1996).

No controlled large-scale grazing experiments have been established along the Dutch mainland coast with artificial marshes. However, three good examples can be found along the German coast. The first site is located at Friedrichskoog in Lower Saxony. It developed after 1854 and was long-term sheep grazed. The experiment was established in 1988 to study the effects of different stocking rates on soil and vegetation (Kiehl *et al.* 1996; Kiehl 1997). The stocking rate was expressed in sheep-units, i.e. adult sheep including their lambs (1 sheep-unit equals 2.8 sheep). The area was heavily grazed 'as a golf course' by 3.4 sheep-units/ha at the end of the grazing season. The control area with 3.4 sheep-units/ha was compared with 1.5 and 1.0 sheep-units/ha and cessation of grazing. At the start of the experiment this salt marsh harboured mainly *Festuca rubra* community, at the lower marsh *Puccinellia maritima* community, and at the intertidal flats *Spartina anglica* and *Salicornia* spp. communities were found (Kiehl 1997). The vegetation revealed a relatively small coverage of the *Elymus athericus* community after the cessation of grazing 11 years after the start of the experiment (Bakker *et al.* 2003a).

Apart from the above large-scale patterns, the Friedrichskoog experiment also revealed different micropatterns in the vegetation with the various stocking rates seven years after the start of the experiment. The micropatterns were formed by a mosaic of short and tall *Festuca rubra* stands on a scale of square decimetres in transects of 2 m × 10 m. In the most intensively grazed and the abandoned paddocks, no micropattern was found. The vegetation in the transects was homogeneously short or tall, respectively. However, micropatterns occurred in the three intermediately grazed paddocks with the highest spatial diversity in the 1.5 sheep-units/ha (Berg *et al.* 1997).

The second site is located at Sönke-Nissen-Koog in Schleswig Holstein. It developed after 1924 and was long-term sheep-grazed. The experimental treatments were established at the same time and had the same layout as at the Friedrichskoog site. At the start of the experiment this salt marsh mainly harboured the *Puccinellia maritima* community with locally some *Festuca rubra* and *Elymus athericus* communities, and with *Spartina anglica* and *Salicornia* spp. near the intertidal flats (Kiehl 1997). The marsh showed a large coverage of the *Elymus athericus* community after the cessation of grazing. The community covered smaller areas at the lower stocking rates, 11 years after the start of the experiment (Bakker *et al.* 2003a).

The third site is in the Ley Bucht (Germany). The site was cattle-grazed since its formation after 1950. The site was established as an experiment in 1980. The area with 2 cattle/ha was compared with areas with stocking rates of 1 and 0.5 cattle/ha and cessation of grazing. The zonation included *Elymus repens/Elymus athericus* and *Festuca rubra* communities close to the seawall, the *Agrostis stolonifera* community at the transition, the *Puccinellia maritima* community at the lower marsh, and *Spartina anglica* and *Salicornia* spp. communities near the intertidal flats. Eight years after the cessation of grazing, the *Elymus athericus* community covered large areas at the higher salt marsh and one spot at the lower marsh. It hardly occurred at the other grazing regimes (Bakker *et al.* 2003a). The *Elymus athericus* community quickly spread over both the higher and the lower marsh, and covered nearly the entire gradient 20 years after the cessation of grazing, at the expense of the *Festuca rubra* and the *Agrostis stolonifera* communities, and the *Puccinellia maritima* community, respectively. Also the 0.5 cattle/ha regime revealed a spread of the *Elymus athericus* community 15 years after the start of the experiment at both the higher and the lower marsh, but to a lesser degree than at the abandoned area.

Both artificial and the natural back-barrier salt marshes tend to transform into a dominance of the *Elymus athericus* community after the cessation of livestock grazing within 10–30 years, as could be predicted from succession without livestock grazing. However, a correlation between the number of years of exclusion of livestock grazing and the spreading of *Elymus athericus* is not always found. The salt-marsh sites that do not follow this rule seem to have a low sediment (nitrogen) input (Schröder *et al.* 2002). In these sites exclusion of livestock grazing did not result in a dominance of *Elymus athericus* within 30 years. A complication may be that because of the low sediment input these sites are building a sedimentation deficit due to continuous sea-level rise, and hence are becoming wetter. This may be an extra factor preventing the establishment of *Elymus*. Another conclusion is that grazing with low stocking rates cannot prevent the spread of *Elymus athericus*, but only retards the spread. In contrast to intensive grazing and no grazing at all, intermediate grazing can create small-scale patterns in the vegetation.

10.9 Exclusion of large herbivores: effects on invertebrates

On the natural mainland salt marsh in Mont Saint-Michel Bay (France), the invasive species *Elymus athericus* outcompetes *Atriplex portulacoides*. Apart from changes in plant communities, this results in changes in invertebrate communities, particularly spiders. The invasion of *Elymus athericus* led to an increase in the overall species richness. Causes may be the formation of a dense, tall sward which allows colonization of web-spinning species such as *Argiope bruennichi*, *Neoscona adianta* and *Larinioides cornutus*. The building of a deep litter layer favours nocturnal wanderers (Gnaphosids, Clubionids), ambush hunters (Thomisids) and litter-sensitive sheet-weavers. Non-coastal species such as the ground-living nocturnal *Pachygnatha degeeri* and the halophilic sheet-web spinning *Arctosa fulvolineata* increased. However, the dominant halophilic species *Pardosa purbeckensis* was strongly negatively affected by the invasion of *Elymus athericus* (Pétillon *et al.* 2005). Some halophilic ground beetle species were more abundant in grazed than in abandoned sites and vice versa. In general, no effect of management on species richness was found for ground beetles. Generally, spiders seem to be more dependent on vegetation and litter structure than ground beetles (Pétillon *et al.* 2008).

The aforementioned experiment in the artificial marsh of the Ley Bucht aimed to study the effect of various stocking rates on the invertebrate fauna (Andresen *et al.* 1990). For invertebrates, it may not only be the plant species composition that is important. Non-flowering *Asters* were found only at the higher salt marsh within the highest stocking rate. The canopy height of the understorey was higher in the abandoned site than in the grazed sites (Andresen *et al.* 1990). In the third year of cessation of grazing, positive effects for several invertebrate groups were recorded for Collembola, Aranea, Amphipoda, Coleoptera and Diptera. This was attributed to the accumulation of litter, increase of flowering plants and hence availability of pollen and nectar and therefore higher aboveground biomass for leaf- and stem-dwelling species (Irmler and Heydemann 1986). *Erigone longipalpis*, a halophilic species, is the most important spider species in the *Puccinellia maritima* community. Other species occur mainly in the *Festuca rubra* community and cannot be considered halophilic, namely *Oedothorax retusus, Pardosa agrestis* and *Pachygnata clerki*. Whereas *Erigone longipalpis* still occurred in high abundance in the *Festuca rubra* community at the start of the experiment in 1980, it has since 1982 moved to the lower *Puccinellia maritima* and *Salicornia* spp. communities in the abandoned site. The other species spread into the lower salt marsh at different rates (Andresen *et al.* 1990). A distinct zonation of the invertebrate communities was observed in the first three years of the experiment. The community diversity was highest in the abandoned site, since communities of the higher marsh spread into the lower marsh. In 1988, however, the community of the higher marsh had spread over the entire elevation gradient, and completely replaced the communities of the lower marsh in the abandoned site. Hence, eventually the community diversity was lowest at the abandoned site. However, the number of species became highest at the abandoned site, partly as a result of immigration from adjacent grassland. But the main reason was that many species are damaged by grazing (Irmler and Heydemann 1986). The authors especially stress the damaging effects of grazing on many arthropod communities.

Two and three years after the start of the aforementioned grazing experiment in Friedrichskoog and Sönke-Nissen-Koog, invertebrates were monitored. Mainly herbivorous and flower visitors were positively affected by cessation of sheep grazing and the resulting flowering of *Aster tripolium* and *Plantago maritima*. A minor part of the herbivorous fauna profits from enhanced plant growth in moderately grazed sites. Typical soil dwellers benefit from grazing owing to greater amounts of bare soil (Meyer *et al.* 1995).

In general, the community structure changes from a dominance of detritivores to a dominance of herbivores after the cessation of sheep grazing, and after the cessation of cattle grazing. The number of species and individuals increases shortly after the cessation of grazing, but after a longer period of cessation of grazing typical halophilic species may decrease. For the time being it is not possible to discuss top-down or bottom-up concepts with respect to the interaction between vegetation and invertebrates. Food web studies could help in this discussion and are currently being carried out.

10.10 Exclusion of large herbivores: effects on birds

10.10.1 Migrating birds

In order to evaluate the importance of livestock grazing for habitat use by geese in the Wadden Sea, a large-scale inventory was made. Sixty-three transects were established, subdivided over 38 sites. Only those sites with a stable and clearly defined management regime for at least six preceding years were included. Management was subdivided into 'long-term ungrazed'(> 10 years), 'short-term ungrazed' (6–10 years), 'lightly grazed' (low stocking rates, i.e. ≤ 4.5 sheep/ha or ≤ 1 cow/ha), and 'intensively grazed' (i.e. with high stocking rate). Only marshes with sufficiently large surface area (> 5 ha), large enough for a flock of geese to land on, were included. The sites were distributed over the entire Danish ($n = 11$), German ($n = 17$) and Dutch ($n = 10$) Wadden Sea. Twenty-two sites harboured transects with at least two different grazing regimes under similar abiotic conditions. Seventeen sites with paired transects were visited twice, once in April and once in May 1999. The transects on back-barrier marshes were, with one exception, visited only by brent geese, whereas most transects on artificial marshes along the mainland coast were utilized by both brent and barnacle geese. For each management regime at each site, one transect was established perpendicular to the seawall and the coastline, along the entire extent of the marsh. Hence, transects were variable in length, ranging from 100 m to 1000 m, and included high-marsh, mid-marsh and lower marsh sections. Twenty plots of 4 m^2 were sampled per transect, and the accumulated number of goose droppings were counted and the plant community was assessed (Bos *et al.* 2005).

The communities of *Elymus athericus*, *Artemisia maritima* and *Atriplex portulacoides* had a significantly taller canopy, but a lower goose dropping density than the communities of *Agrostis stolonifera*, *Festuca rubra* and *Puccinellia maritima*. Dropping density at the transect level declined with decreasing livestock grazing regime. However, only the long-term ungrazed regime combined for barrier marshes and artificial marshes had significantly lower dropping densities than the other regimes (Fig. 10.7). These results are valid for May, the end of the staging period for both goose species. In April, goose-dropping densities at the transect level did not differ between grazing regimes. There were no significant differences in dropping densities by geese between transects grazed by sheep or cattle (Bos *et al.* 2005). We conclude that the long-term exclusion of livestock on salt marshes will result in a decline in utilization of these areas by spring-staging geese.

10.10.2 Breeding birds

The effects of excluding livestock grazing on breeding birds cannot be studied in small-scale exclosure experiments as for plants and invertebrates. Also a comparative study in the entire Wadden Sea, as for migrating birds, has not been carried out so far. We derive our knowledge from a small number of studies describing differences in ungrazed and differently grazed marshes. At the natural marsh on Schiermonnikoog, including some low dunes, the breeding population was monitored in 1973 and 1978. The 83 ha of marsh ungrazed since 1958 harboured maximally 31 species with in total 850–1000 breeding pairs, the 77 ha continuously cattle-grazed marsh hosted maximally 25 species with in total 550–600 breeding pairs. In 1978, the grazed marsh harboured 133 breeding territories for oystercatcher, 10 for lapwing and 71 for redshank, whereas the grazed marsh harboured 85, five and 48 territories, respectively (Van Dijk and Bakker 1980).

Studies on the relationship between management and vegetation, and the occurrence of breeding birds have been summarized by Koffijberg (in press). Most studies have been carried out on artificial marshes in Germany (Hälterlein 1998; Eskildsen *et al.* 2000; Hälterlein *et al.* 2003; Oltmanns 2003; Schrader 2003; Thyen and Exo 2003, 2005; Thyen 2005). They reveal a trend that relaxation of formerly heavily grazing regimes results in an increase in species richness, particularly due to a species group shift from waders, gulls and terns towards ducks and songbirds. Another trend is the decrease of avocet (*Recurvirostra avosetta*), great ringed plover (*Charadrius hiaticula*), Kentish plover (*Charadrius alexandrinus*), common tern (*Sterna hirundo*) and Arctic tern (*Sterna paradisaea*) after the cessation of grazing and subsequent vegetation succession. A problem in these studies is that the results represent snapshots, describing 'pioneer situations' a few years after transition of management, and do not include the long-term effects of cessation of grazing.

For some species more detailed information is available. Increased grazing negatively affects the number of redshanks. This was attributed to the destructive effects of trampling of nests and hatchlings, whereas changes in the vegetation composition were considered less important (Schultz 1987). However, in salt marshes in Great Britain the occurrence of redshank densities were positively related to the extent of the *Elymus athericus* community. This relation could be explained by the variation in vegetation structure. Cattle-grazed plots, with

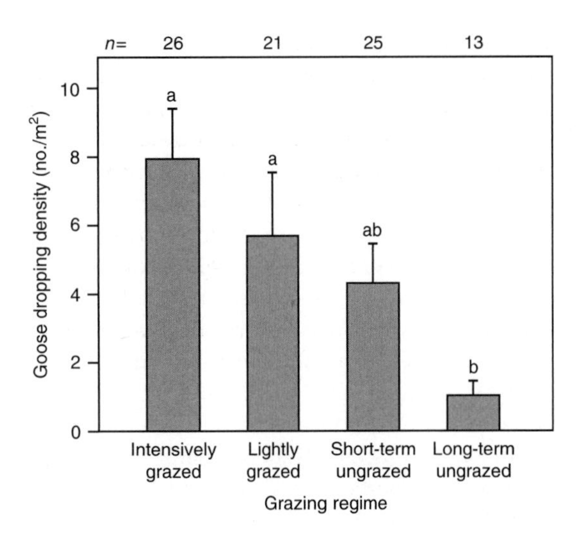

Figure 10.7 Average goose grazing pressure at the transect level in relation to livestock grazing regime for all transects that were paired within the same site. Bars that do not share the same letter differ significantly ($P < 0.05$). After Bos *et al.* (2005).

Elymus athericus covering up to 30%, supported the most structurally diverse vegetation and the highest breeding densities. In contrast, ungrazed plots of similar habitat contained tall, uniform vegetation and supported significantly lower breeding densities (Norris *et al.* 1997). The period of abandonment was not indicated. However, a survey on 77 saltmarsh sites in Great Britain revealed that breeding redshank densities were lowest on heavily grazed marshes and tended to be highest on lightly or ungrazed marshes (Norris *et al.* 1998). Redshanks breeding on salt marshes partly feed on nearby intertidal flats and build their nests hidden among vegetation of intermediate height, avoiding areas with low cover or with very tall vegetation (Cramp and Simmons 1983). In the Dollard (The Netherlands) cattle-grazed salt marsh, densities of redshanks were approximately two breeding pairs per hectare at a grazing regime of ~200 cattle-days/ha in 1984, and decreased to less than one breeding pair per hectare in 1998. Within the same period cattle grazing was reduced to ~50 animal-days/ha. The redshanks preferentially breed in the *Elytrigia repens* community, and in the less preferred short-grass stands with *Festuca rubra*, *Agrostis stolonifera* and *Puccinellia maritima*. Especially the latter stands were partly replaced by bare soil and secondary pioneer community of *Salicornia* spp. and *Suaeda maritima*, which was, however, attributed to increasing numbers of spring-staging barnacle geese and not to decreased cattle grazing (Esselink 2000).

We have to conclude that the effects of cessation of livestock grazing on breeding birds need further study. From the results so far, we suppose an initially positive, but in the long term negative, effect.

10.11 Ageing of salt marshes and implications for management

As long as the area of salt marshes increases, marshes will feature the successional series of pioneer, young and older mature marshes. When these extension processes stabilize eventually, only mature marshes will be found. This happens at back-barrier marshes that do not expand. It also happens along the mainland coast where the present area is maintained, and no further expansion into the intertidal flats takes place. In the past, it was economically feasible to embank marshes, and start new sedimentation fields (Esselink 2000). Nowadays, it is no longer economically feasible for many farmers to graze livestock at the marshes. The combination of decrease in the pioneer zone, and hence maturation of the marshes, and abandonment of livestock grazing results in the encroachment of *Elymus athericus* on artificial marshes (Dijkema 2007).

What are the implications for management (often livestock grazing) in view of these ageing processes of salt marshes? According to the 'wilderness concept' (a contradiction in itself for an artificial marsh), the solution with respect to the question 'to graze or not to graze' (Bakker *et al.* 2003a) is easy: the management option will be 'no grazing'. This will undoubtedly result in a loss of biodiversity at the local scale. However, at the scale of the entire Wadden Sea, it should be a preferred option for the marshes that have never been grazed by livestock such as the eastern parts of Terschelling (The Netherlands), Schiermonnikoog (The Netherlands), Ameland (The Netherlands) and Spiekeroog (Germany). In the long run, these areas will demonstrate whether there is a world beyond *Elymus athericus*.

According to the 'biodiversity concept' the answer to the question 'to graze or not to graze' will be: define the biodiversity target at a distinct scale, and decide to what extent livestock grazing as a management tool may help to reach the biodiversity target.

It is known that no grazing results in a low diversity for plants and less favourable feeding conditions for hares and spring-staging geese. High-intensity livestock grazing is a good option for spring-staging geese. Low-intensity grazing renders a pattern of intensively grazed short swards and lightly or no-grazed taller patches of vegetation. The difference with respect to the options *no grazing* or *intensive grazing* seems the patchiness and the spatial scale. However, our knowledge of the consequences of such a mosaic for the diversity of breeding birds and invertebrates is fragmentary.

Another option with respect to grazing is rotational grazing. Livestock grazing can be abandoned after a period of intensive livestock grazing. The result will be flowering of the plants and the possibility of replenishing the soil seed bank. Flowers and taller stems will attract invertebrates, which can be the prey items for breeding birds. Before *Elymus athericus* invades, the intensive grazing regime should be re-installed. The results of such a *rotational grazing regime* have not been monitored so far. Salt-marsh communities and their management will profit from large-scale and long-term experiments in which the interactions of plants, invertebrates and birds are studied.

In summary, in order to have a full display of salt-marsh communities, including many species of plants, vertebrates and invertebrates, the best management option is to have variety in the structure of the vegetation. This can be achieved by variation in grazing management, both in space and time.

PART V

FUTURE DIRECTIONS

CHAPTER 11

Evolutionary processes in community ecology

Jacintha Ellers

11.1 Introduction

Community ecology examines the distribution, abundance, demography and interactions among populations of coexisting species at a particular site or in a specific area. In past decades, it has been extremely successful in describing and explaining patterns of species diversity, food web structure, invasion, etc. However, at the same time community ecology has been under fire for lack of two important attributes. First, community ecology is short of general, mechanistic principles leading to quantitative predictions (Lawton 1999). Mechanistic approaches in ecology try to functionally link traits of individuals to higher level processes such as multispecies interactions and community structure. Yet in an effort to reduce the inherent complexity of biological communities, most studies have grouped species according to their trophic positions while disregarding differences in traits within these functional groups. Such simplification of community structure facilitates the study of ecological community properties, but it overlooks the species-specific contribution to multiple species interactions in the community. Analysis of functional traits is one way to quantitatively predict the impact of local species loss or biological invasions on ecosystems. Second, community ecology has always assumed homogeneous populations that are impervious to evolutionary change, thereby excluding any evolutionary community dynamics resulting from selection on traits of individuals. For a long time, it was assumed that the relatively slow timescale of evolutionary changes rendered an evolutionary perspective unnecessary. However, evolutionary biology has produced compelling evidence that strong selection pressures and fast (co-) evolution are commonplace in nature (Thompson 2005). Hence, local adaptation of populations to community context may significantly affect community functioning. A growing number of community ecologists have come to realize the value of an evolutionary perspective for addressing ecological questions.

11.1.1 Bridging the gap between evolutionary biology and community ecology

Several recent papers have explored the possibilities for a synthesis between community ecology and evolutionary biology (Agrawal *et al.* 2007; Johnson and Stinchcombe 2007). Evolutionary biology studies how ecological factors regulate genetic variation and evolution, and can thus provide community ecology with a more mechanistic, quantitative insight into how genetic and phenotypic diversity can shape community processes. On the other hand, most evolutionary studies perform experiments with the single species in isolation, or only in direct interaction with another species such as studies on coevolution. Including community composition as an ecological context for evolution may increase our understanding of the maintenance of genetic and phenotypic variation in the field.

However, part of the difficulty in bringing together the community perspective with the evolutionary perspective is that the yard-sticks used to measure community performance differ from those relevant to evolutionary biology. Community properties most often used include productivity, carbon storage, nutrient acquisition or decomposition rate, all of which are related to community functioning. Others reflect community properties desirable for nature management such as species diversity, community stability and resilience. Although such properties can certainly derive from species composition and individual performance, they are not the primary targets of natural selection. Rather, from an evolutionary point of view, they are by-products of the competitive and mutualistic interactions between individuals under selection to enhance their own fitness. This discrepancy in views has led community biologists to discard individual variation and selection in favour of a more phenomenological approach, whereas it has caused evolutionary biologists to turn away from community ecology because of its descriptive nature. The value of a synthesis between community ecology and evolutionary biology will hence critically depend on the potential for common standards.

In this chapter, I aim to outline a newly emerging field of research: evolutionary community ecology. I will argue that considering variation in traits of individuals and species is essential to understand and predict community functioning and composition, and, vice versa, that community composition is a key component of the selective forces determining genetic and phenotypic variation at the individual level. Bridging the gaps between community and evolutionary ecology can be mutually beneficial. In fact, we may even need an integrative approach in order to face fast changing environmental conditions such as global warming and urbanization, which pose ecological as well as evolutionary challenges.

I will first review the main principles regarding genetic and phenotypic variation from evolutionary biology, including maintenance of genetic diversity, measuring genetic variation and phenotypic plasticity. I will emphasize parallels between genetic diversity and species diversity, and population and community ecology, respectively. I will show as a proof of principle that community properties result from variation and selection at the level of individual organisms. Subsequently, I will address how variation in one species mediates interaction at higher trophic levels or at lower trophic levels. I will focus specifically on two aspects of evolutionary community ecology: (1) how genetic diversity and phenotypic plasticity of single species influence the composition of communities and (2) how species diversity and composition in a community influence the genetic diversity and realized phenotype of single species.

This chapter will not go into research on macroevolutionary timescales such as community phylogenetics and co-diversification of communities. Even though evolutionary history can moderate community assembly, I will focus on the shorter timescale interactions as these are the most important ones to cope with the ever-changing environmental dynamics.

11.2 Evolutionary biology: mechanisms for genetic and phenotypic change

Evolutionary ecologists study the cause and effect of variation in fitness-related traits of individuals such as longevity, fecundity and feeding rate in their natural environment. They lean heavily on the robust framework provided by quantitative and population genetics for quantitative prediction of evolutionary change. The two main topics that can be identified are maintenance of genetic variation and the population benefits of variation. In other words, how does natural selection act on phenotypes while still preserving genetic diversity, and does variation in phenotype affect performance of an individual and its associated population? Understanding these processes at the population level may allow us to extrapolate the consequences of genetic diversity to community level and may illustrate its effect on community properties.

11.2.1 Benefits and maintenance of genetic diversity at the population level

The population-level consequences of genetic variation have been the domain of evolutionary biologists for decades. Genetic variation can be generated through sexual reproduction and the associated recombination during meiosis. Alternatively, mutations, insertions or deletions can change the genetic make-up of individuals. The majority of mutations and new genetic combinations are believed to be deleterious, either because they render an enzyme or promoter region non-functional or because they break up co-adapted gene complexes. Natural selection will quickly purge such variation from the population. But even without selection a significant proportion of genetic variation is lost each generation through the random effect of genetic drift. Especially in small populations, random drift can quickly reduce genetic variation. Only if genetic variation has a clear adaptive value will it be maintained in the population, because the fitness benefits associated with genetic diversity in a population will decrease extinction risk.

Two principal mechanisms of reducing extinction risk have been put forward. First, the tangled bank hypothesis proposes that individuals with different genotypes may each use a slightly different niche and therefore together are able to extract more food from their environment than genetically identical individuals (Bell 1991; Barrett *et al.* 2005). In addition, different genotypes may compete less because they explore different microniches (Antonovics 1978). The second hypothesis to explain the advantage of genetic diversity is the Red Queen hypothesis, which argues that genetic diversity reduces the risk of infection or attack by natural enemies (Jaenike 1978). Empirical studies have provided ample support for the benefits of genetic diversity within populations, including positive correlations with increased resistance to disturbance by grazing (Hughes and Stachowicz 2004), increased productivity and foraging rate (Mattila and Seeley 2007) and enhanced settling success (Gamfeldt *et al.* 2005).

However, few studies have actually determined the underlying mechanism of fitness enhancements. The tangled bank hypothesis assumes that different genotypes show positive complementarity, because microniche differentiation causes them to compete less with other genotypes than with their own genotype. In a study by Reusch *et al.* (2005) on a coastal community dominated by the seagrass species *Zostera marina*, experimental plots with a high genotypic diversity produced a higher shoot number and more biomass than genetically impoverished communities or monocultures of genotypes (Fig. 11.1). They were able to attribute the effect of genotypic diversity to positive interactions between genotypes, particularly facilitation. Some genotypes that performed poorly in monoculture had a proportionally strongly reduced mortality in genotypic mixtures (Reusch *et al.* 2005). Other studies, specifically testing the advantage of genetic complementarity using clonal diversity, also found effects. Semlitsch *et al.* (1997) found significant differences in the life history traits among different clonal groups in *Rana esculenta* and an increased proportion of metamorphosed larvae in clonal mixtures of frogs compared with a clone reared alone. In another study, genetically diverse groups of clones were demonstrated to be better invaders than genetically uniform groups of invaders. The better performance of the genetically diverse group of invaders was attributed to competitive release experienced by individuals in genetically diverse populations (Tagg *et al.* 2005).

Experimental evidence for improved population performance by reduced infection risk in genetically diverse populations is rare. Schmid (1994) found evidence that genetic diversity can influence mildew infection levels in *Solidago altissima*. Infection rate affected individual performance with increased height and biomass of less infected plants, but mean plant performance per plot was not correlated with genetic diversity levels. The few studies so far indeed provide evidence for genetic complementarity as a significant mechanism maintaining genetic variation.

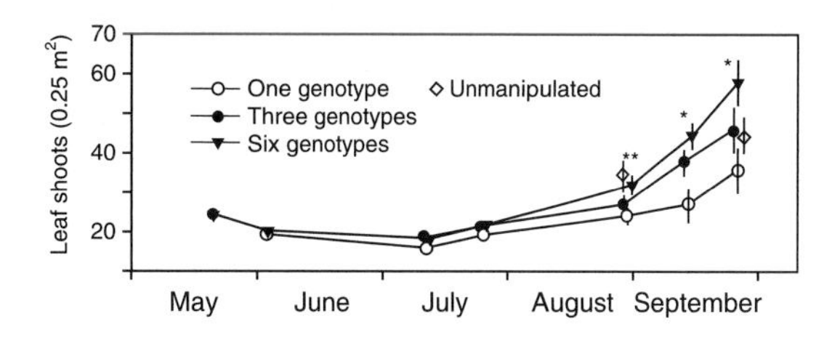

Figure 11.1 Comparison of mean leaf shoot density (SE) of *Zostera marina* among one-, three- and six-genotype treatments after climate perturbation, and natural shoot density at the experimental site. From Reusch *et al.* (2005). Copyright 2005 National Academy of Sciences, USA.

11.2.2 The source and nature of genetic variation

Before discussing any further the possible benefits of genetic diversity, it is worth considering the question of what is genetic variation? In many studies on the effect of genetic diversity on population or community functioning, genetic variation is taken to be the number of genotypes present. However, no measure of the degree of genetic dissimilarity is given. Although chances are negligible that two genotypes are exactly identical, that still leaves a wide range of genetic dissimilarities possible. Genome-wide screening methods such as amplified fragment length polymorphisms (AFLPs) or the use of large numbers of microsatellites or single nucleotide polymorphisms (SNPs) may give a more or less reliable measure of overall genetic diversity because they encompass the entire genome. Even then, overall genetic diversity in itself might not be sufficient to enhance population evolutionary potential. Genetic variation matters only if it translates into phenotypic variation upon which natural selection can act. If the rise of molecular biology has taught us one thing, it is that the relationship between genetic and phenotypic diversity is not a straightforward one. Previously, the most important distinction in genetic variation was between coding variation and non-coding variation such as neutral markers. Genetic variation in coding regions leads to amino acid substitutions in functional proteins, and therefore can contribute to the relative fitness of the genotype in the population. However, the long-held assumption that non-coding or neutral markers are not part of the process of transcription and translation, and hence are not subject to natural selection, has been challenged by advances in molecular biology. A much more complex picture of the molecular genetic organization structure now recognizes variation in coding regions, promoter regions, transcription factor binding sites, regulatory genes, etc.

The neutral molecular markers often used to quantify the genetic variability of populations are at best poorly related to variation in quantitative traits (Merila and Crnokrak 2001; Reed and Frankham 2001; McKay and Latta 2002). Reed and Frankham (2001) carried out a meta-analysis across 71 studies to determine the mean correlation between molecular and quantitative measures of genetic variation. Although molecular measures are commonly used as a proxy for quantitative genetic variation, the observed correlation between the two was weak and not significant ($r = 0.217$). This analysis shows the risk in using molecular measures of genetic diversity to predict population performance or other population properties, unless there is direct evidence that the markers used are indicative of the underlying quantitative genetic variation (Merila and Crnokrak 2001). In all other cases quantitative variation is better measured directly.

11.2.3 The relationship between genetic and phenotypic diversity

Another reason to apply caution in the extrapolation from genetic diversity to population performance is the existence of phenotypic plasticity. Phenotypic plasticity is the property of a given genotype to produce different physiological or morphological phenotypes in response to changing environmental conditions. Differences in phenotypic plasticity can be quantified by measuring a reaction norm (Via *et al.* 1995), which describes the change in a trait across environments (Fig. 11.2a). Reaction norms with a steep slope indicate strong trait sensitivity to the environmental factor, i.e. strong phenotypic plasticity, whereas a flat reaction norm denotes weak phenotypic plasticity.

Phenotypic plasticity may account for much of the phenotypic variation in populations and communities and is thought to play an important role in adaptation to spatio-temporal heterogeneity in environmental conditions. Induced phenotypic responses are a successful conditional strategy to cope with fluctuating conditions such as temperature, particularly when there are costs involved in the induction of the response. For example, the induction of heat shock proteins protects the organism from damage through misfolded proteins due to heat shock exposure. However, there are substantial energetic and metabolic costs involved, due to repression of standard cell activity after exposure to heat shock (Krebs and Holbrook 2001). Constitutive expression of this heat shock response

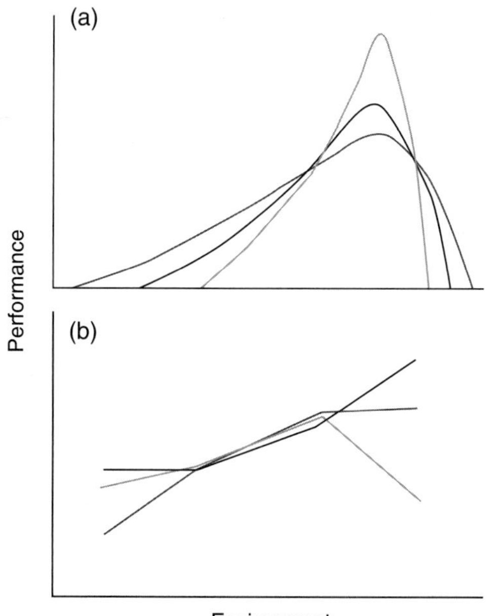

Figure 11.2 (a) Schematic representation of different reaction norms, showing continuous reaction norms that differ in the degree of sensitivity of the trait (slope of the reaction norm) and in the maximum performance (height of the peaks), but not in optimal value of the environmental variable. (b) Canalized reaction norms show the same phenotype under average environmental conditions but deviate in performance at either extreme of the range of environmental conditions.

is therefore undesirable. Environmental conditions triggering phenotypic plasticity explicitly include the biotic environment as well. Well-known examples of phenotypic plasticity induced by species interactions are the production of defensive plant compounds in response to herbivory and the induction of anti-predator defences.

Genotypes may differ in their response to environmental change, i.e. have different levels of phenotypic plasticity. In other words, genotypes that have similar phenotypes under one environmental condition may diverge in their performance when the environment changes, or vice versa. A common phenomenon causing seemingly monomorphic populations is canalization: genotypes have all evolved to identical optimal trait values under the most common environmental condition (Schlichting and Pigliucci 1998). Exposure to more extreme conditions, however, often reveals hidden variation for such traits (Fig. 11.2b). For example, now that global change causes extreme temperatures to become more common, genetic variation hitherto unexposed to selection will become apparent and important for survival. Overwhelming evidence now exists that the degree of phenotypic plasticity is genetically determined (Scheiner and Lyman 1989; Loeschke *et al.* 1999) and that natural selection can lead to local adaptation in plasticity levels (Liefting and Ellers 2008). There is a growing appreciation of the potential of phenotypic plasticity to modify species interactions.

11.3 Proof of principle: community properties result from genetic identity and selection at the level of individual organisms

How do population genetic processes extend to the community level? Population genetic theory could equally well apply to communities if genes have extended phenotypes (Dawkins 1982; Whitham *et al.* 2003). Genes with extended phenotypes affect not only the individual carrying the genes but also the performance of associated species in the community. For example, we can think of genetic differences in secondary chemical compounds that affect plant defensive capability against foliar herbivores (Havill and Raffa 2000; Harvey *et al* 2003). Although *interspecific* differences in allelochemical composition of host plants have long been recognized as an important factor structuring communities (Dungey *et al.* 2000; Sznajder and Harvey 2003; Wimp *et al.* 2005), the potential effect of *intraspecific* differences on community composition and functioning has been acknowledged only relatively recently. The performance and abundance of herbivores can be expected to be influenced in a similar way by intraspecific variation in host plants, provided that the magnitude of genetic differences is large enough to detect extended phenotype effects. Other examples of traits with extended phenotypes are induction of morphological anti-predator defences in many amphibians, or variation in thermal tolerance of algae in coral communities which is related to bleaching mortality (Fabricius *et al.* 2004).

The extended consequences of intraspecific genetic differences have received ample attention over recent years. Several recent studies show that plant genotype significantly affects various community properties, such as disease infection (Roscher *et al.* 2007), arthropod abundance and diversity (Stiling and Rossi 1996; Johnson and Agrawal 2005; Crawford *et al.* 2007) and decomposition rate (Madritch *et al.* 2006). For instance, genetic identity of the evening primrose (*Oenothera biennis*) explained more than 40% of the variation in arthropod diversity (Johnson and Agrawal 2005). Genotypic differences accounted for more variation in arthropod community structure than did environmental variation, indicating the relative importance of genetic factors. Not only species interactions are influenced by plant genotype. Madritch *et al.* (2006) showed that community properties such as decomposition and nutrient release from aspen litter are determined by genotype identity. Litter from different aspen genotypes differed significantly in carbon and nitrogen release, most likely because litter chemistries varied across genotypes (Madritch *et al.* 2006).

Most studies have focused on genetic differences in plants because plant diversity is thought to shape the composition and dynamics of animal communities. Only some studies have looked at the effect of genetic differences at other trophic levels than primary producers. For instance, plant productivity itself is to a large extent controlled by association with arbuscular mycorrhizal fungi (AMF). Experiments with genetically different isolates of the arbuscular mycorrhizal fungus *Glomus intraradices* show that AMF genotype can affect plant growth response and the extension of the fungal mycelium (Munkvold *et al.* 2004; Koch *et al.* 2006). The effect of genetic identity of *Glomus intraradices* on plant growth ranged from enhanced plant growth to no growth benefit or even a reduced growth, also depending on the environmental conditions (Koch *et al.* 2006). Associations with different individuals in the AMF population will present plants with either costs or benefits, potentially mediating the outcome of competition between plants.

Although most studies address the importance of genotype identity, effects of different phenotypes may also mediate community processes. Harvey *et al.* (2003) examined the effect of differences in a herbivore's diet on growth and development of its primary parasitoid and secondary hyperparasitoid. Hyperparasitoid body mass and survival were significantly larger if the herbivore had fed on plants containing lower concentrations of glucosinolates, a chemical defence compound. Hence, phenotypic plasticity caused by the nutritional differences in herbivore diet may affect performance across several trophic levels.

11.4 Effects of genetic and phenotypic diversity on community composition and species diversity

In recent years more and more researchers have identified the influence of genetic diversity on community functioning as an emerging frontier in ecology (Whitham *et al.* 2003; Agrawal *et al.* 2007). Yet it is not sufficient to merely observe a correlation between genetic diversity and community properties. To contribute to the understanding of general mechanistic principles of community ecology, studies need to assess the relative importance of ecological and genetic factors, and address the mechanism underlying these correlations.

Individual organisms in communities can contribute in two ways to the genetic diversity of the community. First, each species represents a discrete genetic entity, and individuals belonging to different species cause an associated rise in genetic diversity within the community. The effect of species diversity on community stability and function has been discussed extensively elsewhere (Hooper *et al.* 2005), so I will touch upon this only briefly. A consensus is emerging that, if plant species richness increases, community productivity as well as stability increases, although this effect levels off at relatively low species numbers. In the context of this chapter, I will mainly focus on the other way individual organisms may contribute to community genetic diversity – if individuals represent different genotypes within species.

11.4.1 Effects of genetic diversity on community functioning

Composition and functioning of communities is thought to be strongly shaped by plant species diversity, hence the effect of genetic variation in the dominant plant species is the best studied. Since plant genotypes can exploit slightly different resources, analogous to population-level effects of within-species genetic diversity, the prediction is that genetic diversity enhances productivity or resource efficiency of the plant, thereby increasing the total availability of resources in the community. The most straightforward experimental approach is to manipulate genotypic diversity in the community and compare indicators of community performance, such as plant productivity, carbon storage, nutrient acquisition or decomposition rate. A flurry of experimental work has now shown that increased plant genotypic diversity can explain a significant proportion of the community properties, including an increased net primary productivity (Crutsinger *et al.* 2006), higher fruit production (Johnson *et al.* 2006), higher resistance to invasion (De Meester *et al.* 2007) and accelerated decomposition rate (Madritch *et al.* 2006). Particularly in species-poor communities the effects of within-species diversity have close parallels to benefits at population level.

It is unlikely that the positive effects of plant genetic diversity on community functioning are solely due to sampling or selection effects. The sampling effect can be observed if diverse mixtures have a higher chance of containing highly productive genotypes (Huston 1997). The positive effect of genotypic diversity was not due to greater chances of obtaining mixtures with more productive genotypes in diverse communities (Crutsinger *et al.* 2006). Similar to the benefits of genetic diversity at population level, the complementarity principle seems to be the major determinant of the increased fitness. Complementarity may indicate that mixed genotypes facilitate one another (Hector *et al.* 1999; Mulder *et al.* 2001), or niche differentiation among different genotypes causes the available resources to be used more completely (Loreau and Hector 2001).

A major limitation of the former studies is that nearly all have focused on plant genetic diversity in aboveground communities. A badly needed next step is to test the generality of the effect of within-species genetic diversity, by including higher trophic levels such as herbivores. Also, little is known about how genetic diversity of soil organisms affects rates of decomposition and nutrient availability. Soil processes, in particular, appear to be influenced primarily by the functional characteristics of dominant species rather than by the number of species present (Heemsbergen *et al.* 2004). As a final remark, genetic diversity is mostly defined as the number of genotypes included in the community. The actual extent of genetic differentiation between genotypes, however, is not known, nor is the level of phenotypic differentiation or functional dissimilarity among genotypes included. Given the weakness of the relation between genetic diversity and variation in quantitative traits, the present conclusion that genotypic diversity has community benefits through complementarity still greatly lacks detailed understanding.

11.4.2 Diversity begets diversity?

Can within-species variation also enhance species diversity of the associated animal communities? Different genotypes of the dominant plant species may favour different species in competitive and trophic interactions, leading to a mosaic of spatially varying selection pressures with a distinct set of associated arthropod species. This is known as the 'diversity begets diversity' hypothesis (Whittaker 1975; Vellend and Geber 2005). On the other hand, genetic diversity enhances productivity and resource efficiency, such that the competitive strength of genetically diverse species increases and may lead to competitive exclusion of other species from the community (Vellend and Geber 2005). Although such a hypothesis would predict that genetic diversity lowers community species diversity, it does not account for the positive effect of the increase in available energy on the number of herbivores and predator species that can be sustained, the so-called 'more individuals hypothesis' (Srivastava and Lawton 1998). The relative strength of species exclusion versus energy availability determines the sign of the relationship between genetic diversity and species richness.

In fact, experimental studies present overwhelming evidence that plant genotypic diversity is positively correlated with arthropod abundance (Reusch *et al.* 2005; Crawford *et al* 2007) and arthropod diversity (Dungey *et al.* 2000; Johnson and

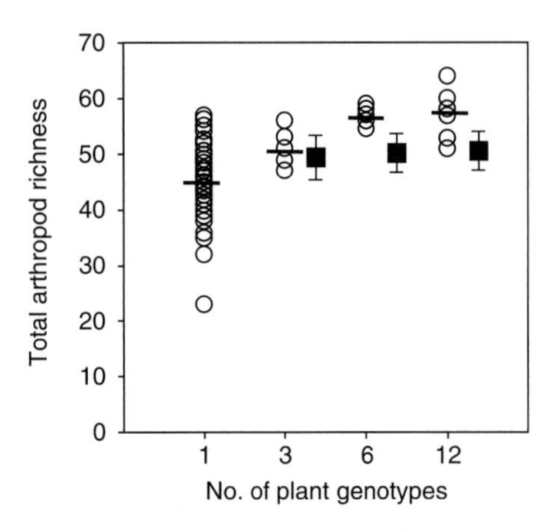

Figure 11.3 The relationship between genotypic diversity of *Solidago altissima* and total arthropod species richness in one-, three-, six- and 12-genotype treatments. Circles indicate plot-level observations, and horizontal lines indicate treatment means. Squares indicate the number of arthropod species predicted by simple additive models. Error bars indicate 95% confidence interval. From Crutsinger *et al.* (2006). Reprinted with permission from AAAS.

Agrawal 2005, Wimp *et al.* 2005; Crutsinger *et al.* 2006; Johnson *et al.* 2006). Experimentally manipulated plots with 12 randomly selected genotypes of *Solidago altissima* contained on average 27% more arthropod species than monocultures (Crutsinger *et al.* 2006). Both herbivorous and predatory species benefited significantly from increased plant genotypic diversity, even more strongly than predicted by the cumulative species richness from each genotype grown in monocultures (Fig. 11.3; Crutsinger *et al.* 2006). In a more detailed study on the same study system, the 12-genotype plots were also shown to contain 80% more galls than the one-genotype plots. Since galls are a preferred habitat for a community of secondary users and their predators, *Solidago altissima* genotypic diversity alters community structure of its associated arthropods through the increased abundance of galls (Crawford *et al.* 2007).

As described above, the competitive exclusion hypothesis predicts that genetically diverse populations will have increased resource efficiency and therefore outcompete populations of genetically impoverished species. This hypothesis addresses the effect of genetic diversity within a trophic level rather than across it. Because it was developed to explain the resistance of communities to species invasions, the main drawback of the competitive exclusion hypothesis is that it considers genetic variation only in the keystone species of a community. Although a number of invasibility studies support the hypothesis (De Meester *et al.* 2007), it appears invalid to explain species coexistence if all species in a community show genetic variation. Booth and Grime (2003) found enhanced coexistence of competing plant species in the most genetically diverse communities. To a large extent this effect could be attributed to genotypic differences in the initial population, but, especially in genetically impoverished communities, genotype by environment interactions determined the structure of the resulting plant communities (Whitlock *et al.* 2007) Given the paucity of empirical studies addressing this issue, a provisional conclusion should be that genetic diversity within species reduces the rate at which species diversity declines.

11.4.3 Phenotypic diversity is also important for community diversity and composition

Although some studies have been able to attribute the positive effects of increased genetic diversity to facilitation (Reusch *et al.* 2005; Crutsinger *et al.* 2006), it has remained unresolved whether and how non-additive effects among genotypes can be predicted. The incidence of facilitative or inhibitory interactions has been hypothesized to depend on functional dissimilarity, defined as the degree of dissimilarity in traits that affect community processes. This hypothesis is confirmed only at species level. Microcosm experiments with soil ecosystems and aquatic grazer communities demonstrated that the effects of community composition on key ecosystem processes can be predicted by measuring the functional dissimilarity in the effect of individual species on community processes. (Heemsbergen *et al.* 2004; Wojdak and Mittelbach 2007). Facilitative interactions occurred in species mixtures with high functional dissimilarity or low niche overlap, independent of species number. If the findings at species level have general applicability, in genetically diverse populations the occurrence of facilitative or inhibitory interactions would depend on the degree of phenotypic dissimilarity

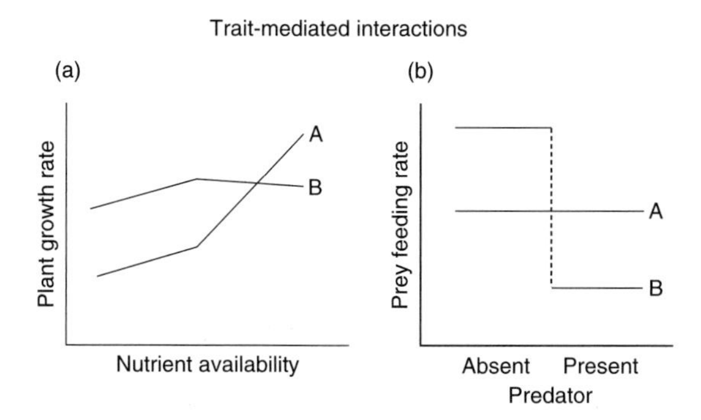

Figure 11.4 The role of phenotypic plasticity in trait-mediated interactions illustrated by two hypothetical examples. (a) A switch in competitive dominance under increased nutrient availability. Plant species A and B have a differential growth rate response to nutrient availability indicated by the steepness of the reaction norm. Under low nutrient availability, species B has the highest growth rate, but species A is better able to profit from high nutrient availability. Hence competitive dominance switches when nutrient availability is high. (b) Predator presence changes competitive interactions of two species. Species B has a higher feeding rate when predators are absent, but is also more sensitive to predation than species A. Therefore, when predators are present, species B has to lower its feeding rate to avoid predation, while species A can continue to feed at the same rate and now outcompetes species B.

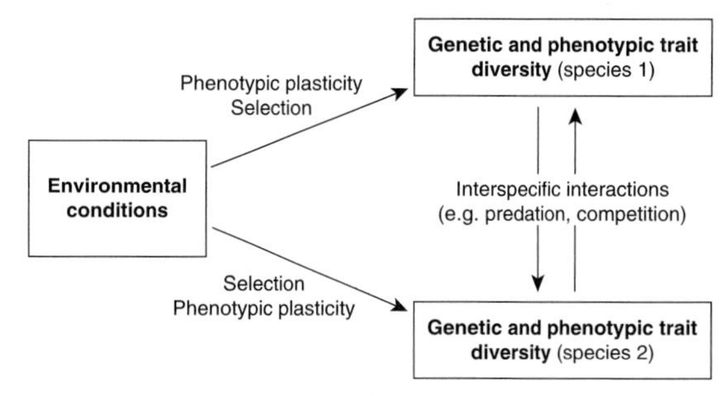

Figure 11.5 The three-way interaction between two species and their abiotic environment (G × G × E). Environmental conditions induce selection or phenotypic plasticity in traits in component species, but not necessarily of the same strength. Performance traits of species may also be mediated by direct interactions such as competition or physical interactions. For illustrational purposes, only two species are shown here but any number can be included.

among genotypes. However, these experiments have yet to be performed.

Some researchers refute a mechanistic approach to community ecology on the basis of the inherent complexity of multiple species interactions, and state that, owing to non-additivity of interactions, the dynamics of communities cannot be predicted from the individual characteristics of the component species (Werner 1992; Sih *et al.* 1998). However, these early studies did not take the effect of phenotypic plasticity into account. If interacting species differ in their response to environmental conditions, environmental change can alter the sign and magnitude of interactions among species (Fig. 11.4), for which Abrams (1995) coined the term 'trait-mediated interactions'. In addition to trait-mediated interactions, interacting species may directly modify each others' gene expression or phenotype by chemical communication, as in the case of induced defences. Condition dependence of interspecific interactions transforms the standard genotype by environment (G × E) interaction into a three-way interaction between two species and the environment (G × G × E; Fig. 11.5). Trait-mediated interactions appear to be common in nature; for instance, the presence of predators may influence foraging behaviour of prey and decrease prey growth rate (Peacor and Werner 2000; Prasad and Snyder 2006). Although identifying how each species alters its traits in the presence of others may be a daunting task, Relyea and Yurewicz (2002) show that this approach yields accurate qualitative predictions on the outcome of mesocosm experiments. More simply measured species traits such as body size do not influence community system functioning in the long run, and seem to be merely effects of initial experimental conditions (Long and Morin 2005). Including the details of such phenotypically plastic responses in theoretical models can affect community dynamics and stabilize tritrophic systems (Bolker *et al.* 2003; Verschoor *et al.* 2004).

11.4.4 Phenotypic plasticity and invasive success

Phenotypically plastic responses have also been implied in the context of invasion success. Invasive species can dramatically change community composition owing to their negative effects on indigenous biodiversity (Mack *et al.* 2000). Phenotypic plasticity may increase invasive success if an individual's plastic response to environmental change allows it to expand ecological niche breadth (Baker 1965; Richards *et al.* 2006). Invasive species are therefore predicted to be more plastic than their native community members. Richards *et al.* (2006) describe three ways in which phenotypic plasticity may contribute to invasive success: a jack of all trades, able to maintain fitness in unfavourable environments; a master of some, able to profit in favourable environments; and a jack-and-master, which combines both. Several studies have demonstrated that the shape of the reaction norm differs between native and invasive species; however, the form of changes and the type of traits showing differences in plasticity levels seem to be idiosyncratic (Chown *et al.* 2007; Muth and Pigliucci 2007).

11.5 Effect of community composition on the genetic and phenotypic diversity of single species

So far, I have discussed only the possible consequences of individual diversity on higher level processes, and the benefits of adopting a more evolutionary perspective for community ecologists. The opposite side of this integrative approach is obviously that evolutionary biologists should become aware of the community context in which evolution takes place (Fig. 11.5). Performance of individual species is typically mediated by interspecific interactions, and the identity of interacting species can determine individual response to changing conditions (Davis *et al.* 1998; Jiang and Morin 2004). Natural selection experiments often leave a large proportion of evolutionary change unexplained, perhaps because fine-scale environmental variation due to identity of competitors is not considered. Several experiments have now revealed that plant response to competition is mediated by interspecific and intraspecific variation in community composition (e.g. Vavrek 1998; Callaway and Aschehoug 2000, Fridley *et al.* 2007).

To predict how community diversity may affect genetic diversity of single species, we can basically just reverse the causation in the two hypotheses discussed in the previous section. In this context the 'diversity begets diversity' hypothesis proposes that species diversity can act as a source of diversifying selection on single species, because each interacting species represents a different competitive environment. If different genotypes of the focal species have differential competitive responses, a positive effect of community diversity on within-species genetic diversity is predicted. In fact, such fine-scale environmental variation in selection pressures suggests the occurrence of local co-adaptation between neighbours. Empirical evidence of such interactions is found for *Taraxacum officinale* genotypes, which exhibited differential response in root biomass, total biomass and leaf area depending on the species identity of the competitor (*Plantago major*, *Poa pratensis* or *Trifolium pratense*). Some genotypes performed best with a specific competitor, but others behaved like generalists with nearly equal performance with all competitors (Vavrek 1998). Similar results, at an even more detailed level, demonstrate that performance depends on the genetic identity of both interacting species (Fridley *et al.* 2007). These studies provide strong support for the diversity begets diversity hypothesis, and give good reason for more empirical emphasis on this topic. Although it is obviously not feasible to include such detailed genetic analysis as a standard protocol in community ecology, it will increase our appreciation of the relevance of genetic diversity in multispecies assemblages and the importance of spatially fluctuating biotic environments for the maintenance of genetic variation.

The alternative hypothesis on the effect of community composition on genetic diversity of single species proposes that a diverse community constrains the ability of the focal species to exploit different niches in the environment. Therefore, community diversity should act as a source of stabilizing selection, reducing the variation within species. To my knowledge, no studies have explicitly measured the degree of genetic variation of a focal species in communities with manipulated community composition. We may tentatively reject this hypothesis because in experimental communities with a fixed number of species with either high or low genotypic diversity, each species retains more genotypes when reared with highly diverse heterospecifics (Whitlock *et al.* 2007). This is opposite to the prediction of the hypothesis, but evidently more work is needed to confirm these results.

A specific prediction derived from this hypothesis is that an increase in community diversity through invasion by a new species will lead to a reduction of genetic variation in the native species within that community. So far, changes in genetic variation as a result of community invasion have been assessed in only one study system. Hild and co-workers studied genetic variability in three native grass species after invasion of Russian knapweed (*Acroptilon repens*) and found a significant reduction in genetic variability after invasion in one native species, but not in the two others (Mealor *et al.* 2004). Also, genetic similarity of both grass species was smaller between invaded and non-invaded communities than within community type, indicating that there is natural selection in response to community invasion (Mealor and Hild 2006). Consistent with this finding, an increasing number of studies shows that natives do show phenotypic changes after exotics have been introduced (reviewed in Strauss *et al.* 2006; Strayer *et al.* 2006) and these studies may provide indirect evidence on changes in genetic diversity of native species. Adaptation to exotics occurs often in the form of shifts in single traits due to directional selection, potentially decreasing genetic variation in these traits. On the other hand, in the case of plant invasions, native phytophagous insects can even be exposed to divergent selection if the induced evolutionary change involves adaptation to a novel host plant, which would enhance genetic variation. At present, we lack sufficiently detailed studies to draw any clear conclusions on this hypothesis.

11.6 Future directions

The empirical evidence presented in this chapter overwhelmingly supports the reciprocal relationship between community ecology and evolutionary biology. Genetic diversity and phenotypic plasticity of component species can shape community structure and composition. Similarly, interspecific interactions and the species composition of communities can influence evolutionary change of species. However, we are only at the beginning of the difficult process of integrating evolutionary principles with the community level. There is an obvious need for a greater number of field studies to show the generality of the results discussed above, especially from the part of the evolutionary biologists. Also, the range of taxa involved in these studies needs to be expanded, because the majority of studies concern plants. Because for some communities measuring genetic and phenotypic diversity for all species will be a prohibitively large undertaking, a careful choice of study systems as well as inventive use of molecular tools is necessary.

If an integrated approach succeeds in adding significant explanatory and predictive power, it may develop into a new discipline of evolutionary community ecology. A true integrative approach to evolutionary community ecology can address a much wider range of issues than just the reciprocal effects of genetic and species diversity that are currently the main focus. Looking from an evolutionary perspective, a crucial question is if, and to what extent, genetically diverse species persist longer in a community than genetically uniform species. This is particularly relevant if one considers plant species that can propagate clonally, or animals that reproduce parthenogenetically. Another essential question is the relationship between community composition and evolutionary change of species within communities. To what extent can community composition drive selection on traits and can divergent selection caused by differences in community composition perhaps initiate ecological speciation? On the other hand, multispecies interactions involved in diffuse coevolution within communities may constrain the possibilities for evolutionary change of trait values, but favour the development of plastic phenotypes.

A multitude of key questions can also be addressed from a community ecological point of view. One question I would regard as crucial is whether species at particular trophic positions in a community are more likely to be genetically diverse? If genetic diversity is maintained by fine-scale environmental variation in interspecific interactions, the degree of connectiveness of species may govern their genetic diversity to a large extent. A second major question is the relationship between phenotypic plasticity and community diversity. Phenotypic plasticity often allows individuals to maintain fitness homeostasis over a much broader range of conditions. Can phenotypically plastic species obtain a broader niche in the community than species that have a fixed phenotype? If so, does plasticity give them a competitive advantage over competing species and is the effect of phenotypic plasticity equivalent to the effect of intraspecific genetic diversity?

It is clear to most of us that evolutionary processes in community ecology can no longer be ignored. Whether the emerging field of evolutionary community ecology will reach its full and exciting potential will depend on the willingness of researchers to look beyond the boundaries of their own discipline. We need to stop seeing the inclusion of genetic diversity or community context as a dictated nuisance, but rather explore its usefulness in obtaining a better understanding of pure and applied problems in biology.

Chapter 12

Emergence of complex food web structure in community evolution models

Nicolas Loeuille and Michel Loreau

12.1 A difficult choice between dynamics and complexity?

A food web is defined as the set of species linked by trophic interactions in a given ecological community. As such, it contains only a subset of the many possible types of ecological interactions and it is a very simplified representation of natural communities. In spite of this simplification, food webs appear to be highly complex networks, if only because any natural system contains several hundreds of species, most of them preying upon or being preyed upon by many others (e.g. Polis 1991). Many food web data sets are now available (Baird and Ulanowicz 1989; Warren 1989; Hall and Raffaelli 1991; Martinez 1991; Polis 1991; Goldwasser and Roughgarden 1993).

It is possible to divide such data sets into two broad categories. The first category will be called 'binary' data sets. Binary data sets simply list species in the food web and the trophic interactions among these species. They do not contain any information in terms of species abundances or trophic interaction strength. Food web theory that deals with binary data sets is primarily interested in:

- comparing food web networks with other types of networks such as protein, genetic, social, neuronal and communication networks (Barabasi and Albert 1999; Amaral and Ottino 2004; Milo *et al.* 2004; Grimm *et al.* 2005; Proulx *et al.* 2005)
- from this comparison, determining properties that are specific to food webs as compared with other types of networks – for example, the fact that food webs are small worlds (Martinez *et al.* 1999; Montoya *et al.* 2006), that they are built in compartments (Pimm 1979; Krause *et al.* 2003) and that they contain many loops (Polis 1991; Neutel *et al.* 2002), a lot of omnivores (Polis 1991), etc.
- finding simple models that would be able to reproduce these features; models such as the Cascade model (Cohen *et al.* 1990; Solow and Beet 1998), the Niche model (Williams and Martinez 2000) and the Nested Hierarchy model (Cattin *et al.* 2004) have been relatively successful in reproducing some of the patterns observed in these binary data sets.

Binary approaches to food webs have been used to draw conclusions about community structure (e.g. food web stability: Pimm 1979; Krause *et al.* 2003) or conservation issues (fragility of food webs to species removal: Dunne *et al.* 2002). In spite of these results, drawing conclusions from binary data sets, or from models that are built on them, to broad ecological issues has proved to be very controversial. Binary approaches have a number of shortcomings:

- descriptors used in binary approaches are highly dependent on species lumping (Solow and Beet 1998) and on the resolution of the data set (Winemiller 1990; Martinez 1991)
- properties measured on binary data sets do not describe the ecological properties of the community satisfactorily; for example, Paine (1980) criticized the use of connectance as derived from these data sets

- binary approaches generally describe foods webs at a given time, while food webs prove to be highly variable in time (Paine 1988)
- because they do not measure species abundances and interaction strength (Cohen *et al.* 1993a; Berlow *et al.* 2004), they are unable to deal with conservation issues (mostly based on species abundance) or functional aspects of ecosystems (such as energy and nutrient fluxes).

An obvious alternative to these high-diversity, static approaches is to describe the dynamics and evolution of species in small food web modules. This approach has been recently reviewed extensively by Fussmann *et al.* (2007). Theoretical studies that follow this approach often consider coevolution of two species (e.g. Levin and Udovic 1977; Saloniemi 1993; Abrams and Matsuda 1997; Loeuille *et al.* 2002; Dercole *et al.* 2006) or evolution of food web modules that contain a restricted number of species (Vermeij 1987; Abrams 1991, 1993; Abrams and Chen 2002; Yamauchi and Yamamura 2005). These models provide interesting insights into species coexistence (Yamauchi and Yamamura 2005) the strength of bottom-up or top-down controls (Loeuille and Loreau 2004), the conditions for the maintenance of intra-guild predation or omnivory (Krivan and Eisner 2003), the conditions for the stability of food web modules (Abrams and Matsuda 1997; Loeuille *et al.* 2002; Yamauchi and Yamamura 2005; Dercole *et al.* 2006), etc. It is unclear, though, how such mechanisms derived from a small number of species may be extended to natural ecosystems that are much more speciose and complex.

Thus, theory is abundant either when dealing with large systems but without dynamics or quantitative information, or when dealing with small dynamical systems in which populations and interactions are explicitly described. The remaining challenge is to develop frameworks that are able to deal with dynamical systems that contain a large number of species and that are able to account satisfactorily for the binary and quantitative aspects of food webs. This is a long-standing issue since Polis (1991) already stressed 16 years ago that theory (Pimm 1982; Pimm and Rice 1987; Cohen *et al.* 1990) was insufficient to tackle the complexity of natural systems.

One possible solution is the use of community assembly models, in which species are drawn from a predetermined regional pool (Post and Pimm 1983; Taylor 1988; Morton and Law 1997; Steiner and Leibold 2004). This type of model has provided useful information on the conditions for the maintenance of large, stable communities. An obvious limitation of these models is that, even when the pool of species is large, it is unable to account for novelties that arise through evolution, and that are potentially infinite. This shortcoming has been addressed by the recent development of evolutionary food web models (Caldarelli *et al.* 1998; Drossel *et al.* 2001; Christensen *et al.* 2002; Anderson and Jensen 2005; Loeuille and Loreau 2005; Ito and Ikegami 2006; Rossberg *et al.* 2006).

The Webworld model (Caldarelli *et al.* 1998; Drossel *et al.* 2001), for example, is based on a large number of traits that may mutate. Traits may be present or absent; thus, species are coded by vectors of 0s and 1s of a predefined length so that the set of species is still finite. An alternative is to base evolutionary models on a few key traits (Loeuille and Loreau 2005; Ito and Ikegami 2006), among which there are trade-offs that are either known or inferred from physiological or morphological constraints. An obvious candidate in the case of trophic interactions is body size. Body size has been suggested to play an important role in the structure of food webs (Cohen *et al.* 1993b, 2003; Neubert *et al.* 2000; Jennings *et al.* 2002a, b; Woodward and Hildrew 2002; Emmerson and Raffaelli 2004; Williams *et al.* 2004, Crumrine 2005). Confirmation of this importance has come from measures showing the tight relationship between the relative difference in body size between predators and prey and the strength of their interaction (King 2002; Jennings *et al.* 2002b; Emmerson and Raffaelli 2004). Body size has also been shown to be of importance for many other life-history traits (Kleiber 1961; Peters 1983; Byström *et al.* 2004; Jetz *et al.* 2004; Savage *et al.* 2004; Reich *et al.* 2006).

In this chapter, we summarize some of the properties of a community evolution model that is entirely based on the evolution of body size. The model shows that, starting with only one morph characterized by its body size, it is possible to obtain stable, complex food webs out of repeated adaptive radiations. The results of the model are then compared with those of other evolutionary food web models to give an overview of possible uses of community evolutionary approaches in community ecology.

12.2 Community evolution models: mechanisms, predictions and possible tests

Community evolution models let entire communities emerge from the basic evolutionary processes of mutation and selection. These models start with one or a small number of species, and new morphs emerge out of repeated mutations. When a mutant is introduced in the system, the selection process comes into play to determine whether it is able to survive or not. The mutant may not survive:

- if its fitness when rare is lower relative to the fitness of its parent
- if its fitness when rare is larger, but demographic stochasticity prevents its invasion; this second possibility is not to be neglected – as mutants are initially rare, demographic stochasticity largely constrains the potential for their invasion.

If a mutant invades the community, several scenarios can follow this invasion:

- The most likely scenario is the extinction of the parent (resident), with the better adapted mutant simply replacing its parent.
- It is also possible that the mutant and the resident coexist. This occurs because fitness may be frequency dependent, i.e. while the mutant's fitness is initially larger (since it invades), this advantage of the mutant against the resident is lost when its frequency increases in the population. When this coexistence occurs, the evolutionary process increases the total diversity of the community.
- Another species of the community goes extinct. This may occur independently of the coexistence or replacement process described in the two previous paragraphs. Because the invasion of the mutant modifies the fitness of other species of the community, it is possible that one or several extinctions follow this invasion.

Community evolution models are in some ways very close to classical community assembly models, since these also contain invasion and selection processes. The main difference between the two types of models lies in the details of the invasion process. In community assembly models, species are introduced from an existing regional species pool (e.g. Post and Pimm 1983; Taylor 1988; Morton and Law 1997; Steiner and Leibold 2004). For this reason, the introduced species do not have to be functionally similar to species already present in the community. Trade-offs between species traits are generally not considered. The timing of the invasion is not constrained, and the diversity of the local species assemblage is bounded by the total number of species present in the regional species pool. By contrast, community evolution models have harsher invasion constraints. The timing of mutation depends on the number of newborn individuals and the probability of mutation per individual. Furthermore, mutations are supposed to have a small phenotypic effect, which means that the characteristics of the mutants are strongly correlated with the phenotypic trait of one of the existing species. Finally, when phenotypic effects are explicitly identified, it is possible to link them mechanistically to physiological or ecological benefits and costs. Therefore, such community evolution models account explicitly for evolutionary trade-offs, while the traits of invading species in community assembly models are often unconstrained, leaving open the question of how such traits emerge in the first place.

12.2.1 One or many traits?

Community evolution modelling is a rapidly growing branch of evolutionary ecology (Caldarelli *et al.* 1998; Drossel *et al.* 2001; Anderson and Jensen 2005; Loeuille and Loreau 2005; Ito and Ikegami 2006; Rossberg *et al.* 2006; Ito and Dieckmann 2007; Lewis and Law 2007). An important choice that governs the characteristics of these models concerns the number of traits and their identity. Although it is obvious that the ecology of species depends on many traits, the number of traits considered is traded off against the biological realism introduced by these traits.

12.2.1.1 Models in which species are defined by many traits

The first community evolution model, named the Webworld model (Caldarelli *et al.* 1998; Drossel *et al.* 2001), had a large number of traits. In this model, each species has a given number L of features (phenotypic traits) picked out of a pool of K traits that constrain the demography of the species and its interactions with other members of the community. A $K \times K$ matrix $[m_{i,j}]$ describes the efficiency of each species' trait against other species' traits. The sum of the matrix elements over the traits possessed by two interacting species yields the strength of their trophic interaction.

A second model inspired by the Webworld model is the Matching model, conceived by Rossberg *et al.* (2006). In this model, each species is characterized by a vector that determines its attack rate and a vector that determines its vulnerability. These vectors contain n components that describe the presence or absence of the trait for the species considered. The interaction strength between two species depends on the matching between the attack traits of one and the vulnerability traits of the other.

Finally, the Tangled Nature model (Christensen *et al.* 2002; Anderson and Jensen 2005) assumes that species interactions are determined by L loci, with two alleles for each locus (noted 1 and 0). The interaction between two species then depends on the allelic composition of the two species. The coupling between two species characterized by their genome is described by a non-symmetrical matrix, whose terms are non-zero with some predefined probability, and then drawn out of a uniform distribution in a predefined interval $[-c,c]$. Contrary to the other two above-mentioned models, the Tangled Nature model is not restricted to trophic interactions *a priori* and may incorporate any kind of interaction.

Both the Webworld and the Matching models have been tested against empirical data (Caldarelli *et al.* 1998; Rossberg *et al.* 2006). They are both successful at reproducing a number of food web structural patterns. They are also particularly useful in addressing the degree of generalism of predators.

12.2.1.2 Models with a limited number of traits

Body size is a key species trait that food web theory has often considered explicitly. Empirical data show that trophic interactions are heavily constrained by body size (Jennings *et al.* 2002b; Emmerson and Raffaelli 2004). In 90% of trophic interactions, the predator is larger than the prey (Warren and Lawton 1987; Cohen 1989). Interaction strength strongly depends on the relative difference between prey and predator body sizes. One of the first models of food web structure, the Cascade model (Cohen *et al.* 1990; Solow and Beet 1998), relies on body size. Besides its effects on species interactions, body size also influences basal metabolic rate and many life-history and physiological traits (Kleiber 1961; Peters 1983; Byström *et al.* 2004; Jetz *et al.* 2004; Savage *et al.* 2004; Reich *et al.* 2006).

An example of a community evolution model based on body size is the model we built (Loeuille and Loreau 2005). In this model, body size affects a number of species traits:

• It determines demographic parameters. A species' fecundity and mortality are supposed to be directly linked to its mass-specific metabolic rate, a fact that is supported by empirical data (Kleiber 1961; Peters 1983). The model assumes that:

$$\begin{aligned} f(x) &= f_0 x^{-0.25} \\ m(x) &= m_0 x^{-0.25} \end{aligned} \qquad (12.1)$$

where x is the species' body size, f is its production efficiency, i.e. the percentage of the nutrient it gets that is allocated to growth and reproduction, and m is its mortality rate. Note that the model uses body mass as a proxy for body size, as is usual in allometric theory.

• Body size affects trophic interactions. A given predator whose size is y is able to consume species whose body size x is smaller because of morphological and behavioural constraints (Warren and Lawton 1987; Cohen 1989). On the other hand, the predator may disregard very small prey items, either because they are hard to detect or because they do not bring enough energy when consumed. The strength of the interaction should then be maximum for some intermediate value of x smaller than y, an assumption that is supported by empirical observations (Emmerson and Raffaelli 2004). A possible candidate function matching all these requirements is the Gaussian:

$$\gamma(y-x) = \frac{\gamma_0}{\sigma\sqrt{2\pi}} e^{\frac{-(y-x-d)^2}{\sigma^2}} \quad (12.2)$$

where σ^2 is the variance of the predation rate, and predators of size y forage optimally on prey of size $x = y - d$.

• Finally, differences in body size also constrain competitive interactions, particularly interference competition. Species that have similar body sizes are more likely to exploit their habitat on similar spatial scales. Habitat use being similar (Price 1978; Jetz *et al.* 2004), interference competition is more likely. A possibility is then to model interference competition between two species with body sizes x and y using a step function:

$$\alpha(|x-y|) = \begin{matrix} \alpha_0 \ \ if(|x-y|) < \beta) \\ 0 \ \ else \end{matrix} \quad (12.3)$$

All these effects of body size are summarized in Fig. 12.1.

These allometric components (equations 12.1–12.3) are then incorporated into the dynamical model:

$$\frac{dN_i}{dt} = N_i\Big(f(x_i)\sum_{j=0}^{i-1}\gamma(x_i-x_j)N_j - m(x_i) - \sum_{j-1}^{n}\alpha(|x_i-x_j|)N_j - \sum_{j=i+1}^{n}\gamma(x_j-x_i)N_j\Big) \quad (12.4)$$

Variable N_i corresponds to the biomass of the species i whose body size is x_i. Species are ordered according to their body mass, so that species 1 is smallest and species n is largest. N_0 describes the amount of inorganic resource whose trait is arbitrarily set to 0 for mathematical convenience. The dynamics of this resource includes nutrient inputs noted I, diffusion of nutrients out of the system at a rate e, as well as recycling of a proportion v of the nutrient that is not assimilated during the consumption process or that is released as a result of

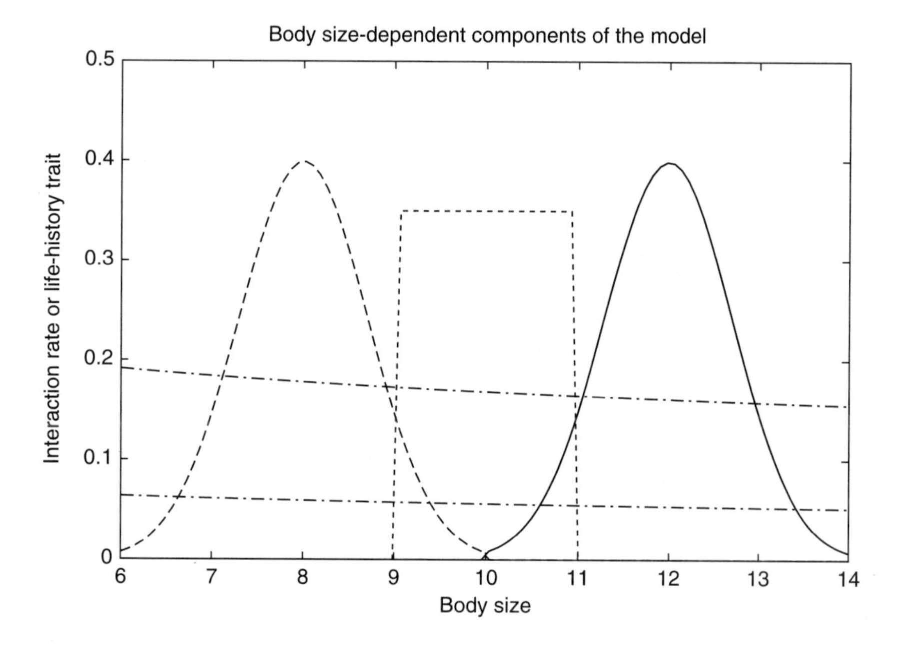

Figure 12.1 Influence of body size on the components of the model. The two dashed-dotted lines show the production rate and mortality rates (equations 12.1). The three other curves detail how interaction rates of a species whose body size is 10 depend on the body size of other species of the community. The solid curve shows the interaction rate with any predators whose body size is included in the interval [10, 14] while the dashed curve shows potential predation rates with a species smaller than itself (equation 12.2). Finally, the dotted step function shows the interference competition rate of the species with species of similar sizes (equation 12.3). Parameters: $m_0 = 0.1$, $f_0 = 0.3$, $\alpha_0 = 0.35$, $\beta = 1$, $\gamma_0 = 1$, $\sigma^2 = 1$.

mortality and excretion. The equation that describes nutrient dynamics is then:

$$\frac{dN_0}{dt} = I - eN_0 - \sum_{i=1}^{n}\gamma(x_i)\ N_iN_0 + \nu N_i\Big(\sum_{i=1}^{n} m(x_i) + \sum_{i=1}^{n}\sum_{j=1}^{n}\alpha(|x_i - x_j|)N_j + \sum_{i=1}^{n}\sum_{j=0}^{i-1}\Big(1 - f(x_i)\Big)\ \gamma(x_i - x_j)N_j\Big) \qquad (12.5)$$

Each simulation starts with a single species N_1, which consumes the inorganic nutrient N_0. At each time step, mutation may occur with a probability μN_i for each species (but the inorganic nutrient does not mutate), where μ is the mutation rate per unit biomass. If a mutation occurs, a mutant is introduced, whose trait is drawn at random in a uniform interval centred on the trait of the parent. When a mutant is introduced, its biomass is set equal to the threshold biomass below which a species goes extinct and is removed from the system.

There are other models based on few species traits. For instance, Ito and Ikegami (2006) used a continuous version of the Webworld model to include two traits for each species, one that describes the species as a prey, and the other that describes it as a predator. We focus below on our own model because it provides an intuitive illustration of how evolutionary dynamics may influence food web structure via one clearly defined trait.

12.2.2 Evolutionary emergence of body-size structured food webs

While the model presented in section 12.2.1.2 starts with a single species, the mutation-selection process adds new morphs to the system, so that total

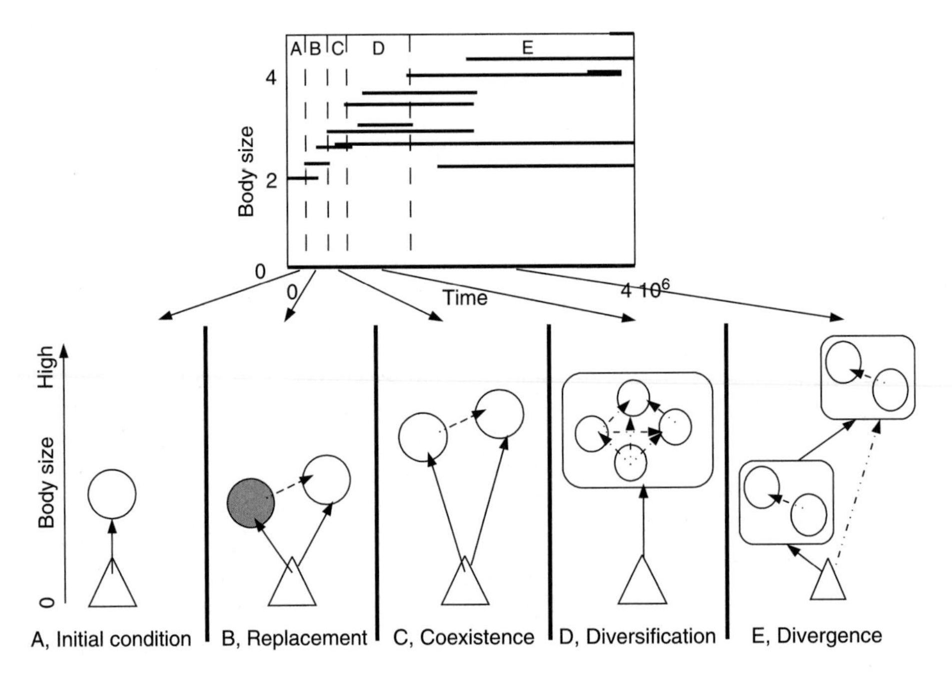

Figure 12.2 First steps of the emergence of a size-structured food web. The main panel shows the trait composition of the community through time, while the lower panel details the different steps of the emergence. The simulation starts with one species that is consuming inorganic nutrient (A). Once in a while, mutants appear (here larger than the resident) and replace their parent (B, in which the grey morph goes to extinction). After several replacements, an evolutionary branching happens, as the mutant and the resident are able to coexist (C). A rapid diversification then occurs in which several morphs are able to coexist (D) but then are selected in differentiated trophic levels (E).

diversity increases through time. This increase in diversity is very fast at the beginning, as the evolutionary process fills (and builds) a niche space that is quite empty at the beginning of the simulation. When mutants invade the system, extinction of the parent species or of other morphs of the community may occur, and after a while total diversity reaches a plateau and compositional turnover becomes small (Fig. 12.2). This plateau is the evolutionary quasi-equilibrium.

The final structure of the food web depends on the parameters of the model. The dimensionality of the food web (total number of morphs and length of the food chain) is mainly limited by energetic parameters such as the nutrient input I and the basal production efficiency f_0. Other characteristics of the food web are sensitive to two parameters:

- The interference competition rate α_0. If there is no interference competition, diversity within a trophic level is reduced and the food web tends to become a food chain. In such cases, the demographic dynamics may become unstable. A small amount of competition (e.g. $\alpha_0 = 0.005$), however, is enough to generate very diverse food webs. At the other end of the spectrum, if the competition rate is very high, individual fitness is mostly determined by competition while selective pressures due to trophic interactions become less important. Under these conditions, having a size that differs at least β from other sizes in the community is the most important condition for a morph to be favoured. As a result, species body sizes become evenly spaced and trophic structure is lost (Fig. 12.3).
- The niche width $nw = \frac{s^2}{d}$, which describes the degree of generalism of predators. The wider a species' niche, the less it is specialized on a given range of body size. Note also that, because the function that describes the niche (equation 12.2) is normalized, when the niche is wider, the maximum consumption rate is smaller. To understand the role of the niche width in the emergence of food web structure, consider the beginning of a simulation in which niches are very narrow. As the inorganic resource has a size 0 and niches are very narrow, morphs whose size is d are strongly favoured because they are the only ones that are capable of taking advantage of the resource efficiently. As a result, evolution will select for body sizes that are close to d. These morphs in turn will provide available energy for morphs whose body size is $2d$. Consequently, evolution generates well-defined body size classes, which also correspond to differentiated trophic levels. By contrast, when niches are wide, the consumption function described by equation 12.2 becomes flatter, so that the consumption advantages described above may be offset by other effects of body size or other components of the model. In these cases, the trophic structure is blurred.

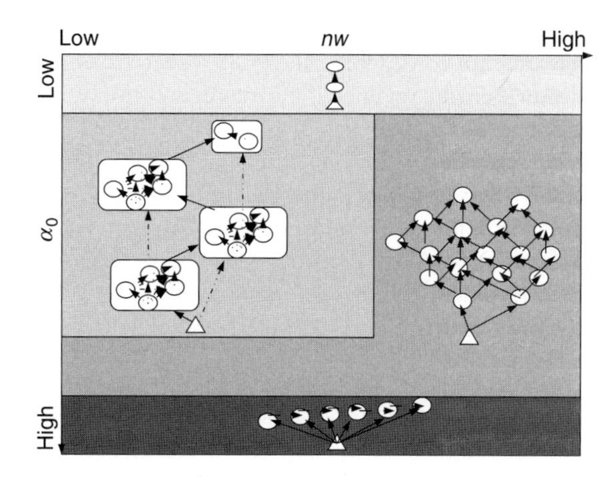

Figure 12.3 Diversity of possible trophic structures emerging from the body-size-based evolutionary model described in section 12.2.1.2. If the interference competition rate α_0 is zero, then a food chain emerges out of the co-evolutionary process. When it is very high, the fitness of the individuals in the community mainly depends on competition, and the trophic structure is organized on one trophic level. In between these two extremes, a wide diversity of outcomes is possible and their structure depends on the niche width parameter *nw*. If niches are narrow, food webs that emerge are structured by an assemblage of distinct trophic levels, but if niches are wide, the trophic structure is blurred as competition and omnivory are ubiquitous in the simulated community.

These effects of niche width and competition strength are illustrated in Fig. 12.3. The interplay of these two parameters is able to produce a complete continuum of trophic structures. Communities that reach an evolutionary quasi-equilibrium may then be used to generate a snapshot describing the shape of the food web. To do this snapshot, all morphs are considered, and the trophic links between them are retained if the interaction strength γ is larger than a threshold value (here, 0.15). The result is a binary food web that describes species and trophic links but ignores quantitative information on biomasses and nutrient fluxes. These simulated food webs can then be compared with empirical data from natural communities (e.g. Warren 1989; Winemiller 1990; Hall and Raffaelli 1991; Martinez 1991; Polis 1991; Havens 1992; Memmott *et al.* 2000). This comparison was done in the following way:

- Food webs were generated for 36 pairs of parameters $\begin{cases} a_0 = \{0, 0.1, 0.2, 0.3, 0.4, 0.5\} \\ nw = \{0.5, 1, 2, 3, 4, 5\} \end{cases}$ and their properties were examined.
- For each property, a surface was drawn in parameter space by interpolating the results of the 36 simulations.
- A least squares fit determined which pair of parameters yielded the community closest to empirical data.

For each empirical data set, it is possible to find parameters that generate a food web whose properties are very similar. For all simulated communities, the properties used for the least squares fit are compared with those of the empirical data in Table 12.1. While the match between the communities produced by the model and the empirical data sets is far from being perfect, it is as good as the match obtained using the best binary food web models, at least for the descriptors listed in Table 12.1 (Loeuille and Loreau 2005). The model introduced here also produces the connectance and total diversity of the community, while these quantities were used as parameters (and therefore left unexplained) in the Niche model as well as in other binary food web models.

12.2.3 Advantages of simple community evolution models

In discussing the advantages of the above model or other simple community evolution models, our aim is not to show that simple models based on one or a few traits are better than more complex ones, but rather to identify their specific contribution to understanding food webs.

12.2.3.1 Comparison with other community evolution models

The main advantages of community evolution models based on a restricted and clearly identified set of traits are a better understanding of the role of evolutionary constraints (trade-offs) and a greater ability to test their predictions.

Models that use a large number of traits do not identify these traits explicitly. The influence of these traits on species interactions and demography is usually determined using a matrix whose elements are drawn at random (see section 12.1). Therefore, traits are not linked mechanistically to the biology of the species. No benefits or costs of the phenotypic traits are explicit. In community evolution models, community properties emerge spontaneously from the evolutionary dynamics, so that a complete understanding of these evolutionary dynamics is required to discuss thoroughly the possible mechanisms producing these properties. In the examples detailed in section 12.2.1.2, an explicit link is made between body size and the biology of species. Of course, such knowledge involves additional hypotheses on trade-offs producing the selective pressures acting on the phenotypic trait. But in the case of body size, these trade-offs are well known because body size has been the focus of a lot of work in ecology and physiology (Kleiber 1961; Peters 1983; Brown 2004). It is then possible to use our model as a tool to understand which allometric components of the model are responsible for the observed community structure. For instance, it is possible to turn off the effects of body size on the life-history parameters *f* and *m* and examine the consequences of the allometric components of competitive and trophic interactions, independently of the effects of body size on life-history parameters.

12.2.3.2 Comparison with binary qualitative models

A large part of food web theory concerns food web topology in tight connection with empirical data. These models use binary data, i.e. species and links are either present or absent but are not quantified.

Table 12.1 Comparison between the characteristics of empirical data sets and those of communities emerging from the model described in Section 12.2.1.2 that are the closest when the interference competition rate α_0 varies from 0 to 0.5 and the niche width *nw* varies from 0.5 to 5

Value	Error	SMI	CD	YE	CB	BBL	LRL	SP
Connectance	99	0.26 (0.12)	0.26 (0.31)	0.23 (0.061)	0.24 (0.072)	0.2 (0.17)	0.15 (0.12)	0.27 (0.32)
Chain length	26	4.15 (4.2)	6.2 (7.18)	3.79 (4.89)	4 (2.77)	4.14 (2.55)	4.6 (N.A.)	5.44 (4.81)
Omnivore	19	54 (60)	76 (79)	42 (53)	46 (38)	48 (36)	47 (N.A.)	74 (60)
Top	22	17 (17)	5 (0)	24 (38)	21 (28)	4 (0)	4 (0)	5 (4)
Intermediate	7	63 (69)	80 (90)	55 (53)	60 (62)	68 (68)	82 (87)	79 (92)
Bottom	75	20 (14)	15 (10)	21 (9)	19 (10)	28 (32)	14 (13)	16 (4)

SMI, St Martin Island; CD, Coachella Dessert; YE, Y than Estuary; CB, Chesapeake Bay; BBL, Bridge Brook Lake; LRL, Little Rock Lake; SP, Skipwidth Pond.In each instance, the characteristics of the simulated community are given, while the empirical data is given in parentheses. Matching of the two is comparable to the ones obtained using the Niche model. Other parameters are fixed: I, 10; e, 0.1; v, 0.5; μ_0, 1; d, 2; f_0, 0.3; m_0, 0.1; β=0.25.

Models of this kind use community properties (usually the total diversity of the system as well as its connectance) to determine other community properties. Species and links are distributed among species using rules that are different among models. The Cascade model (Cohen *et al.* 1990; Solow and Beet 1998), the Niche model (Williams and Martinez 2000) and the Nested Hierarchy model (Cattin *et al.* 2004) are examples of such models. All of them are able to match a number of topological descriptors of the empirical data sets satisfactorily.

Compared with these binary models, community evolution models have the advantage that they can provide quantitative information such as interaction strength and species abundances (Loeuille and Loreau 2006). Moreover, since they let community structure emerge from the evolutionary process, they provide the whole dynamics that leads to this structure, not just a snapshot of it (Caldarelli *et al.* 1998; Drossel *et al.* 2001; Loeuille and Loreau 2005; Ito and Ikegami 2006; Rossberg *et al.* 2006). In the case of models that are based on one or a few traits (such as the body-size model presented above), parameters are also measured at the individual level, so that all the community topologies emerge out of processes defined at a lower level. For this reason, these models are able to assess quite accurately how the dynamics really lead to the observed structure. In contrast, binary models are parametrized using community properties (species diversity and connectance). Consequently, they simply use large-scale patterns to infer other large-scale patterns, but whether the internal dynamics of the system can lead to these patterns or not is left unknown.

12.2.3.3 Testing predictions

One of the major caveats of food web theory is the proper test of models. Although the study of topological features such as those listed in Table 12.1 may lead to rejection of a model if the latter fails to reproduce them, the ability of a model to reproduce these topological features is insufficient to accept it. For instance, the Cascade model (Cohen 1989; Solow and Beet 1998), the Niche model (Martinez *et al.* 1999), the Nested Hierarchy model (Cattin *et al.* 2004), the model presented here (Loeuille and Loreau 2005 2006) and the Matching model (Rossberg *et al.* 2006) all provide a good fit to these data, although their assumptions and mechanisms are quite different. Community evolution models, however, provide dynamical features, which may be used for additional tests of model predictions (provided that empirical data on the dynamics of food webs is also available).

Community evolution models also produce additional quantitative predictions that can be tested. For instance, at any given time of the evolutionary process, it is possible to get the distributions of species abundances and interaction strengths in the system. Nutrient and energy flows can also be quantified in the simulated communities. These quantitative predictions can be compared with corresponding empirical data or with existing theories that deal with energy constraints in natural ecosystems (e.g. Quince *et al.* 2005; Loeuille and Loreau 2006; Rossberg *et al.* 2008).

When models are based on clearly identified traits, it is also possible to use empirical information on these traits to assess the quality of the model. For instance, using the model presented in section 12.2.1.2, it is possible to get the density and body size of each species. It is then possible to use these additional pieces of information to test the model. The food web data for Tuesday Lake incorporate these pieces of information (Cohen *et al.* 2003).

An obvious limit to quantitative tests is the quantity and reliability of empirical data (Winemiller 1990; Hall and Raffaelli 1991; Martinez 1991; Havens 1992; Krause *et al.* 2003). Topological measures already depend quite strongly on the sampling effort and on the aggregation of species in functional groups (or tropho-species). Quantitative data are hard to get and require new standards to make them comparable across different ecosystems (Berlow *et al.* 2004). Another problem is the short-term variability of quantitative descriptors (Baird and Ulanowicz 1989; Winemiller 1990; Polis 1991). Measures of energy fluxes or biomasses are highly variable depending on the season, while long-term averages require a large sampling effort and long-term funding. Under the assumption that food webs are at equilibrium, it is possible to infer some quantities using only partial information (Christian and Luczkovich 1999; Trites *et al.* 1999;

Neira *et al.* 2004; Neira and Arancibia 2004; Sànchez and Olaso 2004). The applicability of such an equilibrium hypothesis, however, is debatable, as evidence of short-term variability and long-term changes accumulates. For all these reasons, although quantitative tests of community evolution models are desirable and theoretically possible, they have not been performed so far.

12.3 Community evolution models and community ecology

In addition to predictions on food web structure, community evolution models can provide interesting insights into many other topics of interest to community ecology. A few of these insights are discussed below, but the possibilities of such extensions depend greatly on the particular assumptions of the models.

12.3.1 Community evolution models and the diversity–stability debate

As seen in Section 12.2.2, community evolution models allow the emergence of diverse communities. The model detailed in section 12.2.1.2 gives rise to food webs that can maintain several hundreds of morphs (Loeuille and Loreau 2005). Similar diversity may be obtained using the Webworld model (Caldarelli *et al.* 1998; Drossel *et al.* 2001) or the Matching model (Rossberg *et al.* 2006, 2008). Remarkably enough, our model as well as the Webworld model generate communities in which population dynamics are quite stable in spite of the large diversity that emerges.

This is an important contribution of these models, since the relationship between diversity and stability has puzzled ecologists for decades. Since May (1973) demonstrated that increased diversity means an increased likelihood that the system may be unstable, ecologists have been looking for mechanisms that could explain the stable assemblages of species that constitute ecosystems. While functional complementarity between species may provide a basis for the ability of ecosystems to maintain a stable overall functioning and resist disturbances (the insurance hypothesis: Yachi and Loreau 1999; Loreau *et al.* 2003), the mechanisms behind the stability of population dynamics in systems that contain a large number of species are still very much an open question. Compared with community evolution models, community assembly models often show more unstable dynamics (e.g. large extinction cascades or cyclic trajectories; Steiner and Leibold 2004). Results of community evolution models suggest that the stability of the food webs that emerge during the evolutionary process is linked to the evolutionary process itself. Adaptation may be one of the bases for the reconciliation of diversity and stability.

In food web models that deal with a restricted number of species, it is noteworthy that the functional response of consumers plays an important role in the stability of population dynamics. Strong instabilities can be produced as non-linearities, such as Holling type II functional responses, are included (Gross *et al.* 2004). In our abovementioned evolutionary model, it is noteworthy that even the incorporation of type II functional responses did not lead to unstable dynamics, or that such dynamics were only transient (results not shown). As in ours, the initial version of the Webworld model used type I functional responses (Caldarelli *et al.* 1998). An updated version of the model uses a functional response determined by optimal foraging of predators (Drossel *et al.* 2001). Both models generate stable species assemblages. Although these results are still limited in scope and other functional responses should be tested before definitive conclusions can be made, these results suggest that stable communities can be obtained when adaptation takes place, regardless of the functional response used.

One of the possible reasons for the stability of complex systems is low interaction strength. If a community contains only species that interact strongly with one another, it is unstable. But stability may be obtained if a large proportion of the interactions are weak (Kokkoris *et al.* 1999, 2002; McCann 2000; Neutel *et al.* 2002). Interestingly, the model presented here possesses a large number of weak interactions (Loeuille and Loreau 2005; see also Emmerson and Raffaelli 2004). Thus, the evolutionary process may favour the maintenance of weak interactions, thereby enabling stable population and community dynamics. The same phenomenon seems to be responsible for the stability of the food webs produced by the Webworld model (Quince *et al.* 2005).

12.3.2 Effects of perturbations on natural communities

Understanding the effects of sustained press perturbations on natural communities is increasingly important as the rapid growth of human populations disrupts natural ecosystems. Unfortunately, tools to assess the effects of such perturbations are few, especially on a long timescale.

Yodzis (2000) found that the uncertainty of the effects of changes in one population on the rest of the food web was high in the Benguela ecosystem because of a large number of indirect demographic effects. In addition to these difficulties, recent studies have shown that evolution of species may occur on a short timescale (Reznick *et al.* 1997; Hendry *et al.* 2000; Huey *et al.* 2000; Heath *et al.* 2003; Reale *et al.* 2003; Hairston *et al.* 2005). Thus, changes in life history and species interaction traits because of evolutionary changes may not be negligible in perturbed ecosystems.

Although our model as well as the other community evolution models discussed here are too simplified to provide detailed realistic predictions, they may provide interesting and testable insights into the evolutionary and population dynamical effects of perturbations. Understanding the influence of evolution on species extinctions would be particularly valuable because this issue has hardly been explored. We can decompose the evolutionary effects on species extinctions due to anthropogenic perturbations in three categories:

- Evolution of species following a perturbation. This evolution may help them to respond to the perturbation. For instance, evolution or phenotypic plasticity has helped some species to track global changes (Wing *et al.* 2005; Balanya *et al.* 2006; Franks *et al.* 2007; Sherry *et al.* 2007).
- The extinction probability of species that interact with the species experiencing the perturbation most strongly is modified because of the latter's evolution (evolutionary murder: Dercole *et al.* 2006).
- The extinction probability of species that interact with the species experiencing the perturbation most strongly is modified because of its evolution in response to changes in the latter's density or trait.

On all these issues, community evolution models are able to provide first answers.

To illustrate this, consider a model based on a trait influenced by the perturbation. For instance, the model introduced in section 12.2.1.2 is based on body size. One of the most common perturbations experienced by animal populations is harvesting by humans, which very often depends on body size. For instance, trophy hunting is preferentially directed towards individuals with a large body size, and has already been shown to have evolutionary effects on bighorn rams (Coltman *et al.* 2003). It may also be linked to the size of ornaments (as in the case of rams), but even then it has a selective effect on body size because the latter is correlated with the size of ornaments (Kodric-Brown *et al.* 2006). In fisheries, harvesting is also heavier on large-sized fish (Pauly *et al.* 1998).

In size-structured food webs, the effects of harvesting on large-sized organisms can be assessed directly. These effects include (1) demographic effects, since population dynamics in the model presented in section 12.2.1.2 depend explicitly on body size (equation 12.4), and (2) evolutionary effects, through correlated modifications of the fitness landscapes of the species composing the community. Selective harvesting of large body sizes means that top predators are more likely to be the target of harvesting, a situation that is well documented in fisheries (Pauly *et al.* 1998). Harvesting predators can disturb the food web through top-down effects. These demographic effects include:

- primary extinctions, as the target species may disappear from the system
- secondary extinctions, if the extinction or decline in population size of the harvested species produces extinctions of other, non-targeted species in the web. In the instance of harvesting predators, this may happen when the disappearance or decrease of the predator population generates negative effects on its prey populations (keystone predator *sensu* Paine 1966).

When top predators are harvested in the model of section 12.2.1.2 a surviving predator's mutant whose body size is smaller may be favoured because it has a lower probability of being harvested. This, in turn, modifies the size refuges of its prey (equation 12.2), so that prey that were protected from strong trophic pressures may now decline or go extinct. Such evolutionary extinctions are theoretically possible and observed in community evolution models, but very little is known about their implications in terms of conservation.

Finally, evolution may rescue some species. Evolution of a harvested species may allow it to adapt fast enough to escape extinction. Even if this is not the case, indirect evolutionary effects of harvesting as described above may provide the necessary conditions for the appearance of new morphs. As evolution possibly creates new extinctions but also new species, the net effect of evolution on the total diversity of the system is not obvious. An analysis of these issues with the model presented in section 12.2.1.2 is currently under way.

12.3.3 Models with identified traits: other possible applications

Part of community ecology relies on traits whose importance has been established in many empirical or experimental studies. The same traits could be used in evolutionary food web models. It would then be possible to make an explicit link between evolutionary dynamics in food webs and other areas of community ecology that are usually discussed without any evolutionary considerations.

Empirical and experimental observations show that the stoichiometry of consumer and resource species influences their interaction (Loladze and Kuang 2000; Grover 2003). For instance, stoichiometric effects are one of the possible explanations for the prevalence of omnivory in nature (Matsumura *et al.* 2004). Stoichiometry also influences the whole structure of food webs (Turner *et al.* 1998; Schade *et al.* 2003). Much is known about elemental ratios, from both a physiological and an ecological point of view, so that trade-offs driving the evolution of elemental ratios can be derived from this knowledge. Therefore elemental ratios could be incorporated in evolutionary food web models. Some work along these lines is already under way. Hopefully, it will then be possible to predict community patterns related to ecological stoichiometry, such as the differences between elemental ratios at different trophic levels, differences in their variance, the prevalence of the Redfield ratio in ecosystems.

Evolution of dispersal and habitat preference also largely determines community organization. Integration of spatial effects in the structure of communities is a rapidly expanding theme of community ecology. A particularly useful framework that has been developed recently is the metacommunity concept, which describes a set of local communities connected by dispersal of individuals among patches (Leibold *et al.* 2004). Studies of the interaction between evolution and dispersal in these metacommunities has already begun (Urban 2006, Rossberg *et al.* 2008; Loeuille and Leibold 2008). However, the integration of spatial components in community evolution is not properly done yet (but see Rossberg *et al.* 2008). Incorporating the evolution of dispersal or habitat choice (Gyllenberg and Metz 2001; Metz and Gyllenberg 2001; Kisdi 2002) would allow evolutionary food web models to link to metacommunity theory, but such an extension is very costly in terms of complexity and few insights are yet available.

The strongest link currently available between evolutionary food web models and other areas of community ecology is with the allometric theory of ecology (reviewed in Brown 2004). This theory uses the relationship between body size and various physiological or life-history traits (metabolism, production rate, etc.) to make various predictions on species biomass and nutrient fluxes in ecosystems. Allometric theory is often successful in describing macro-scale patterns of community structure and ecosystem functioning. However, it usually deals with snapshot pictures of communities. It does not account for the dynamical processes that generate the structure itself, although it often invokes coevolution of species as a mechanism (Damuth 1981; Maiorana and Van Valen 1990; Marquet *et al.* 1995; Brown 2004). As a result, community evolution models relying on body size are complementary to allometric theory. First, models based on body size such as the one detailed in section 12.2.1.2 rely on some similar assumptions. For instance, our model contains the influence of body mass on individual production and mortality rates (equations 12.1), two components of allometric theory. Second, community evolution models account explicitly for the coevolutionary dynamical process that is supposed to underlie the patterns revealed by allometric theory.

Consider one of the main results of allometric theory, i.e. the distribution of species abundances as a function of body size. Damuth (1981) showed with empirical data that the density D of a given species is related to its mean body mass, noted x by the relationship $D = kx^{-0.75}$. Since the mean metabolic rate M of an individual is linked to its body size by the relationship $M = k'x^{0.75}$ (Kleiber 1961), the total amount of resources E consumed by a given species in the system should be $E = MD = kk'x^0$, i.e. the energy consumed by a species is independent of its body mass. This prediction is called the energetic equivalence rule (Damuth 1981; Nee *et al.* 1991). Although the mechanism that is supposed to lead to this equal partitioning of resources among species is somewhat vague, coevolution of species that share a same set of resources has been invoked (Damuth 1981; Maiorana and Van Valen 1990). This influential rule has been tested using empirical data with both successes (Damuth 1981, 1991, 1993; Marquet *et al.* 1990; Nee *et al.* 1991; Long and Morin 2005) and failures (Brown and Maurer 1986; Greenwood *et al.* 1996; Cyr 2000; Cohen *et al.* 2003; Russo *et al.* 2003) Although it was initially derived for species within a single trophic level, it was later extended by others to systems that contain multiple trophic levels. Allometric theory then predicts that the exponent that links density and body size is -1, so that $D=kx^{-1}$ (Brown and Gillooly 2003).

Interestingly, the model presented in section 12.2.1.2 contains some components that are similar to the ingredients used in Damuth's energetic equivalence rule. The allometric relationships used for production and mortality rates are inferred from individual metabolism, and the model simulates species coevolution on shared resources, the mechanism that was proposed for the emergence of the perfect sharing of resources between community members. Therefore, it is possible to test this mechanism (keeping in mind, of course, the limits of the model's assumptions) and to see for which parameters, if any, the predicted links between population density or energy use and body size are observed. The results show that population density is a decreasing function of body mass, but the exponent of the relationship depends on the strength of competitive interactions and on the niche width of consumers, so that coevolution does not lead to an equal partitioning of energy among species (Loeuille and Loreau 2006). This example illustrates how community evolution models may give additional insights to the allometric theory of ecology. Such models can include allometric components when they consider body size as an evolving trait. Because they consider dynamical components of populations instead of focusing on the equilibrium communities, they may also be used to test mechanisms assumed to explain allometric patterns.

12.4 Conclusions, and possible extensions of community evolution models

Community evolution models make three major contributions to community and ecosystem ecology. First, they extend classical pairwise coevolutionary models to large, complex ecosystems, with new results. Take the example of how evolution, or coevolution, affects population dynamics. In small communities, some studies show that evolution or coevolution may have stabilizing effects (Pimentel 1961; Saloniemi 1993; van Baalen and Sabelis 1993) while others suggest the contrary (Abrams and Matsuda 1997; Yoshida *et al.* 2003). As we have pointed out in section 12.3.1, the results seem to be less ambiguous in more complex community evolution models, in which evolution tends to produce large assemblages of species that are stable on a demographic timescale.

Second, they provide, for the first time, insights into the evolutionary emergence of entire food webs or ecosystems. Classical evolutionary models have mostly considered evolution or coevolution of pre-existing species. In community evolution models, species themselves emerge spontaneously from the evolutionary dynamics of the system.

Third, they provide new perspectives on food web and community properties, and potentially a more complete understanding of the mechanisms that generate them. We provided several examples of such applications in sections 12.2.2 and 12.2.3. Community evolution models are capable of giving as good a match to binary data sets as classical food web models such as the Cascade and Niche models. But, additionally, they provide the dynamics of food web structuring whereas other models are only able to reproduce empirical data at a given time. Finally, community evolution models describe species interactions based on individual-level traits, so that community properties are emergent properties of processes that take place at a smaller scale. As a consequence, the mechanisms underlying emerging structures are much clearer than in the case of the Niche or Cascade models, which use large-scale patterns, such as species diversity and connectance, to predict other large-scale patterns, but cannot account for species diversity and connectance in the first place.

12.4.1 Possible extensions of community evolution models

As discussed in section 12.3.3 community evolution models can include other traits than body size. Whatever other traits are chosen, however, body size seems a natural candidate for a primary trait. Body size has well-documented effects on many life-history traits and trophic interactions in all taxonomic groups, on both plants and animals. It has been suggested as a good proxy for a species' trophic level, and has been used abundantly in both static food web models and the new community evolution models.

Although the importance of body size is undisputed, species interactions are the product of several traits. Therefore, a straightforward extension of these models would be to include one or several other traits to better account for species interactions. Some of the good candidates, such as elemental ratios, habitat choice and dispersal rates, are discussed in section 12.3.3. In addition to these, another important trait is niche width, which encapsulates a species' ability to consume a more or less large array of prey species. In the model presented in section 12.2.1.2, we made the simplifying assumption that niche width is constant among species and does not evolve. We are currently working to add evolution of niche width in this model.

Another possible extension of the model is the incorporation of other types of interactions. Current community evolution models account for trophic interactions, and sometimes interference competition. There is increasing evidence that other types of interactions, such as mutualism and parasitism, play an important role in the structure and dynamics of natural communities (e.g. Callaway *et al.* 2002; Lafferty *et al.* 2006; Michalet *et al.* 2006). Networks of mutualistic interactions are now documented, and some recent studies suggest a possible role of evolution in constraining their structure (Jordano *et al.* 2003; Vázquez and Aizen 2004; Bascompte *et al.* 2006). The biomass of parasites is sometimes comparable to the biomass of predators, so that nutrient flows involved in parasitic interactions may no longer be neglected (Lafferty *et al.* 2006). The main problem with the inclusion of such interactions in community evolution models is to find traits that can be linked to them unambiguously in the same way as body size is for trophic interactions. Goudard and Loreau (2007) recently proposed a first community assembly model that includes all types of species interactions. Their model could be extended to include evolutionary dynamics.

12.4.2 Empirical and experimental implications of community evolution models

When community evolution models are based on well-defined traits, it is possible to include physiological or genetic information on these traits. The benefits and costs of these traits are then assessed from empirical or experimental knowledge, and the evolutionary trade-offs that constrain them are built as assumptions into the models. This is both a blessing and a curse. The advantage is the possibility to play with the various fitness components to determine how each trait influences emerging patterns. On the other hand, evolutionary trade-offs are notoriously difficult to obtain, and their shape strongly influences the results of the evolutionary dynamics (de Mazancourt and Dieckmann 2004; Loeuille and Loreau 2004).

In the case of body size, many observations exist, so that costs and benefits can be determined relatively safely. Things are less obvious for the other traits that were proposed as possible extensions to existing community evolution models (section 12.4.1). Elemental ratios are typically linked to the growth rate of individuals (Justic *et al.* 1995; Kooijman 1998; Makino *et al.* 2003; Klausmeier *et al.* 2004; Frost *et al.* 2006). Similarly, predators modulate their attack rates between their different prey depending on prey stoichiometry (Loladze and Kuang 2000; Grover 2003). Thus, life-history and species interactions are dependent upon elemental ratios, but the exact shape of this dependence is not well known.

Habitat choice and dispersal probably affect interaction strength too. For instance, habitat choice by a predator may be driven by prey palatability, so that interaction strength is increased. By a symmetric argument, it may be assumed that dispersal or habitat choice by prey can reduce interaction strength. Habitat choice involves costs linked to the uncertainty of finding a suitable place and increased mortality while moving, in addition to the energy spent.

Finally, evolution of niche traits implies a trade-off between the maximum consumption rate and niche width. Note that this trade-off is already included in equation 12.2. When niche width increases, for example because σ^2 is increased, then the maximum interaction rate is decreased because the function γ is normalized (i.e. its integral is constant and equal to γ_0). But niche width might also influence other traits that determine the species' life-history or their trophic interactions. These indirect costs and benefits are less documented.

Thus, including other traits hinges on the empirical knowledge we have of their associated trade-offs. To determine these trade-offs, controlled experiments in common garden are promising tools. Such experiments have already yielded interesting results on the costs of anti-herbivore defences in plants (Mauricio 1998; Strauss *et al.* 2002). Such studies are required for other traits so that their effects on life-history and ecological interactions are better represented in models.

Other empirical needs include the development of quantitative data. Quantitative data sets exist (Baird and Ulanowicz 1989; Winemiller 1990; de Ruiter *et al.* 1995; Christian and Luczkovich 1999; Trites *et al.* 1999; Yodzis 2000; Neira and Arancibia 2004; Neira *et al.* 2004; Sànchez and Olaso 2004; Williams *et al.* 2004; Tewfik *et al.* 2005), but several problems remain:

- There is a need for new standards for these quantitative data (Cohen *et al.* 1993a; Berlow *et al.* 2004). Some studies use density to describe species abundances while others use biomass. Some use energy flows for measuring interaction strength, others use the frequency of the interaction, still others use the effect of predator removal, etc. Because of this lack of standards, quantitative data sets are very heterogeneous, making it difficult to test some predictions of community evolution.
- There is a need for longer term studies. Quantitative data typically show a high variability in species abundances and interaction strength, for instance through seasonal variations. Long-term trends, however, might show less variability. This means that the quantification of food web properties should be performed over several years. Projects that describe food webs should be funded on a long-term basis, as requested by Cohen *et al.* (1993a).
- There is a need for better assessment of some critical hypotheses underlying quantitative food web data. Because *in situ* measurements are very costly, both in money and in time, many indirect methods have been used, such as an extensive use of bibliographical or gut content data and reliance on equilibrium assumptions to infer some of the data set using partial information (e.g. using the ECOPATH software). Errors involved in these methods should be carefully quantified and error bars included in food web quantification, as should possible errors of direct observations.

The ideal data sets to test evolutionary food web models contains species abundances, interaction strengths and detailed knowledge of the traits described in the model under standardized conditions. Of course, getting such data is very difficult, perhaps sometimes even impossible. But linking model predictions and empirical data will be an indispensable step to fully assess the scope and potential of recent theoretical advances.

Chapter 13

Mutualisms and community organization

David Kothamasi, E. Toby Kiers and Marcel G.A. van der Heijden

13.1 Introduction

When the Titanic attempted to manoeuvre past the infamous iceberg, the captains did not realize that the majority of the challenge was hidden. Current knowledge about species interactions in community ecology is comparable to the visible tip of a huge iceberg – the majority of species interactions remain invisible and unknown. In the past, ecologists have focused on negative interactions such as competition, predation and parasitism to explain the organization of communities because these interactions could be integrated with Darwin's theory of natural selection with relative ease. Mutualisms, positive interactions between two or more species that support each other's fitness, were difficult to reconcile with the struggle for existence implied by natural selection. Wilkinson and Sherratt (2001) emphasize that one reason for the under-representation of mutualisms in ecology may be a historical absence of useful models. For example, when testing cooperative interactions, the classic Lotka–Volterra models produce 'silly results' as a result of unconstrained positive feedbacks. However, even Lotka–Volterra models have unveiled some fundamental characteristics of persistent mutualisms, especially in obligate partnerships (Vandermeer and Boucher 1978).

Mutualistic interactions between species are ubiquitous in ecosystems and involve organisms from every kingdom. Habitats ranging from deserts, to tropical rainforests to coral reefs are dominated by species that depend on mutualists (Bronstein 2001). While these interactions have been described as mathematically unstable, key evolutionary events such as the origin of the eukaryotic cell, invasion of land by plants and the radiation of the angiosperms are linked to mutualism (Bronstein 2001; Bronstein *et al.* 2006; Morris and Blackwood 2007). Indeed, the complex situations that an organism is likely to encounter in an ecosystem will often favour cooperation over competition (Cohen 1998).

Mutualisms can enhance diversity and influence community organization by mechanisms involving habitat modifications, acquiring food sources, dispersal and protection. These interactions promote coexistence of competing species through positive feedbacks on abundances in a manner that is similar to the negative feedbacks of predation that mediate coexistence of competing prey species by reducing the probability of competitive exclusion. For instance, mutualistic arbuscular mycorrhizal fungi increase the competitive ability of plants that are otherwise inferior competitors (Schmitt and Holbrook 2003). Disruptions of mutualistic interactions cause dramatic declines in population sizes of plants and even shifts in plant community composition (Riera *et al.* 2002). Threats to critical mutualists, for instance pollinators (Memmott *et al.* 2007), can potentially endanger the evolutionary persistence of the plants that depend on them and consequently cause significant perturbations in community function and organization.

We focus here on the role that mutualists play in plant community organization. We begin by tracing the evolution of mutualisms from initial conflicts that evolved through trade-offs into cooperation that bestowed selective advantages on the mutualistic partners. We then discuss the role of mutualisms in community organization and conclude by evaluating performance in two prominent examples of mutualistic interactions: the legume–rhizobia and plant–mycorrhiza interactions.

13.2 Conflicts, cooperation and evolution of mutualisms

Cooperative behaviours that benefit both the actor and the recipient(s) of the behaviour are termed 'mutually beneficial' (West *et al.* 2007b). Even though interspecific mutualisms are often viewed in the context of reciprocal exploitations, they nonetheless provide net benefits to each partner (Herre *et al.* 1999). Mutualistic interactions may be direct or indirect (West *et al.* 2007a; Fig. 13.1). In direct mutualisms, the cooperating species interact physically, whereas in indirect mutualisms cooperating species benefit from each other's presence, but there is no direct contact. Direct mutualisms can include symbiotic interactions, which are defined as intimate interactions among different species. In symbiotic interactions the benefits exchanged can be classified in four different types: nutritional, supply of energy, protection and transport mutualisms.

One useful approach to understanding mutualisms is via an economic framework that defines the value of the benefits and costs exchanged by the partners. A biological trading price determined by the balance between supply and demand for the benefits being exchanged locates the interaction along the mutualism–parasitism continuum. When a price is favourable for both partners the interaction moves to the mutualistic end of the continuum, but if the price is favourable for one species and not for the other the interaction becomes parasitic (Schwartz and Hoeksema 1998; Hoeksema and Bruna 2000; Hoeksema and Schwartz 2006). Such trade-based models are useful for conceptualizing partnerships, but they tend to gloss over the in-

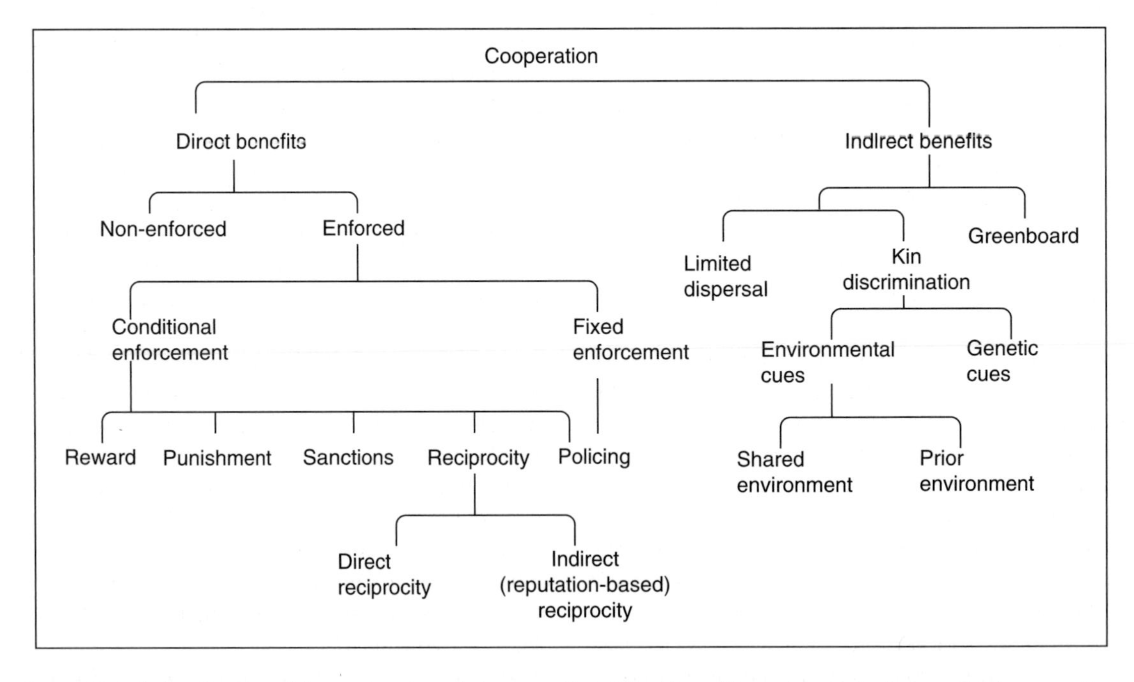

Figure 13.1 A classification of the explanations for cooperation. Direct benefits explain mutually beneficial cooperation, whereas indirect benefits explain altruistic cooperation. Within these two fundamental categories, the different mechanisms can be classified in various ways. These possibilities are not mutually exclusive; for example, a single act of cooperation could have both direct and indirect fitness benefits, and interactions with relatives could be maintained by both limited dispersal and kin discrimination. Reproduced from West *et al.* (2007a) with permission from Elsevier.

credibly complex physiological mechanisms that allow mutualisms to function. In reality, the decision to cooperate, and with whom to cooperate, needs to be understood in the context of their unique physiological constraints (Kiers and van der Heijden 2006).

To gain a broad perspective of mutualisms at the community level, it is important to understand how biotic and abiotic factors influence transitions from parasitism to mutualism (Johnson *et al.* 1997; Saikkonen *et al.* 1998; Pellmyr and Leebens-Mack 1999). Biotic factors, such as intraspecific competition, number of partners available or other alternatives available to a partner (Yamamura 1993; Lipsitch *et al.* 1995; Herre *et al.* 1999) may determine the value conferred by a given mutualistic strategy. If there are viable alternatives, cooperation may be destabilized. Abiotic factors will play a role as well. For example, in the acacia–ant mutualism, nutrient-rich habitat (created by termites) plays a significant role for the coexistence of different mutualist guilds (Palmer 2003). Such mutualist guilds may function along a continuum of cooperation and parasitism (Stanton *et al.* 1999).

Switching from parasitic to mutualistic lifestyles may occur in interactions between plants and fungal endophytes. The fungus *Colletotrichum magna* causes anthracnose in cucurbit plants, but mutant isolates of this fungus can exhibit mutualistic effects (Kogel *et al.* 2006). The recognition molecules involved in both mutualistic and parasitic interactions appear to be similar. However, it is not clear to what extent friendly recognition overbalances unfriendly recognition. The mutual interactions between fungal endophytes and plants are an outcome of balance under environmental, physiological and genetic control that results in fitness benefits for both partners (Kogel *et al.* 2006).

Mutualisms are often characterized by an apparent asymmetry (one partner extracts significantly more benefits than the other) in benefits received by the two sides of the interaction (Kawakita and Kato 2004). For instance, many plant species are pollinator generalists that recruit the services of diverse potential pollinator species. Several pollinators are also generalists that may visit many plants. When the asymmetry becomes exacerbated, a previously mutualistic interaction can become parasitic, and some interactions have been shown to slide back and forth between the two (Hoeksema and Bruna 2000). Plant–mycorrhiza associations that are generally at the mutualistic end of the continuum can become parasitic depending on ecological conditions, suggesting that mycorrhizas may have begun as parasites dependent on host plant carbon, and subsequently evolved into mutualists by offsetting carbon costs to the plants by providing net gains through uptake of other nutrients (Johnson *et al.* 1997). Trade models have identified three factors important in determining whether a potential partner becomes parasitic or remains mutualistic under asymmetrical conditions: (1) relative differences between the partners in their resource acquisition abilities; (2) relative differences between the partners in their resource requirements; and (3) variation in the shape of resource acquisition trade-offs (Hoeksema and Schwartz 2003). If the environmental conditions change, such as an increase in the productivity of the environment, interactions are predicted to become less beneficial (Kiers *et al.* 2002; Thrall *et al.* 2007).

There is much interest in the identity of the factors that align the interests of cooperating partners so that the relationship remains mutually beneficial and evolutionarily stable (Herre *et al.* 1999; West *et al.* 2002a; Sachs *et al.* 2004; Moran 2007). Mode of transmission may be important in aligning partner interests. Mutualisms established by vertical transmission of symbionts from parent to offspring will tend to favour cooperation. Conversely, high horizontal transmission among different individuals of the host and competition with other endosymbionts within a host may push towards parasitism (Herre *et al.* 1999), because the symbiosis has to be formed anew each time the symbionts meet, providing opportunities for parasites (or less beneficial symbionts) to invade the system. Wilkinson and Sherrat (2001), however, argue that, depending on the conditions, horizontal transmission can also lead to successful mutualisms. Other criteria for aligned interest include (1) genotypic uniformity of symbionts within individual hosts, (2) spatial structure of populations (but see West *et al.* 2001) leading to repeated interactions and (3) restricted options outside the host (Herre *et al.* 1999; Sachs *et al.* 2004). Additionally, interests may be aligned through control mechanisms, such as host-mediated punishment, in which those that fail to cooperate suffer from sanctions, leading to decreased fitness (Denison 2000; West *et al.* 2002a, b; Kiers *et al.* 2003). Moreover, the exchange of luxury resources may stabilize many mutualistic associations. For instance, in mycorrhizal associations plants may trade surplus carbon for excess fungal nutrients, such as phosphorus or nitrogen (Kiers and van der Heijden 2006).

The predicament faced by any pair of species engaged in new mutualisms is how to initiate cooperation from a previously non-cooperative state and then maintain a stable interaction. This is the classic prisoner's dilemma problem: although individuals can benefit from mutual cooperation, each one can do even better by exploiting the cooperative efforts of others (Axelrod and Hamilton 1981). The payoff to a mutualist is in terms of the effect on its fitness. Regardless of what the other partner does, the selfish choice of cheating yields a higher payoff than cooperation, if there are no punishment mechanisms in place. However, if both the partners cheat, costs for both partners are more than if both had cooperated (Axelrod and Hamilton 1981). If the interaction is iterated repeatedly, then the best strategy can be to cooperate with other individuals who cooperate too (Nowak and Sigmund 1992). The investment decision of a mutualist depends on the payoff received in the previous iteration. Biologically, this means that healthy organisms have more to offer their partners and the amount invested evolves as mutations periodically arise (Doebeli and Knowlton 1998). Successful mutualists should be better competitors (if they are rewarded for their mutualistic behaviour), and eventually establish in the community as the now combined abilities of the partners in acquiring resources will reduce competition from

Figure 13.2 *Butea monosperma*, an important legume from a deciduous forest ecosystem of India, is a keystone mutualist forming mutualistic associations with some animal species for pollinator services. *Butea monosperma* also forms belowground mutualisms with rhizobia and arbuscular mycorrhizal fungi. Inset, a rose ringed parakeet (*Psittacula longicauda*) foraging on the flowers of *Butea monosperma*; the parakeet is a nectar robber and does not provide any reciprocal benefit to the plant. See plate 4.

ecologically similar species (Boucher *et al.* 1982). However, ultimately, partners can achieve stable mutualisms through the enforcement of cooperative behaviour by actively rewarding cooperation and punishing cheating (West *et al.* 2002a). Mutualisms will remain stable if benefits for both partners exceed costs; otherwise the interaction is predicted to shift to parasitism (Holland *et al.* 2004).

13.2.1 Mutualism can also develop without evolution

While an extensive body of literature describes the coevolution of mutualists (Anstett *et al.* 1997; Kato *et al.* 2003; Machado *et al.* 2005; Mehdiabadi *et al.* 2006; Moran 2006), the different traits of different species can give rise to mutualistic associations *de novo* even between organisms that are not coevolved. The association may subsequently be stabilized through natural selection acting on each species (Moran 2007). For example, invasive ant species may provide some dispersal services to native plant species that previously relied on native mutualistic ant species (Lach 2003). The facultative nature of many mutualisms also indicates that their formation is not always a result of coevolution. An established mutualistic association is open to parasitic exploitation by either partner or a third species that might profit from the benefits provided while not give anything in return (Boucher *et al.* 1982). Several birds and small animals forage for nectar on *Butea monosperma* (Fig. 13.2), a leguminous tree from India. While the purple sunbird (*Nectarinia asiatica*) and the three striped squirrel (*Funambulus tristriatus*) provide reciprocal pollination services, the rose ringed parakeet (*Psittacula longicauda*) is a cheater that benefits by foraging on the flowers but provides no reciprocal benefits to the plant (Tandon *et al.* 2003). These examples show that mutualistic interactions are complex and are subject to continuous evaluation and selection by the environment.

13.3 Mutualisms in community organization

Mutualisms provide partner species with novel options for adjusting to changing physical and biotic environments. They can be pivotal in affecting the organization, structure and function of communities. Mutualisms commonly support the key species that define entire ecosystems and can play important roles in moving energy and nutrients across ecosystem borders (Hay *et al.* 2004). Mutualists, through affecting the fitness of their partners, can have a strong influence on community organization. Modern angiosperms are estimated to comprise ~250 000 species and an estimated 70–90% of these recruit animal mutualists for pollination services (Heywood 1993; Kearns *et al.* 1998; Fontaine *et al.* 2006). Up to 80% of all terrestrial plants are believed to form symbiotic associations with mycorrhizas for nutrient acquisition and mycorrhizas are believed to have hastened the invasion of land by plants (Malloch *et al.* 1980; Simon *et al.* 1993; Smith and Read 1997). Below we will discuss a number of mutualisms that are known to influence plants and plant community composition. We will start with plant–pollination and plant–protector mutualisms. In the two subsequent sections we will then discuss two prominent examples of mutualistic interactions, the legume–rhizobia and plant–mycorrhiza interactions.

13.3.1 Plant–pollinator interactions

Plant–pollinator interactions have been hypothesized to coevolve towards an increasing degree of specialization (Stebbins 1970). However, recent empirical studies are demonstrating a scenario in which most plants exploit a wide and diverse range of pollinators (Sahli and Conner 2006; Gomez *et al.* 2007). Generalization may benefit pollinators foraging on several plant species as they can acquire multiple resources, such as pollen, nectar, mates or prey, each provided at different plants (Ghazoul 2006). Increased diversity may be one consequence of plant and pollinator seeking generalization in mutualist partners. Generalizations provide mutualist partners with 'competitor-free space' (Ghazoul 2006). Pollinators displaced from preferred flowers by aggressive competitors have the option of returning to their preferred flowers if they temporarily relocate to alternative species, not visited by the competitively dominant pollinators. By providing a competitor-free space, less rewarding flowers benefit from hosting the displaced pollinators. For instance, non-rewarding orchids are known to benefit from displaced pollinators by being located close to rewarding flowers (Johnson *et al.* 2003; Ghazoul 2006). Generalizations in pollination mutualisms facilitate the pollination of several coexisting species and contribute to coexistence and diversity. The plant–pollinator interaction continuum can range from full specialist to full generalist. Such pollinator functional diversity enhances diversity in plant communities (Fontaine *et al.* 2006).

13.3.2 Plant–protector mutualism

Mutualisms involving protection may have the effect of decreasing species diversity in a community. An estimated 20–30% of all grass species host endophytic symbionts (Leuchtmann 1992). *Neotyphodium coenophialum,* a fungal endophyte of the invasive grass species *Lolium arundinaceum,* protects the host from herbivores and pathogens through the production of toxic alkaloids (Finkes *et al.* 2006). *Neotyphodium coenophialum* also alters the detritivore composition in the rhizosphere, suggesting possible influences on ecosystem processes like decomposition and nutrient turnover (Lemons *et al.* 2005). Because of the unpalatability rendered by the endophyte, the competitive ability of *Lolium arundinaceum* is increased and this grass is able to displace other competing species from the community, thereby bringing about a reduction in species diversity. In this way, a mutualist associated with a dominant plant species enhances the competitive ability of its host, leading to lower biomass of the competing species and reduction in the productivity of the community (Rudgers *et al.* 2005). The strong influence of fungal endophytes on plants and herbivores may cascade to other trophic levels. The *Lolium–Neotyphodium* mutualism has been reported to influence composition of spider communities by causing reductions in prey populations and changes in plant assemblages that affected web building (Rudgers and Clay 2005; Finkes *et al.* 2006).

Another example of plant-protection metabolism is that several plant species form mutualistic associations with ants. The ants protect the plants against herbivores while the plants provide food and/or housing in return. For example, the whistling acacia (*Acacia drepanolobium*) has modified thorns called stipular spines that house stinging ants. The ants deter herbivores by swarming out of their nests and attacking an intruder at the smallest movement. Without the protection by ants, plants like the whistling acacia might be eliminated by herbivores such as giraffes, for whom thorns are no great deterrent. However, recent experiments in which large herbivores were excluded from feeding on *Acacia* trees demonstrated that, in the absence of defoliation threats from large herbivores, *Acacia depranolobium* reduced investments in extrafloral nectaries and modified stipular thorns for feeding and housing symbiotic ants. The reduction in rewards to ants caused a shift in competitive dominance within plant–ant community from nectar-dependent symbiotic ant species to an antagonistic ant species. This shift in ant community and the resultant breakdown of ant–*Acacia* mutualism could lead to potentially negative consequences for *Acacia* growth and survival (Palmer *et al.* 2008).

In a few cases, protective ants might actually castrate their hosts and reduce the plant's fitness (Stanton *et al.* 1999). *Cordia alliodora* forms symbiotic associations with ants for protection against insect herbivores. Ant associates such as *Azteca pittieri* gave significant benefits to the plants by protecting them against insect herbivores, while other associates such as *Cephalotes setulifer* (which do not provide any benefits) actually formed a drain on the plant resources (Tillberg 2004).

An example of perturbation in ecological processes demonstrates the important role mutualisms have in community organization. Habitat fragmentation results in disruptions in the foraging range of pollinators and reductions in pollination services (Kearns *et al.* 1998). This has implications for the fitness and genetic profile of plant populations. In fragmented plant populations, pollinators increase geitonogamy by visiting a higher proportion of flowers on individual plants, leading to increased genetic drift and inbreeding depression (Kearns *et al.* 1998). Shifts in pollinator assemblages can result in parallel changes in plant communities. In Britain a declining insect pollinator population caused a parallel decline in insect-pollinated plants and an increase in species reliant on abiotic factors for pollination. Similarly, in The Netherlands, a declining bee population has caused a decline in plants exclusively pollinated by bees (Beismeijer *et al.* 2006).

Invasions by alien species have the potential to disrupt mutualistic interactions within native communities and can also modify successional trajectories. Impacts of alien pollinators have yet to be conclusively demonstrated. However, it is thought that alien pollinators might either increase pollen transfer among plants, thereby increasing plant fitness, or might have very low visitation rates and consequently might not (or negatively) affect reproductive success (Traveset and Richardson 2006). Alien pollinators are known to displace native pollinators. Specialist plants dependent on the displaced pollinators are also likely to be displaced, resulting in changes in plant species composition. Alien pollinators might bring about changes in community composition by preferentially visiting plants that might not be preferred by native pollinators. The red-whiskered bulbul (*Pycnonotus jocosus*), an invasive species in Mauritius, visits the flowers of the extremely rare endemic *Nesocodon mauritianus* more frequently than do native birds (Traveset and Richardson 2006).

Disruptions of mutualisms may accelerate the decline and extinction of species in a community. The loss of one species could cause the subsequent loss of other species that are directly or indirectly dependent upon them. Disruption of belowground plant mutualists like mycorrhizas and symbiotic bacterial associates can have significant effects on aboveground plant communities (Klironomos 2002; Stinson *et al.* 2006; van der Putten *et al.* 2007). *Alliaria petiolata* (garlic mustard), an invasive species, disrupts arbuscular mycorrhizal associations (Stinson *et al.* 2006). Garlic mustard may consequently affect composition of mature forest communities by favouring plants with low mycorrhizal dependency and repressing the regeneration of canopy trees that are dependent on arbuscular mycorrhizal (AM) fungi symbiosis. While some invasions might alter soil-borne mutualisms and consequently reorganize recipient plant communities, mutualistic associations might also favour invasions (Richardson *et al.* 2000). Although there is little evidence for dominance and competitive exclusion of native species by invasive plants through establishment of new mutualisms, evidence is plentiful for promotion of plant invasions caused by mutualistic interactions with other soil biota (Simberloff and Von Holle 1999; Reinhart and Callaway 2006). Arbuscular mycorrhizas can potentially colonize a broad range of hosts although specificity may exist for growth responses, making it possible for invaders to use the native mycorrhizae of a new region (Callaway *et al.* 2004).

13.3.3 Plant nutrition symbiosis

In the following sections we discuss two of nature's most important mutualistic interactions, the legume–rhizobia and the plant–mycorrhiza symbioses and how these important mutualisms affect community organization. Both rhizobia and mycorrhizal fungi supply limiting nutrients to their plant hosts and are especially important in nutrient-poor ecosystems.

13.3.3.1 Legume–rhizobia symbioses

More than 15 000 species of legumes are involved in symbioses with rhizobia (de Faria *et al.* 1989). Plant symbioses with nitrogen-fixing soil microorganisms play an important role in organizing community structures. Nitrogen-fixing organisms may be free-living or form intimate associations and fix atmospheric nitrogen in specialized structures such as root nodules in many legumes. The legume–rhizobia association involves a formal and physiologically complex symbiosis. Rhizobia are Gram-negative heterotrophic bacteria classified within six genera–*Allorhizobium*, *Azorhizobium*, *Bradyrhizobium*, *Mesorhizobium*, *Rhizobium* and *Sinorhizobium* – that interact with plants of the family Leguminosae, leading to profound physical alterations in both organisms (Pepper 2000).

Although quite an extensive literature exists on the cross-talk between plants and rhizobia and the subsequent formation of the symbiotic nodules, little information has been gathered as to the fate of the rhizobia after this point. However, this information is critical to understanding both the evolutionary and ecological functioning of the symbiosis. What we do know is that, following the formation of nodules in the plant, the bacterial cells undergo a morphological transformation into bacteroids. Rhizobia can fix nitrogen only in this transformed state, which is reached through a terminal development event during which bacteroids may lose the ability to reproduce (Zhou *et al.* 1985). Loss of reproductive viability is generally prevalent in the nodules of the indeterminate types, such as nodules found on peas and alfalfa. In such cases it is believed that the rhizosphere populations are replenished by the undifferentiated infection threads (Denison 2000). In determinate nodules, such as in soybean, reproductive viability of the transformed bacteroids is high (Sutton and Paterson 1980), and it is thought that bacteroids remain viable after the nodule senesces. Exceptions to these generalizations exist, and more research is needed to understand the processes that control reproductive viability of rhizobia in nodules.

One of the most intriguing aspects of the symbiosis is how the relationship has persisted for millions of years. Rhizobia expend an incredible amount of their energy in fixing nitrogen and supplying it to their host (Gutschick 1981). This is thought to incur considerable costs in terms of reproductive fitness for the rhizobia (Denison and Kiers 2004a). It is true that, by supplying its host with nitrogen, an individual rhizobium enhances host photosynthesis, thereby potentially increasing the rhizobium's own access to photosynthate. However, we also know that plants are typically infected by more than one strain of rhizobia (Hagen and Hamerick 1996; West *et al.* 2002b and references within). This means that rhizobia that supply their host with nitrogen may indirectly benefit competing strains of rhizobia infecting the same individual plant. As the number of strains per plant increases, evolutionary theory predicts a rise in symbiont parasitism (Smith and Szathmary 1995). The situation is analogous to the classic tragedy of the commons problem from human economics (Hardin 1968). The tragedy is that 'free-rider' rhizobia, those that cheat by extracting carbohydrates from the host while fixing little to no nitrogen, are predicted to spread at the expense of efficient nitrogen-fixing strains (Denison 2000; Kiers *et al.* 2002; Denison *et al.* 2003). The problem of hosting multiple partners is a recurring theme in rhizosphere mutualisms (Kiers and Denison 2008). The question is, why do rhizobia expend resources on fixing nitrogen for the benefit of their host plant (perhaps also indirectly benefiting competing rhizobial strains) when they could use those resources for their own reproduction (Denison 2000; West *et al.* 2002a, b)?

Bever and Simms (2000) proposed a model in which nitrogen-fixing bacteroids in the nodules would convey reproductive benefits to their free-living kin in the rhizosphere. They hypothesized that, because rhizobia reproduce largely by asexual fission, genetically identical kin of the bacteroids inside the nodule could be found directly outside the nodule, when there was little mixing in the soil. Thus, although a bacteroid may have sacrificed its own reproduction in the process of fixing nitrogen, the transformation could be evolutionarily advantageous if the benefits from enhanced exudates of the host could increase the fitness of its related kin in the rhizosphere. This was hypothesized to offset the loss in reproductive capacity of the bacteroid (Bever and Simms 2000).

There are three flaws with this reasoning. First, the bacteroid form is not a reproductive dead-end for all rhizobia, as explained above. Their hypothesis would apply only to those rhizobia in indeterminate nodules. Second, this hypothesis is inconsistent with the way natural selection operates in nature. The rhizosphere surrounding the host plant root may be colonized by kin, but it is also colonized by rhizobia of many different lineages, some of which may not even be fixing nitrogen (Denison and Kiers 2004b). Yet these 'cheaters', even though unrelated to those in the bacteroid, will also benefit by the increases in exudates from the altruistic sacrifice of the transformed bacterium. Third, even if strong spatial structuring in the soil (no mixing) increased the chances that the exudates would be directed to kin, increased spatial structure also increases competition between relatives, making competition more local. These contrasting effects tend to balance each other out (West *et al.* 2002b). Therefore, spatial structuring of soil populations will not necessarily favour greater mutualism (see Denison and Kiers 2004a).

In contrast, preferentially allocating resources directly to cooperative rhizobial strains represents a stable evolutionary strategy (West *et al.* 2002a, b). If host plants are able to discriminate among nodules, supporting those that are fixing more nitrogen with increased resources, then cooperation will tend to be favoured. This has been termed the 'sanctions' hypothesis and suggests that the reproductive success of the rhizobium strain is contingent on the strain's ability to export nitrate to the host (Denison 2000; West *et al.* 2002a, b; Kiers *et al.* 2003). It has been found that legume hosts will impose fitness-limiting sanctions ('punishment') to rhizobial strains based on the actual nitrogen-fixing benefits the strain provides (Kiers *et al.* 2003; Simms *et al.* 2006). The mechanism of punishment is thought to involve a decreased oxygen supply to nodules and the severity of this punishment varies, depending on the extent of cheating (Kiers *et al.* 2006).

The nitrogen-fixing symbiosis can confer incredible advantages to both the host plant and the rhizobial symbiont. For rhizobia, participating in the symbiosis can dramatically enhance reproductive output. A single rhizobium cell that infects a soybean root can produce up to 10^{10} descendents from a single large nodule (Denison and Kiers 2004b). For plants, the extra nitrogen from fixation can facilitate their rapid spread across the landscape. While free-living nitrogen fixation association may result in approximately 0.1–25 kg/ha/year, symbiotic nitrogen can have fixation rates between 100 and 300 kg/ha/year (Pepper 2000). Given that the rhizobia–legume symbioses can increase soil nitrogen fourfold compared with free-living nitrogen fixers, the impact of this symbiosis on ecosystem functioning is significant.

Some of the world's most troublesome invaders of natural ecosystems have been nitrogen-fixing legumes and actinorrhizal species such as *Acacia, Albizzia, Prosopis* and *Myrica faya* (Richardson *et al.* 2000). It is believed that invading legumes are able to alter community composition through their modifying influence on soil nitrogen levels. Plant species of most terrestrial ecosystems are adapted to low-nitrogen soils (Rice *et al.* 2004). Nitrogen-fixing legumes can cause a subtle, but continuous, increase in nitrogen pools and fluxes in nitrogen-limited ecosystems (Olde Venterink *et al.* 2002).

Changes in soil nutrient profiles will have direct effects on the competitive success of species leading to dramatic alterations in community composition (Bobbink *et al.* 1998; Sala *et al.* 2000; Scherer-Lorenzen *et al.* 2007). One clear example is invasion by the actinorrhizal *Myrica faya* in a nitrogen-limited forest in Hawaii. *Myrica faya* caused an influx of 18 kg nitrogen/ha/year whereas the nitrogen amounts in the Hawaiian forest soils prior to the introduction of *Myrica faya* was only 5.5 kg nitrogen/ha/year (Vitousek *et al* 1987; Vitousek and Walker 1989). Here, owing to a substantial increase in nitrogen inputs, *Myrica faya* could successfully displace the native *Metrosideros polymorpha*, leading to significant community reorganization.

Successful colonization by an alien legume species will depend largely on its ability to find a compatible rhizobial symbiont. Competitive exclusion of species adapted to low nutrient levels by faster growing nitrophilic species is independent of whether it is an alien invading species or a native species (Scherer-Lorenzen *et al.* 2007). Native legume species, however, might be successful in regulating the community structures through their influence on nitrogen cycling. The long-term effects of nitrogen fixation on ecosystem functioning and plant species composition are largely unknown (Scherer-Lorenzen *et al.* 2007). Nitrogen fixation not only influences primary productivity, but may also have cascading effects on successional patterns, community composition and disturbance regimes (Rice *et al.* 2004). Scherer-Lorenzen *et al.* (2007) suggest that the increased nitrogen availability that nitrogen-fixing plants provide might be an important pathway by which nitrogen-fixing invaders alter community structures.

A model developed by Parker (2001) predicts how a legume–rhizobia partnership can facilitate the invasion of an ecosystem. The model sets a legume invader to compete with a resident non-mutualist plant (non-legume) in an nitrogen-limited environment. In the absence of mutualist rhizobia, both the legume and the non-legume compete for nitrogen in the soil according to Lotka–Volterra competition dynamics. In the absence of the rhizobia, legumes are assumed to be inferior competitors. Both plants can coexist as long as nitrogen is available. When nitrogen supplies are exhausted, the non-legume excludes the inferior legume from the community. However, when mutualist rhizobia are introduced into the soil, they provide a benefit to the legume host by making fixed nitrogen available to the host. As nitrogen is depleted, the non-legumes are excluded and the legumes establish on the basis of a regular nitrogen supply provided by symbiotic rhizobia (Fig. 13.3). Ensuring the establishment of legumes and the exclusion of non-legumes may have a selective advantage for the bacterial symbionts as this would ensure a ready supply of photosynthetic carbon from the legume hosts. However, suitable rhizobia are not always available (see Parker *et al.* 2006). van der Heijden *et al.* (2006a) have demonstrated in microcosm experiments that legume biomass decreases in the absence of rhizobia. In these experiments, the non-legume biomass was largely unaffected by rhizobia. While the non-legumes mainly acquired nitrogen from the soil, the legumes obtained nitrogen fixed from the atmosphere. This suggests that, in nitrogen-limited environments, rhizobia could selectively favour the establishment of their host legumes. As the legumes become abundant, soil nitrogen levels increase due to nitrogen fixation. Grasses benefit from the increased nitrogen availability and outcompete the legumes until nitrogen availability decreases so that legumes become more abundant, starting the cycle again. This fluctuation in species dominance can have important consequences for stability of plant communities (Schwinning and Parsons 1996).

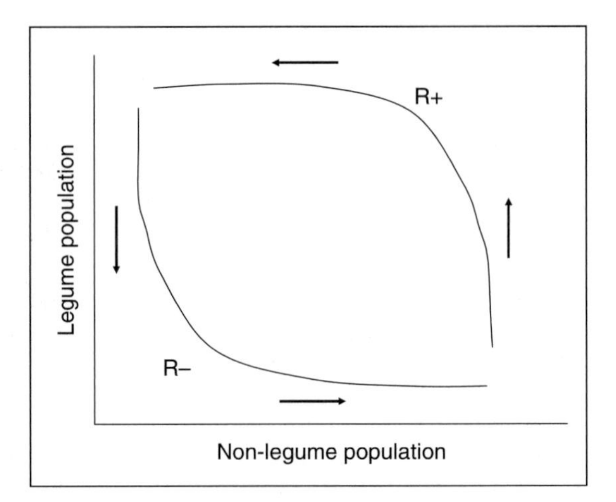

Figure 13.3 Competition between a mutualistic legume and a non-mutualistic plant. The legumes are inferior competitors for nitrogen extraction and in the absence of the rhizobia mutualist (R–), the legume population is eventually excluded. In the presence of the rhizobia mutualist (R+), the legume population is able to obtain nitrogen and excludes the competing non-legume population.

Extensive literature is available on the legume–rhizobia symbioses and the effects of these symbioses on plant growth and nitrogen fluxes into the rhizosphere. However, little attention has been paid to the impacts of these symbioses on community organization, structure and ecosystem processes. Further studies are needed to understand (1) how these symbioses function in community organizations and (2) the effects of fixed nitrogen on plant species composition at the ecosystem level.

13.3.3.2 Mycorrhizal symbioses

The symbiosis between the majority of plants and mycorrhizal fungi is one of the most abundant and ecologically important symbioses on Earth. Mycorrhizal fungi can provide resistance to disease and drought, and supply a range of limiting nutrients including nitrogen, phosphorus, copper, iron and zinc to the plant in exchange for plant carbon. Mycorrhizal fungi can forage effectively for these nutrients because they usually form an extensive mycelial network of fine hyphae in the soil. It is not unusual to find over 10 m of mycorrhizal hyphae per gram of soil (Leake *et al.* 2004). Moreover, the diameter of mycorrhizal hyphae is up to 10 times smaller than those of plant roots, indicating that hyphae can enter small soil pores that are inaccessible for plants roots. The most abundant and important groups of mycorrhizal fungi are the arbuscular mycorrhizal (AM) fungi, the ecto-mycorrhizal (EM) fungi and the ericoid mycorrhizal (ERM) fungi. AM fungi are abundant in grassland, savanna and tropical forests and associate with many grasses, herbs, tropical trees and shrubs (Read and Perez-Moreno 2003). EM fungi associate with about 6000 tree species and are abundant in temperate and boreal forests and in some tropical forests (Alexander and Lee 2005). Ericoid mycorrhizal fungi are most abundant in heathland, where they associate with members of the Ericaceae (Smith and Read 1997). The fungi involved in mycorrhizal associations are phylogenetically diverse (James *et al.* 2006), and members of several of the major fungal clades, the Glomeromycota, Ascomycota and Basidiomycota, interact with plant roots

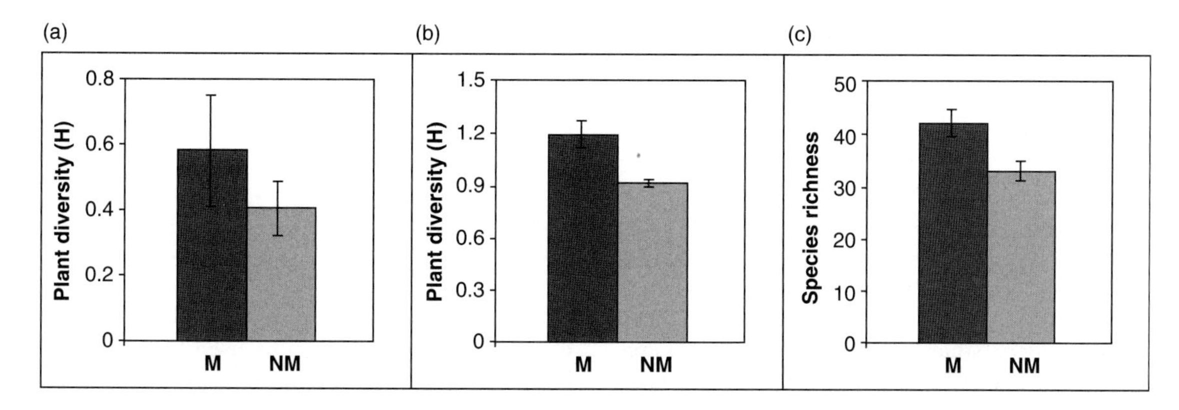

Figure 13.4 The impact of arbuscular mycorrhizal fungi on plant diversity and plant species richness in European grassland. Studies from (a) calcareous grassland in the UK (modified after Grime *et al.* 1987), (b) Swiss calcareous grassland (modified after van der Heijden *et al.* 1998a) and (c) an early successional grassland community in the UK (modified after Gange *et al.* 1993). M, mycorrhizal; NM, non-mycorrhizal.

and form typical fungus–root structures that are referred to as mycorrhizas (Smith and Read 1997). In this section, we largely focus on the role of AM fungi in community organization.

AM fungi alter plant community composition, plant productivity and plant diversity. For instance, AM fungi increase plant diversity in European grassland by as much as 30% (Fig. 13.4). The fungi do this by promoting seedling establishment and enhancing competitive ability of subordinate plant species relative to dominants (Grime *et al.* 1987; van der Heijden *et al.* 2006b). AM fungi may also prevent non-mycorrhizal plants from growing via antagonistic interactions (Francis and Read 1994) and by enhancing the competitive ability of mycorrhizal hosts (see below). Such interactions may also contribute to plant succession (Allen 1991), although this is still poorly understood. In some cases, AM fungi can reduce plant diversity, especially in ecosystems where the dominant plants have a high mycorrhizal dependency and obtain most benefit from AM fungi, such as in tall grass prairie (Hartnett and Wilson 1999) or some annual plant communities in Australia (O'Connor *et al.* 2002). Similarly, in tropical rainforests ecto-mycorrhizal associations may encourage dominance of certain tree species, at the expense of arbuscular mycorrhizal trees that are less able to acquire nutrients and tolerate pathogen attack, thereby reducing species coexistence (Connell and Lowman 1989). Over 20 000 plant species are completely dependent on microbial symbionts (including mycorrhizal fungi) for growth and survival, pointing to the importance of plant–microbe symbiosis as regulators of plant species richness on Earth (van der Heijden *et al.* 2008).

Mycorrhiza-enhanced efficiency of nutrient uptake may be important mechanisms in competition between plants and may affect plant coexistence and community organization (van der Heijden 2002). A competition model proposed by Tilman (1988) has been applied by van der Heijden (2002) to predict the influence of mycorrhiza in a competition between host and non-host plants. The model represents the growth of a species by isoclines that show the amount of growth in relation to the availability, or supply, of limiting resources. The resource supply levels where plant populations stop growing define the zero population growth isoclines. The supply of additional resources by AM fungi could reduce the growth isoclines of a mycorrhizal-dependent plant, thus broadening its potential niche. This is shown for a hypothetical plant species (plant B) in Fig. 13.5. In this case, AM fungi enhance the supply of phosphorus, thereby lowering the isocline of plant B (Fig. 13.5). The supply of nitrogen is not affected by AM fungi in this situation because it is unclear whether AM fungi have a big impact on nitrogen uptake and contrasting observations have been made; hence, the growth isocline moves only down and not to the left. Tilman

(1988) predicts that, if two plant species compete for resources, the one with the lowest zero population growth isocline, i.e. the one with lowest resource requirements, will be competitively superior. Furthermore, the model predicts that two plant species can coexist if their zero population growth isoclines cross. Coexistence could be possible because the growth of each plant species is limited by a different resource. As a consequence intraspecific competition is greater than interspecific competition (Tilman 1988). The reduction of the zero population growth isocline of mycorrhizal-dependent plant species by AM fungi, as explained above, can alter their competitive ability with other plant species. For example, plant B will be outcompeted and replaced by plant A in the absence of AM fungi because the zero growth isocline of plant B is always inside that of plant A (Fig. 13.5). However, if AM fungi are present, the zero growth isocline of plant A is unaltered, but that of plant B is reduced and it crosses the zero growth isocline of plant A, so that coexistence occurs (Fig. 13.5). Several studies have indeed shown that AM fungi alter plant competition (Fitter 1977; Hetrick *et al.* 1989; Allen and Allen 1990; West 1996; Marler *et al.* 1999). Several plant species are able to coexist with other plants only if AM fungi are present, indicating that AM fungi enhance their competitive ability (Grime *et al.* 1987; Hetrick *et al.* 1989; van der Heijden *et al.* 1998a). The model presented in Fig. 13.5 can also be extended to plant communities when the response of individual plants to AM fungi is known, and it can be used to predict the impact of AM fungi on the composition of plant communities (see also Urcelay and Diaz 2003). Moreover, some plant species can also suppress the abundance of AM fungi (Stinson *et al.* 2006). In that situation, competitive abilities between plants would be altered in the opposite direction (as shown in Fig. 13.5).

In many ecosystems mycorrhizal fungi form large mycelial networks, also called 'wood wide webs' (Helgason *et al.* 1998). These networks are extremely fascinating because many plant

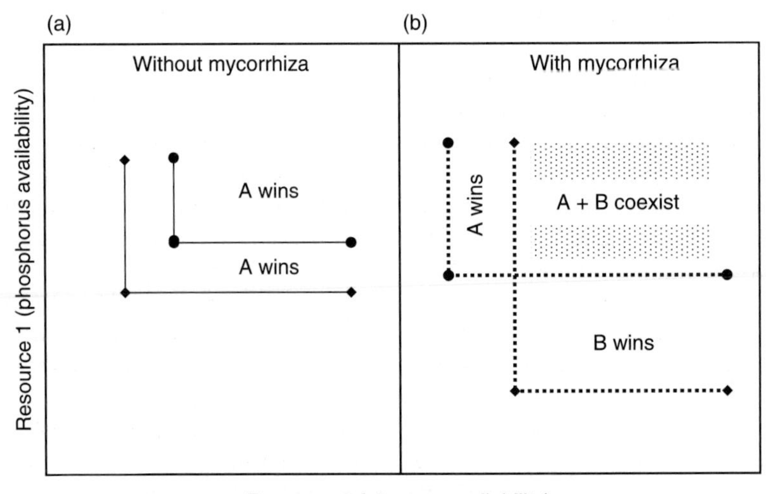

Figure 13.5 Competition between two hypothetical plant species (plant A (◆) and B (●)) without arbuscular mycorrhizal (AM) fungi (solid line) and with AM fungi (dashed line). The lines represent the resource-dependent zero growth isoclines, which show the minimum amount of phosphorus (*y*-axis) and nitrogen (*x*-axis) that is necessary for plant growth. The positions of the isoclines determine the outcome of competition between plants A and B. In the absence of AM fungi (a), plant B is outcompeted by plant A at each resource level because the resource requirement of plant B is higher than that of plant A. Plant B is able to coexist with plant A when AM fungi are present because plant A and B are limited by different resources (b). Reproduced from van der Heijden (2002) with permission from Springer-Verlag.

individuals can be interconnected to these hyphal networks. Such networks not only connect individuals of the same plant species, but, owing to a lack of specificity in many mycorrhizal associations, also individuals of different plant species. Subsequently these networks can act as symbiotic support systems for seedling establishment of different plant species (van der Heijden 2004; Horton and van der Heijden 2008). Some studies even suggest that carbon and nutrients can flow from one plant to another through hyphal networks (Francis and Read 1994; Simard *et al.* 1997; Selosse *et al.* 2006; Whitfield 2007). However, it is still controversial whether significant amounts of carbon and nutrients can be transferred and some authors view carbon transfer as essentially a fungal phenomenon wherein the carbon stays inside the mycelium. For instance, Fitter *et al.* (1998) and Pfeffer *et al.* (2004) observed that carbon transfer to a recipient plant through AM fungal hyphae remains largely or entirely within the mycorrhizal roots, even under conditions that facilitate root to shoot migration. However, some plants do obtain carbon from hyphal networks. There are about 400 species of myco-heterotrophic plants that are completely dependent on carbon and nutrients that they receive from mycorrhizal fungi that they parasitize (Bidartondo *et al.* 2002; Selosse *et al.* 2006). Myco-heterotrophic plants lack chlorophyll (Björkman 1960; Bidartondo *et al.* 2002). Myco-heterotrophic plant lineages have evolved in 11 families of plants in five orders (Leake 1994), suggesting that there is a widespread potential for carbon flow from mycorrhizal fungi to their 'host' plants. Furthermore, several other green plants have a mixed strategy and are thought to acquire carbon through photosynthesis and via fungal links (Tedershoo *et al.* 2007; but see Zimmer *et al.* 2007). Some *Pyrola* species may obtain up to 50% of carbon from fungal hyphae (Tedershoo *et al.* 2007). Moreover, mycorrhizal fungi provide germinating orchid seeds with carbon and nutrients (Cameron *et al.* 2006).

About 250 different AM fungal species have been described (Morton *et al.* 1994). The actual number of AM fungal species is probably much higher because molecular techniques have shown that about 60% of environmental sequences of AM fungi do not match with AM fungi that have been brought into culture (van der Heijden *et al.* 2008). The identity of AM fungi present in a plant community is important because plant species respond differently to different AM fungi (van der Heijden *et al.* 1998a; Maherali and Klironomos 2007) and the diversity of AM fungal communities can affect plant productivity, plant community composition and plant diversity (van der Heijden *et al.* 1998b, 2006b; Vogelsang *et al.* 2006). An intriguing question in mycorrhizal ecology is how uncultured AM fungi contribute to plant diversity and productivity in natural communities. The fact that specialist fungi are most affected by soil perturbations (Helgason *et al.* 2007) might even suggest that our knowledge of how AM fungi affect plant communities is far from complete, especially because this knowledge is based on studies of easily culturable generalist AM fungi. It is also important to mention that AM fungal communities in the soil are completely different from those in plant roots (Hempel *et al.* 2007), opening up many new questions of how members of AM fungal communities interact and support plant growth.

13.4 Conclusions

In nature, organisms from every kingdom are involved in some kind of interspecific mutualism and the mechanisms through which mutualisms influence community organization are likely to be similar across different organismal groups from terrestrial to marine habitats. We focused on mutualisms involving plants, as they occupy a pivotal position in community organization owing to their role as primary energy producers. Mutualisms provide partner species with novel options for adjusting to changing environment and biotic stresses. They may play a critical role in regulating organization, structure and function of communities through activities that regulate the acquisition of resources and ameliorate stresses. Disruption of established mutualisms by habitat fragmentation or through biological invasions has caused cascading shifts in community composition. Mutualisms may mediate the outcome of interspecific interactions such as competition and facilitate the coexistence of competing species. However, in some cases, mutualist interactions may enhance the competitive ability of the dominant species and consequently cause the exclusion of subordinate species. An in-depth understanding of mutualist interactions would allow us to effectively predict the effects of natural perturbation and human interferences in functioning of communities.

Acknowledgements

The authors thank DST (BOYSCAST), Government of India, for funding (D.K.); NWO for funding (E.T. K.); Deepika Sharma for preparing Fig. 13.3; Springer-Verlag for permission to use Fig. 13.5 and portions of text from *Ecological Studies*, vol. 157: *Mycorrhizal Ecology*; and Elsevier for permission to use Fig. 13.1 from *Current Biology*, vol. 17.

CHAPTER 14

Emerging frontiers of community ecology

Peter J. Morin

14.1 Introduction

Speculation about the future direction of any discipline is always problematic, frequently foolhardy, and typically embarrassing to the speculator in retrospect. Predictions about the future of community ecology are no exception. Nonetheless, the contributions in this volume point to a number of important and vital developments in community ecology that may presage important new directions of study. Obviously, these areas constitute only a small sample of the active research frontiers in community ecology. An overview of the chapters in this volume and a number of other recent publications suggests that the following themes may be particularly important in future research.

14.1.1 Spatial ecology

A spatial context determines the outcome of interactions within and among metacommunities, as repeatedly emphasized in Chapters 2, 5, 6, 7 and 9 in this volume. The growing awareness that communities are open systems, which have dynamics that can be strongly influenced by the movement of organisms among subsets of metacommunities, is increasingly clear (Leibold *et al.* 2004). Despite that recognition, theoretical studies far outnumber experimental studies of the consequences of such spatial dynamics. Clever tests of metacommunity theory with amenable experimental systems, such as those of Holyoak and Lawler (1996) and B. Kerr *et al.* (2006), will remain an important future topic of research. It is remarkable that we have so few empirical studies of dynamics in metacommunities, so many years after Huffaker (1958) first demonstrated their likely importance.

14.1.2 Complex dynamics

Even simple models can give rise to remarkably complex dynamics (May 1976). It seems even more likely that complex and often chaotic dynamics will be a feature of systems involving three or more species (de Roos *et al.* 2002; see Chapter 3). Those dynamics can define assembly rules or create alternate stable states for some systems (Scheffer *et al.* 1993; see Chapters 4 and 5).

14.1.3 Size-dependent interactions

Interactions whose strength and consequences depend on the relative sizes of interacting organisms figure prominently in the emerging mechanistic understanding of food web patterns (de Roos *et al.* 2002; Beckerman *et al.* 2006; Petchey *et al.* 2008; see Chapter 5). The many ecological correlates of organism size provide a possible entry into mechanistic ecological theory that links the physiology, energetics and behaviour of individual organisms to their role in communities (Brown *et al.* 2004).

14.1.4 Interactions between topology and dynamics

The many suggestions that aspects of food web topology (May 1973; Pimm and Lawton 1977; de Ruiter *et al.* 1995; Montoya *et al.* 2006) are constrained by dynamics continue to figure prominently in our ways of attempting to understand food web patterns. The various aspects of theory point to the ways that topological patterns can emerge from dynamics (see Chapter 1), and the equally important ways that topology can influence dynamics (see Chapters 1 and 3).

One of the continuing challenges in community ecology is to link the dynamics predicted by models of complex community features, such as food webs, with the dynamics of actual systems. For a number of pragmatic reasons, models are used to draw inferences about the dynamics of complex systems that differ in structure, often because the dynamics of real systems are difficult to observe. This means that the statics, or structure, of natural networks are used to constrain the form of models, which in turn are used to draw inferences about how the dynamics of natural systems might vary as the details of structure vary. To do this, it is usually assumed that natural systems exhibit stable dynamics, and then, given that constraint and others, it is possible to estimate other network properties, such as the strengths of interactions among species in the network (de Ruiter *et al.* 1995; Neutel *et al.* 2002, 2007; Moore *et al.* 2003; Beckerman *et al.* 2006; Petchey *et al.* 2008). In other cases, it is possible to estimate how deletions of species from networks will result in the extinction of other species that depend solely on the deleted species for energy or nutrients (Solé and Montoya 2001; Montoya *et al.* 2006).

The interesting feature of such studies is that in most cases the actual population dynamics of the species in question remain unknown. Partly, this is because dynamics are difficult to measure. Partly, it can also be difficult to selectively remove one species from a network and see how others respond. Nonetheless, it is surprising that we still know the detailed dynamics of only a very few food webs, and these food webs are often relatively simple in structure. This is a fundamental gap in our knowledge of complex ecological networks.

14.1.5 Evolutionary community dynamics

Evolution obviously has the potential to modify interspecific interactions, along the way influencing food web topology, and driving large-scale diversity patterns (Yoshida *et al.* 2003; see Chapters 11 and 12). Nonetheless, evolution remains poorly integrated into community ecology (see Chapter 11). The other important way that evolution may influence and perhaps supersede the role of interactions in determining large-scale patterns of community structure, such as regional diversity gradients, is by determining the diversity and composition of regional species pools from which communities are assembled (Ricklefs 2004, 2008).

It is a source of some discomfort to community ecologists that the causes of both local and global diversity gradients remain uncertain. It is particularly troubling that the most conspicuous large-scale diversity pattern, the latitudinal gradient in species diversity, has no generally accepted explanation, and is instead the subject of continuing debate. In general, it seems possible that both local and regional diversity patterns could be the result of (1) ongoing ecological interactions, including priority effects, (2) historical evolutionary processes, the consequences of speciation and adaptive radiation, and (3) purely neutral processes, including statistical sampling processes such as the mid-domain effect.

At local scales, it seems that a number of factors can conspire to produce a range of diversity patterns. At least four different productivity–diversity patterns have been observed in studies of different systems (Waide *et al.* 1999), including ones that are concave-up, concave-down, increasing, or essentially flat. One explanation proposed for the difference between concave-down and increasing patterns is the scale of study (Chase and Leibold 2002). Concave-down patterns appeared at a local scale (among nearby ponds drawing on the same potential species pool), while increasing patterns appeared at larger spatial scales where the species pool might be expected to increase with productivity. However, other processes can produce a range of productivity diversity patterns, even when drawing on exactly the same species pool. Fukami and Morin (2003) showed that priority effects related to the sequence of species assembly in communities arrayed along a productivity gradient could produce a variety of productivity–diversity patterns, including hump-shaped (concave-down), approximately linear increasing, and concave-up patterns. Because these patterns resulted over the same spatial scale for species drawn from the same species pool of bacterivorous microorganisms, it is clear that factors like history of community colonization can also influence the form of productivity–diversity patterns.

At much larger spatial scales, the causes of diversity patterns remain contentious. The well-known latitudinal gradient in species richness (Pianka 1988) has been attributed to a host of factors, none of which seems to be easily tested. Ricklefs (2004, 2008) has proposed that the causes of these large-scale patterns are unlikely to be revealed by small-scale studies that focus on local ongoing ecological interactions, and that the patterns instead represent the end result of long periods of evolution and diversification within biotas. If so, the most conspicuous community-level pattern in ecology will not be explained by the kinds of ongoing interactions that community ecologists usually dwell on. Other explanations invoke a purely statistical explanation for the apparent peak in species richness along geographical gradients. The mid-domain effect suggested by Colwell and his colleagues (e.g. Colwell and Hurtt 1994; Colwell and Lees 2000) provides a null model explanation for a humped diversity pattern along any geographic gradient, just as a consequence of the way that species ranges will overlap along any gradient. This idea, while elegant in its simplicity, also has its critics (J.T. Kerr *et al.* 2006; Storch *et al.* 2006).

Just as evolution may ultimately provide a viable explanation for large-scale diversity patterns, evolutionary processes have been invoked with increasing frequency to explain phenomena including the population dynamics of interacting species (Yoshida *et al.* 2003) and the structure of food webs (Drossel *et al.* 2001; Loeuille and Loreau 2005; see Chapter 12). Although the need to integrate ecological and evolutionary perspectives has long been recognized, in reality this integration remains tentative and incomplete. A couple of examples point to ways that an understanding of evolutionary processes can provide insights into ecological patterns.

Yoshida *et al.* (2003) used a simple laboratory chemostat system to study the dynamics of herbivorous rotifers feeding on the alga *Chlorella*. Both the rotifers and the algae reproduce clonally, but large differences in dynamics materialized depending on whether algal populations consisted of single or multiple clones (Fig. 14.1). A simple model of the dynamics also predicts that predator–prey oscillations will have a longer period when rotifers are feeding on multiple algal clones, if the clones differ in susceptibility to predation or nutritional value (Fig. 14.1). This study makes the point that the dynamics observed in communities can be modified by dynamic shifts in the genetic composition of prey populations–a feature seldom included in simple models of predator–prey dynamics. Similar results emerge from other chemostat studies of interactions between the bacterium *Escherichia coli* and various types of bacteriophage that act as predators. Although bacteriophages initially have a large effect on bacterial abundances, these effects rapidly diminish as mutant genotypes arise in the bacterial population that are resistant to attack by bacteriophage (Chao *et al.* 1977).

Other possible roles for evolution in creating community patterns are suggested by recent models of evolving predators and prey in food webs. These models (described further in Chapter 12) make few initial assumptions about food web structure, and begin with a single model species that is allowed to evolve in size over time. Organisms that are similar in size are assumed to compete for resources, while those that diverge sufficiently in size come to interact as predators and prey. Depending on the intensity of competition among species and the range of prey size that can be consumed (analogous to niche width), systems evolve to have many of the main attributes of real food webs. Whether food webs display these properties because they are a consequence of evolution within the context of the web, or because webs simply assemble from species that have evolved in various food web contexts to have certain sets of traits, remains uncertain.

14.1.6 Applied community ecology

Several contributions to this volume make it abundantly clear that any boundaries between basic and applied community ecology are artificial and not particularly helpful. Community ecology has much to offer to society, through an enhanced understanding of the mechanisms and consequences of exotic species invasions, sustainable restoration, resource management in multispecies systems and

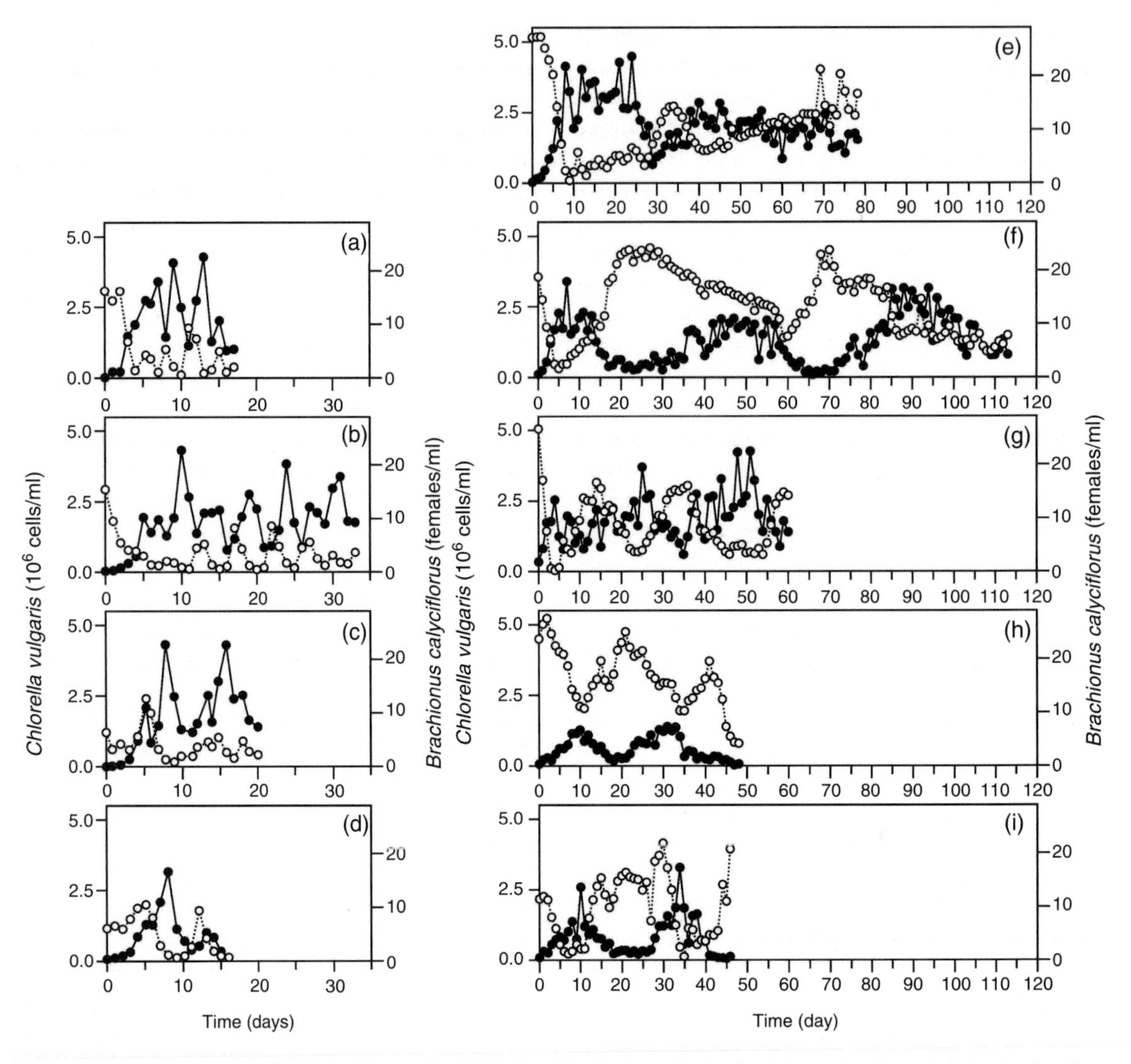

Figure 14.1 Predator–prey dynamics for rotifers (solid dots) feeding on single algal clones (a–d); and dynamics for the same rotifer species feeding on multiple clones of the same algal species (e–i). Note the shift from short period cycles to ones of longer period with an increase in clonal diversity. Used with permission from Yoshida *et al.* (2003).

the reliable maintenance of critical ecosystem services (see Chapters 7, 9 and 10).

14.2 Future directions

Of the various topics emerging from or suggested by the content of the previous chapters in this volume, I want to focus on the need for much additional work in two areas to consolidate what we think we know, or to resolve ongoing controversies. These future directions include the problem of invasions by exotic species as a special case of community assembly processes, where species traits are likely to predict community-level effects, and the need to integrate the full range of ecological interactions into ecological networks–complexity beyond food webs.

14.2.1 Biotic invasions

Ecologists recognize that biological invasions and the biotic homogenization that may consequently result constitute an enormous problem, comparable to concerns over the loss of species to human-caused extinctions (Mack *et al.* 2000; Lockwood *et al.* 2007; van der Weijden *et al.* 2007). The problem is that most of what we know about invasions comes either from observational studies that document the invasions that have occurred, which gives little insight into possible mechanisms promoting invasions, or from small-scale experimental studies that provide insights into mechanisms, but that also suggest patterns that are at variance with large-scale patterns. One example of this is the difference between patterns of invasibility and diversity observed at small and large spatial scales.

Experimental studies conducted at small spatial scales, of the order of a few square metres or less, tend to show that the ability of species to invade established communities declines with the diversity of species already present in those systems (McGrady-Steed *et al.* 1997; Knops *et al.* 1999; Stachowicz *et al.* 1999; Levine 2000; Naeem *et al.* 2000; Symstad 2000; Kennedy *et al.* 2002; Fargione *et al.* 2003). The pattern occurs in terrestrial and aquatic environments, and seems to be quite general. It is also consistent with notions of biotic resistance first articulated by Elton (1958).

Observational studies that consider patterns of invasions at much larger spatial scales tend to show a rather different pattern, although the measure of invasibility used is fundamentally different. The measure of invasibility used is the total number of invasive species that have become established, rather than the performance of any single invading species (such as ability to increase when rare, survival of invading propagules, or some other fitness component of the invaders). These studies rather consistently show that the most diverse communities also house the greatest number of successful invading species (Robinson *et al.* 1995; Planty-Tabacchi *et al.* 1996; Wiser *et al.* 1998; Lonsdale 1999; Stohlgren *et al.* 1999; Levine 2000; Brown and Peet 2003; Stohlgren *et al.* 2003). Obviously, this pattern seems to be inconsistent with the mechanisms of biotic resistance invoked to explain reduced invasion success at small spatial scales. This disparity has resulted in a rather heated discussion about the merits and proper interpretations of studies conducted on different scales (Fridley *et al.* 2007). Obviously, the studies differ in more than just the scale considered. There is the obvious difference between experimental and observational approaches used at these different scales, and the potential to infer incorrect causal pathways in observational studies. There is also a fundamental difference in the operational definitions used to measure invasibility. In one case (small scale), invasibility is measured by the actual performance of experimentally introduced invaders. At large scales, we know only how many invasive species have become established, and we typically do not know how many have failed to become established.

To this end, it would be useful to use model systems to attempt to resolve these differences. Indeed, Mack *et al.* (2000) have written, 'We need to develop innocuous experimental releases of organisms that can be manipulated to explore the enormous range of chance events to which all immigrant populations may be subjected . . . '. We have used such an approach to explore one of the possible factors that could contribute to the apparent positive relation between diversity and invasibility seen at large spatial scales (Jiang and Morin 2004). This approach builds on the suggestion that successful invasion depends on productivity and the availability of resources for invaders (Shea and Chesson 2002).

Jiang and Morin (2004) suggested that variation in productivity among communities could swamp out within community effects to create an apparent, but causally spurious, correlation between diversity and the success of invaders over an array of different experimental communities. The approach first documented that increased productivity generated increased diversity of pre-invasion communities. The success of model invaders added to these established communities also increased with productivity. These two patterns resulted in a positive correlation between 'native' diversity and invasibility (Fig. 14.2). However, when the effects of productivity on 'native' diversity were statistically controlled by partial correlation analysis, the correlation between diversity and invasion success became statistically non-significant. The inference is that productivity, and not 'native' diversity, is the cause of enhanced invisibility, and that more diverse communities are more subject to invasion, not because they are more diverse but because

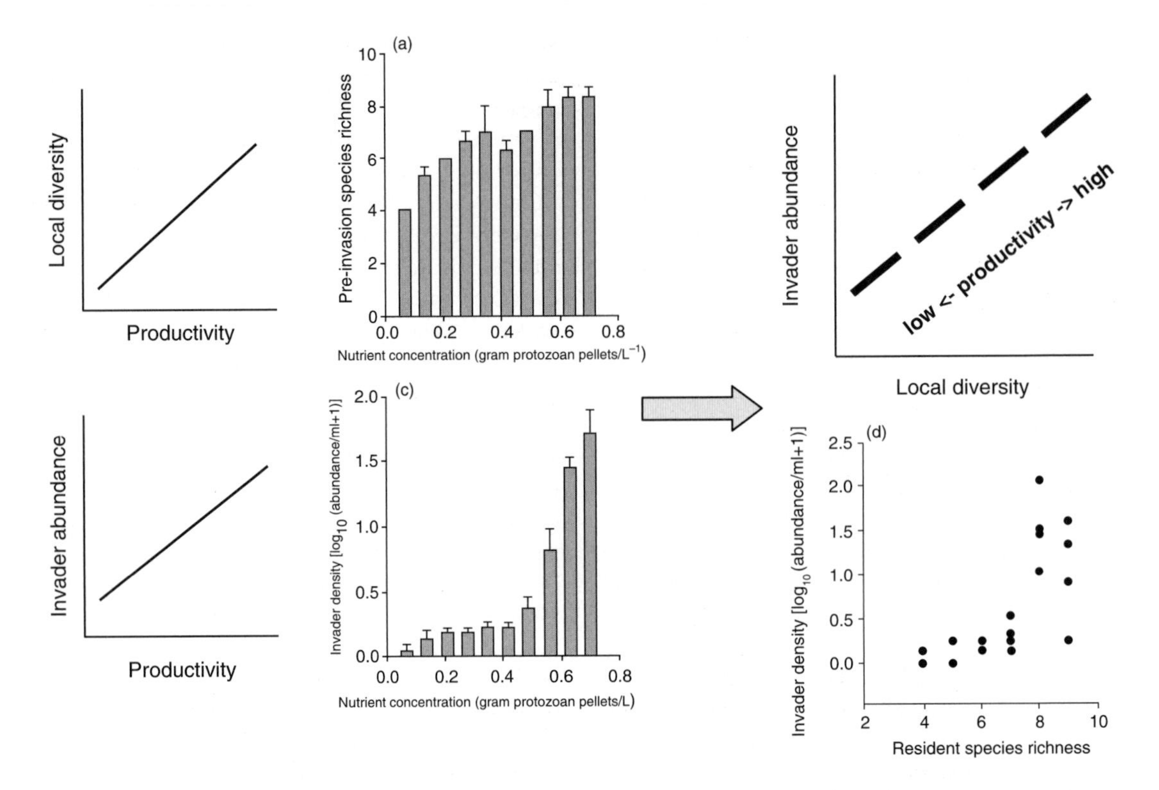

Figure 14.2 Positive relations between productivity and diversity and productivity and invader success in experimental microbial communities (modified from Jiang and Morin 2004). These separate patterns combine to produce an apparent positive correlation between diversity and invader success, but the correlation is spurious and not indicative of a cause–effect relationship.

they are more productive. It would be very interesting to see whether similar patterns would hold in the data used to infer large-scale diversity–invasibility patterns.

Experimental approaches can also be very useful in addressing some of the many different factors that can potentially influence the success of invading species. For example, Lockwood *et al.* (2005) have emphasized the underappreciated role of propagule pressure, the number of invaders arriving at a community per unit time, in affecting invasion success. Using approaches similar to those outlined above, it should be relatively easy to vary either propagule size – the number of invaders arriving per unit time and area – during any invasion event, or the frequency of invasion events to see how these factors influence potential establishment of invaders in communities with different initial properties. Those properties could include diversity, productivity or any of a number of features of interest.

Another potentially interesting approach would be to explore how the trophic position of an invader interacts with diversity at multiple trophic levels in a community. Most inferred effects of local diversity on invasion success invoke some sort of biotic resistance operating within the trophic level that the invader is attempting to enter. The arguments essentially rely on increasing competition within a diverse trophic level, guild or functional group to reduce invasion success. However, differences in diversity on other trophic levels could conceivably have either positive or negative effects. For example, consider the situation depicted in Fig. 14.3, in which an invader on the second trophic level, a herbivore, potentially encounters differences in species diversity of its prey (trophic level 1) or its predators (trophic level 3). Greater prey diversity

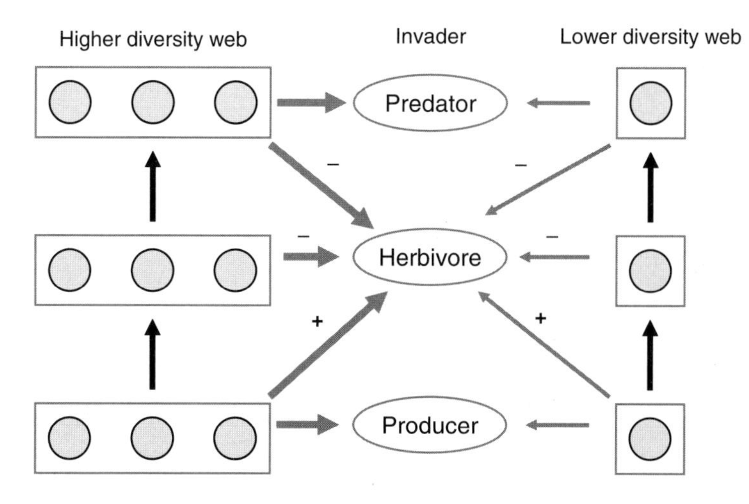

Figure 14.3 Possible differences in the effects of invader trophic level and diversity within different invaded trophic levels on invader success. For simplicity, we focus on effects of diversity in different trophic levels on the success of an invader on the intermediate herbivore level, but similar within and between trophic level effects could influence invaders on other trophic levels. Line thickness corresponds to the strength of positive and negative effects on invader success, shown in grey. Trophic interactions are shown in black.

could enhance invasion by increasing the likelihood that a suitable prey species would support the growth of the invader. Conversely, greater predator diversity could provide another avenue of biotic resistance by increasing the likelihood that a native predator would be able to prevent the establishment of the invader. These possible positive and negative effects of diversity on different trophic levels are just one of the many ways that diversity can potentially influence community structure and functioning. Chapter 8 provides many more examples of diversity-functioning relationships in complex communities.

14.2.2 Interaction networks beyond food webs

The first problem faced in representing the full range of pairwise ecological interactions in community networks is to devise a convention for representing links that correspond with different kinds of interactions: predator–prey, competitive and mutualistic ones. One way to do this is to use different line styles, or colours, to represent different kinds of interactions. Obviously, the other way to do this in a non-graphic fashion is simply to have a matrix of interactions keyed by the signs of the effect of individuals of each species on the other. While easy to do, visual inspection of such tables for community patterns becomes problematic as the tables extend to more than a small number of species. Figure 14.4 shows in a very general form some of the complexities of community interactions that would have to be integrated into anything approaching a complete depiction of an ecological network or a moderately complex community module. Keep in mind that this figure simply represents the kinds of interactions that need to be included, but it in no way completely represents the aspects of complexity shown by the full species richness and patterns of interactions (connectance) in natural systems. For practical computational reasons, such interactions might be best represented as a matrix or table showing the signs and in some cases mechanisms of interactions, rather than as a graph.

There are two obvious additions to the basic interactions depicted in food webs that need to be incorporated in any comprehensive ecological network. As suggested in Chapter 1, networks of competitive interactions should be included, since such interactions can potentially influence the coexistence of species. Describing the network of interactions (who competes with whom) is relatively straightforward, but encoding information about

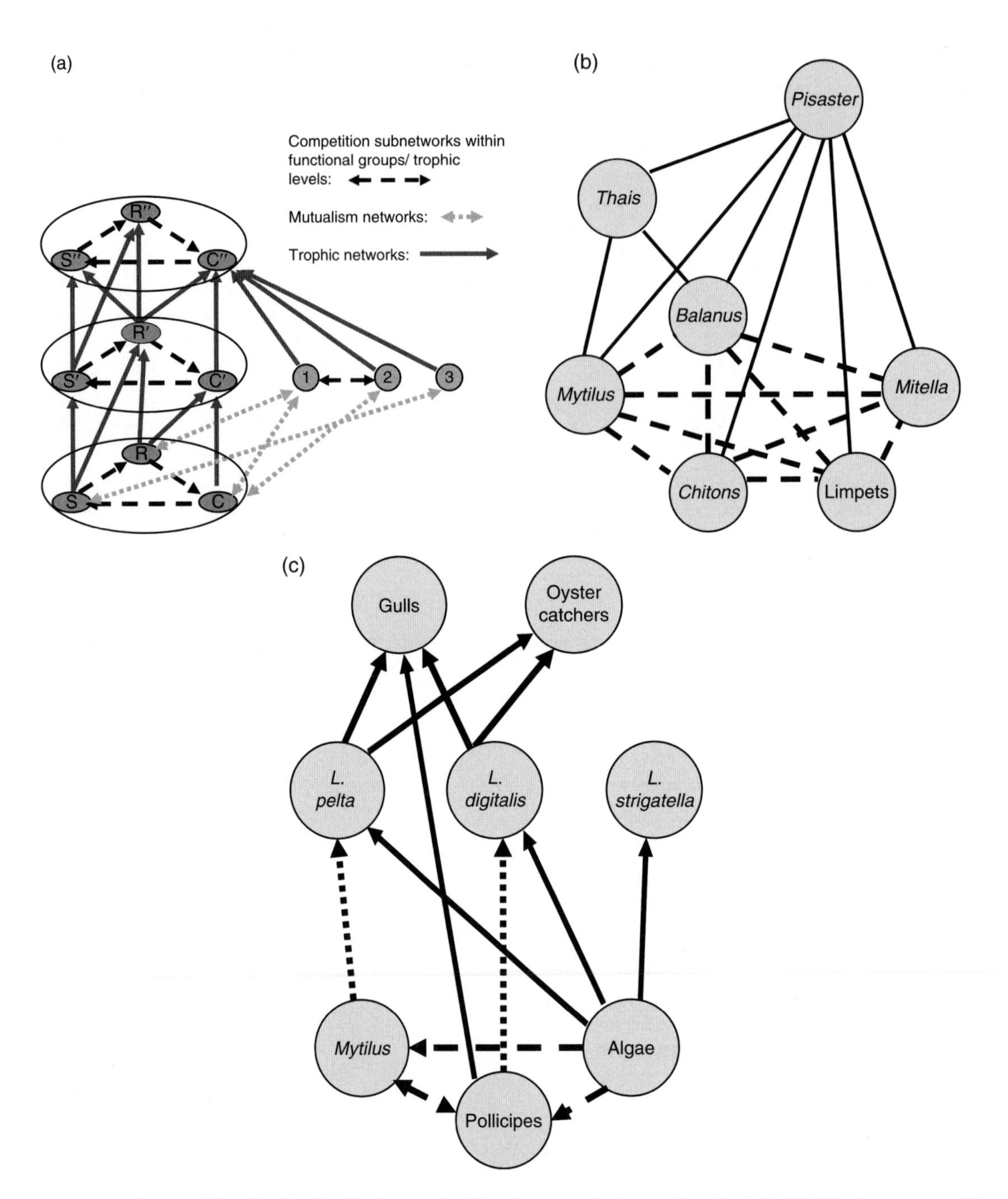

Figure 14.4 What would an idealized unified network of local community interactions look like? (a) Ovals correspond to idealized trophic levels, and species within trophic levels interact via intransitive competitive interactions, indicated by dashed lines. Predator—prey links between trophic levels are indicated by solid lines. Mutualistic interactions are indicated by dotted lines, and mutualists may compete and interact with other species as prey or predators. Examples for small subwebs from the literature are shown in (b) for data from Paine (1966) and in (c) for data from Wootton (1992).

the mechanism of competition, or the competitive hierarchy among these species, is not so easy. The other important feature that is missing from most food webs is the network of mutualistic interactions, such as those described in Chapters 2 and 13. Here, the inclusion of mutualistically linked species is relatively straightforward, as is the possibility that mutualists may compete with one another, or may consume or be consumed by other species. The main limitation is that for most networks we have relatively little information about the structure of competitive or mutualistic interactions.

Once these more interactively complete ecological networks have been realized, we can begin to ask fundamental questions about their topology and functional attributes that we currently cannot do with the fragmented information that we have about networks of predators and prey, competitors and mutualists. As a first pass, it would be interesting to evaluate the relative frequency of consumer, competition and mutualistic links in webs, to assess the potential relative importance of these different interactions in networks of different provenance. Similarly, the relative contribution of links/interactions of different sorts to overall community stability could be inferred by modelling their respective impacts on inferred network stability. There may also be interesting differences in those portions of total interactive connectance associated with three fundamental kinds of interspecific interactions. These are ideas that we can only begin to explore as we obtain more complete information about the range of interactions that occur across a broad range of communities. It also behooves ecologists not to focus exclusively on the topology of these systems, as the real goal here is to explore the extent to which interactions confer structure on the many kinds of communities that fascinate us (Ings *et al.* 2009).

References

Introduction

Begon, M., Harper, J.L. and Townsend, C.R. (1996) *Ecology, Individuals, Populations and Communities*. Blackwell Science, London.

Clements, F.E. (1916) *Plant Succession: Analysis of the Development of Vegetation*. Publication no. 242. Carnegie Institute of Washington, Washington, DC.

Gleason, H.A. (1926) The individualistic concept of the plant association. *Bulletin Torrey Botanical Club*, **53**, 7–26.

Krebs, C.J. (1972) *Ecology, the Experimental Analysis of Distribution and Abundance*. Harper & Row, New York, NY.

Morin, P.J. (1999) *Community Ecology*. Blackwell Science, Oxford.

Ricklefs, R.E. (2008) Disintegration of the ecological community. *The American Naturalist*, **172**, 741–50.

Chapter 1

Abrams, P.A. and Ginzburg, L.R. (2000) The nature of predation: prey dependent, ratio dependent or neither? *Trends in Ecology & Evolution*, **15**, 337–41.

Arsenault, R. and Owen-Smith, N. (2002) Facilitation versus competition in grazing herbivore assemblages. *Oikos*, **97**, 313–18.

Barbarasi, A. (2002) *Linked. The New Science of Networks*. Perseus Publishing, Cambridge, MA.

Bascompte, J., Jordano, P., Melian, C.J. and Olesen, J.M. (2003) The nested assembly of plant-animal mutualistic networks. *Proceedings of the National Academy of Sciences of the United States of America*, **100**, 9383–7.

Bascompte, J., Jordano, P. and Olesen, J.M. (2006) Asymmetric coevolutionary networks facilitate biodiversity maintenance. *Science*, **312**, 431–3.

Beckerman, A., Petchey, O.L. and Warren, P.H. (2006) Foraging biology predicts food web complexity. *Proceedings of the National Academy of Sciences of the United States of America*, **103**, 13745–9.

Bengtsson, J. (1994) Confounding variables and independent observations in comparative analyses of food webs. *Ecology*, **75**, 1282–8.

Beninca, E., Huisman, J., Heerkloss, R., *et al.* (2008) Chaos in a long-term experiment with a plankton community. *Nature*, **451**, 822–5.

Bersier, L.F., Banasek-Richter, C. and Cattin, M.F. (2002) Quantitative descriptors of food-web matrices. *Ecology*, **83**, 2394–407.

Boerlijst, M.C. and Hogeweg, P. (1991) Spiral wave structure in pre-biotic evolution: hypercycles stable against parasites. *Physica D*, **48**, 17–28.

Boucher, D.H. (1985) The idea of mutualism, past and future. In *The Biology of Mutualism* (ed. D.H. Boucher). Oxford University Press, Oxford.

Brooker, R.W., Maestre, F.T., Callaway, R.M., *et al.* (2008) Facilitation in plant communities: the past, the present, and the future. *Journal of Ecology*, **96**, 18–34.

Brose, U., Williams, R.J. and Martinez, N.D. (2006) Allometric scaling enhances stability in complex food webs. *Ecology Letters*, **9**, 1228–36.

Brown, J.H. (1981) Two decades of homage to Santa Rosalia: toward a general theory of diversity. *American Zoologist*, **21**, 877–88.

Cohen, J.E. (1978) *Food Webs and Niche Space*. Princeton University Press, Princeton, NJ.

Cohen, J.E., Jonsson, T. and Carpenter, S.R. (2003) Ecological community description using the food web, species abundance, and body size. *Proceedings of the National Academy of Sciences of the United States of America*, **100**, 1781–6.

Connell, J.H. (1978) Diversity in tropical rain forests and coral reefs. *Science*, **199**, 1302–10.

Czárán, T.L., Hoekstra, R.F. and Pagie, L. (2002) Chemical warfare between microbes promotes biodiversity. *Proceedings of the National Academy of Sciences of the United States of America*, **99**, 786–90.

Dawah, H.A., Hawkins, B.A. and Claridge, M.F. (1995) Structure of the parasitoid communities of grass-feeding chalcid wasps. *Journal of Animal Ecology*, **64**, 708–20.

Dean, A.M. (1983) A simple model of mutualism. *The American Naturalist*, **121**, 409–17.

DeAngelis, D.L. (1975) Stability and connectance in food web models. *Ecology*, **56**, 238–43.

de Ruiter, P.C., Neutel, A.-M. and Moore, J.C. (1998) Biodiversity in soil ecosystems: the role of energy flow and community stability. *Applied Soil Ecology*, **10**, 217–28.

de Ruiter, P.C., Wolters, V. and Moore, J.C. (2005) *Dynamic Food Webs: Multispecies Assemblages, Ecosystem Development, and Environmental Change*. Elsevier-Academic Press, Burlington, VT.

Dunne, J.A., Williams, R.J. and Martinez, N.D. (2002) Food-web structure and network theory: the role of connectance and size. *Proceedings of the National Academy of Sciences of the United States of America*, **99**, 12917–22.

Dykhuizen, D.E. (1998) Santa Rosalia revisited: why are there so many species of bacteria? *Antonie Van Leeuwenhoek International Journal of General and Molecular Microbiology*, **73**, 25–33.

Elton, C. (1927) *Animal Ecology*. Sidgwick & Jackson, London.

Feinsinger, P. (1976) Organization of a tropical guild of nectarivorous birds. *Ecological Monographs*, **46**, 257–91.

Fonseca, C.R. and John, J.L. (1996) Connectance: a role for community allometry. *Oikos*, **77**, 353–8.

Fretwell, S.D. (1977) Regulation of plant communities by food-chains exploiting them. *Perspectives in Biology and Medicine*, **20**, 169–85.

Gardner, M.R. and Ashby, W.R. (1970) Connectance of large dynamic (cybernetic) systems: critical values for stability. *Nature*, **228**, 784.

Hairston, N.G. (1981) An experimental test of a guild: salamander competition. *Ecology*, **62**, 65–72.

Haskell, J.P., Ritchie, M.E. and Olff, H. (2002) Fractal geometry predicts varying body size scaling relationships for mammal and bird home ranges. *Nature*, **418**, 527–30.

Havens, K.E. (1992) Scale and structure in natural food webs. *Science*, **257**, 1107–9.

Havens, K.E. (1993) Effects of scale on food web structure. *Science*, **260**, 243.

Holt, R.D. (1977) Predation, apparent competition, and the structure of prey communities. *Theoretical Population Biology*, **12**, 197–229.

Hubbell, S.P. (2001) *The Unified Neutral Theory of Biodiversity and Biogeography*. Princeton University Press, Princeton, NJ.

Huisman, J. and Olff, H. (1998) Competition and facilitation in multispecies plant-herbivore systems of productive environments. *Ecology Letters*, **1**, 25–9.

Huisman, J. and Weissing, F.J. (1999) Biodiversity of plankton by species oscillations and chaos. *Nature*, **402**, 407–10.

Hutchinson, G.E. (1961) The paradox of the plankton. *The American Naturalist*, **95**, 137–45.

Janzen, D.H. (1966) Coevolution between ants and acacias in Central America. *Evolution*, **20**, 249–75.

Jenkins, B., Kitching, R.L. and Pimm, S.L. (1992) Productivity, disturbance and food web structure at a local spatial scale in experimental container habitats. *Oikos*, **65**, 249–55.

Jordano, P., Bascompte, J. and Olesen, J.M. (2003) Invariant properties in coevolutionary networks of plant-animal interactions. *Ecology Letters*, **6**, 69–81.

Kaunzinger, C.M.K. and Morin, P.J. (1998) Productivity controls food-chain properties in microbial communities. *Nature*, **395**, 495–7.

Kerr, B., Riley, M.A., Feldman, M.W. and Bohannan, B.J.M. (2002) Local dispersal promotes biodiversity in a real-life game of rock-paper-scissors. *Nature*, **418**, 171–4.

Kirkup, B.C. and Riley, M.A. (2004) Antibiotic-mediated antagonism leads to a bacterial game of rock-paper-scissors in vivo. *Nature*, **428**, 412–14.

Lawler, S.P. and Morin, P.J. (1993) Food web architecture and population dynamics in laboratory microcosms of protists. *The American Naturalist*, **141**, 675–86.

Leaper, R. and Huxham, M. (2002) Size constraints in a real food web: predator, parasite and prey body-size relationships. *Oikos*, **99**, 443–56.

Leibold, M.A. (1996) A graphical model of keystone predators in food webs: trophic regulation of abundance, incidence, and diversity patterns in communities. *The American Naturalist*, **147**, 784–812.

Leibold, M.A., Holyoak, M., Mouquet, N., *et al.* (2004) The metacommunity concept: a framework for multi-scale community ecology. *Ecology Letters*, **7**, 601–13.

Lewinsohn, T.M., Prado, P.I., Jordano, P., Bascompte, J. and Olesen, J.M. (2006) Structure in plant-animal interaction assemblages. *Oikos*, **113**, 174–84.

Lindeman, R.L. (1942) The trophic-dynamic aspect of ecology. *Ecology*, **23**, 399–418.

MacArthur, R.H. and Pianka, E.R. (1966) On optimal use of a patchy environment. *The American Naturalist*, **100**, 603–9.

Martinez, N.D. (1991) Artifacts or attributes? Effects of resolution on the Little Rock lake food web. *Ecological Monographs*, **61**, 367–92.

Martinez, N.D. (1992) Constant connectance in community food webs. *The American Naturalist*, **139**, 1208–18.

Martinez, N.D. (1993) Effect of scale on food web structure. *Science*, **260**, 242–3.

Martinez, N.D. (1994) Scale-dependent constraints on food-web structure. *The American Naturalist*, **144**, 935–53.

May, R.M. (1972) Will a large complex system be stable? *Nature*, **238**, 413–14.

May, R.M. (1973) *Stability and Complexity in Model Ecosystems*. Princeton University Press, Princeton, NJ.

McCann, K., Hastings, A. and Huxel, G.R. (1998) Weak trophic interactions and the balance of nature. *Nature*, **395**, 794–8.

Montoya, J.M. and Sole, R.V. (2003) Topological properties of food webs: from real data to community assembly models. *Oikos*, **102**, 614–22.

Murtaugh, P.A. and Kollath, J.P. (1997) Variation of trophic fractions and connectance in food webs. *Ecology*, **78**, 1382–7.

Neutel, A.-M., Heesterbeek, J.A.P. and de Ruiter, P.C. (2002) Stability in real food webs: weak links in long loops. *Science*, **296**, 1120–3.

Nishikawa, K.C. (1985) Competition and the evolution of aggressive-behavior in 2 species of terrestrial salamanders. *Evolution*, **39**, 1282–94.

Ohgushi, T. (2005) Indirect interaction webs: herbivore-induced effects through trait change in plants. *Annual Review of Ecology, Evolution and Systematics*, **36**, 81–105.

Oksanen, L., Fretwell, S.D., Arruda, J. and Niemelä, P. (1981) Exploitation ecosystems in gradients of primary productivity. *The American Naturalist*, **118**, 240–61.

Olesen, J.M., Bascompte, J., Dupont, Y.L. and Jordano, P. (2006) The smallest of all worlds: pollination networks. *Journal of Theoretical Biology*, **240**, 270–6.

Olff, H., Vera, F.W.M., Bokdam, J., *et al.* (1999) Shifting mosaics in grazed woodlands driven by the alternation of plant facilitation and competition. *Plant Biology*, **1**, 127–37.

Otto, S.B., Rall, B.C. and Brose, U. (2007) Allometric degree distributions facilitate food-web stability. *Nature*, **450**, 1226–30.

Pace, M.L., Cole, J.J., Carpenter, S.R. and Kitchell, J.F. (1999) Trophic cascades revealed in diverse ecosystems. *Trends in Ecology & Evolution*, **14**, 483–8.

Paine, R.T. (1966) Food web complexity and species diversity. *The American Naturalist*, **100**, 65–75.

Paine, R.T. (1988) Food webs: road maps of interactions or grist for theoretical development? *Ecology*, **69**, 1648–54.

Petandiou, T. and Ellis, W.N. (1993) Pollinating fauna of a phryganic ecosystem: composition and diversity. *Biodiversity Letters*, **1**, 9–22.

Petchey, O.L., Beckerman, A.P., Riede, J.O. and Warren, P.H. (2008) Size, foraging, and food web structure. *Proceedings of the National Academy of Sciences of the United States of America*, **105**, 4191–6.

Pierce, G.J. and Ollason, J.G. (1987) 8 reasons why optimal foraging theory is a complete waste of time. *Oikos*, **49**, 111–18.

Pimm, S.L. (1980) Properties of food webs. *Ecology*, **61**, 219–25.

Pimm, S.L. (1982) *Food Webs*. Chapman and Hall, London.

Pimm, S.L. (1984) The complexity and stability of ecosystems. *Nature*, **307**, 321–6.

Pimm, S.L. (1991) *The Balance of Nature?* The University of Chicago Press, Chicago, IL.

Pimm, S.L. and Lawton, J.H. (1977) Number of trophic levels in ecological communities. *Nature*, **268**, 329–31.

Pimm, S.L. and Lawton, J.H. (1978) On feeding on more than one trophic level. *Nature*, **275**, 542–4.

Polis, G.A. (1991) Complex trophic interactions in deserts: an empirical critique of food web ecology. *The American Naturalist*, **138**, 123–55.

Polis, G.A. and Winemiller, K., eds. (1996) *Food Webs: Integration of Patterns and Dynamics*. Chapman and Hall, London.

Post, D.M. (2002) The long and short of food-chain length. *Trends in Ecology & Evolution*, **17**, 269–77.

Post, D.M., Pace, M.L. and Hairston Jr, N.G. (2000) Ecosystem size determines food-chain length in lakes. *Nature*, **405**, 1047–9.

Prins, H.H.T. and Olff, H. (1997) Species richness of African grazer assemblages: towards a functional explanation. In *Dynamics of Tropical Communities* (eds D. Newbery, H.H.T. Prins and N.D. Brown), pp. 448–90. Blackwell, Oxford.

Proulx, S.R., Promislow, D.E.L. and Phillips, P.C. (2005) Network thinking in ecology and evolution. *Trends in Ecology & Evolution*, **20**, 345–53.

Reichenbach, T., Mobilia, M. and Frey, E. (2007) Mobility promotes and jeopardizes biodiversity in rock-paper-scissors games. *Nature*, **448**, 1046–9.

Rezende, E.L., Lavabre, J.E., Guimaraes, P.R., *et al.* (2007) Non-random coextinctions in phylogenetically structured mutualistic networks. *Nature*, **448**, 925-U6.

Ringel, M.S., Hu, H.H., Anderson, G. and Ringel, M.S. (1996) The stability and persistence of mutualisms embedded in community interactions. *Theoretical Population Biology*, **50**, 281–97.

Rooney, N., McCann, K., Gellner, G. and Moore, J.C. (2006) Structural asymmetry and the stability of diverse food webs. *Nature*, **442**, 265–9.

Root, R.B. (1967) The niche exploitation pattern of the blue-gray gnatcatcher. *Ecological Monographs*, **37**, 317–50.

Rosenzweig, M.L. (1971) Paradox of enrichment: destabilization of exploitation ecosystems in ecological time. *Science*, **171**, 385–7.

Schmid-Araya, J.M., Schmid, P.E., Robertson, A., *et al.* (2002) Connectance in stream food webs. *Journal of Animal Ecology*, **71**, 1056–62.

Schoener, T.W. (1971) A theory of feeding strategies. *Annual Review of Ecology and Systematics*, **2**, 369–404.

Schoener, T.W. (1983) Field experiments on interspecific competition. *The American Naturalist*, **122**, 240–85.

Schwartz, M.W. and Hoeksema, J.D. (1998) Specialization and resource trade: biological markets as a model of mutualisms. *Ecology*, **79**, 1029–38.

Sinclair, A.R.E. (1979) Dynamics of the Serengeti ecosystem: process and pattern. In *Serengeti: Dynamics of an Ecosystem* (eds A.R.E. Sinclair and S.M. Norton-Griffiths), pp. 1–30. University of Chicago Press, Chicago, IL.

Slobodkin, L.B. (1960) Ecological energy relationships at the population level. *The American Naturalist*, **94**, 213–36.

Sterner, R.W., Bajpai, A. and Adams, T. (1997) The enigma of food chain length: absence of theoretical evidence for dynamic constraints. *Ecology*, **78**, 2258–62.

Stomp, M., Huisman, J., Stal, L.J. and Matthijs, H.C.P. (2007a) Colourful niches of phototrophic microorganisms shaped by vibrations of the water molecule. *ISME Journal*, **1**, 271–82.

Stomp, M., Huisman, J., Voros, L., *et al.* (2007b) Colourful coexistence of red and green picocyanobacteria in lakes and seas. *Ecology Letters*, **10**, 290–8.

Stouffer, D.B., Camacho, J., Guimera, R., *et al.* (2005) Quantitative patterns in the structure of model and empirical food webs. *Ecology*, **86**, 1301–11.

Sugihara, G., Schoenly, K. and Trombla, A. (1989) Scale invariance in food web properties. *Science*, **245**, 48–52.

Teal, J.M. (1962) Energy-flow in salt-marsh ecosystem of Georgia. *Ecology*, **43**, 614–24.

Thompson, J.N., Reichman, O.J., Morin, P.J., *et al.* (2001) Frontiers of ecology. *Bioscience*, **51**, 15–24.

Tilman, D. (1990) Constraints and tradeoffs: toward a predictive theory of competition and succession. *Oikos*, **58**, 3–15.

Torsvik, V., Goksøyr, J. and Daae, F.L. (1990) High diversity in DNA of soil bacteria. *Applied and Environmental Microbiology*, **56**, 782–7.

Ulanowicz, R.E. (1995) *Utricularia*'s secret: the advantage of positive feedback in oligotrophic environments. *Ecological Modelling*, **79**, 49–57.

Ulanowicz, R.E. (1997) *Ecology, the Ascendent Perspective*. Columbia University Press, New York, NY.

van Veen, F.J.K., Müller, C.B., Pell, J.K. and Godfray, H.C.J. (2008) Food web structure of three guilds of natural enemies: predators, parasitoids and pathogens of aphids. *Journal of Animal Ecology*, **77**, 191–200.

Vazquez, D.P., Melian, C.J., Williams, N.M., *et al.* (2007) Species abundance and asymmetric interaction strength in ecological networks. *Oikos*, **116**, 1120–7.

Warren, P.H. (1990) Variation in food web structure: the determinants of connectance. *The American Naturalist*, **136**, 689–700.

Warren, P.H. (1994) Making connections in food webs. *Trends in Ecology & Evolution*, **9**, 136–41.

Waser, N.M., Chittka, L., Price, M.V., *et al.* (1996) Generalization in pollination systems, and why it matters. *Ecology*, **77**, 1043–60.

Winemiller, K.O. (1989) Must connectance decrease with species richness. *The American Naturalist*, **134**, 960–8.

Winemiller, K.O. and Pianka, E.R. (1990) Organization in natural assemblages of desert lizards and tropical fishes. *Ecological Monographs*, **60**, 27–55.

Wolin, C.L. (1985) The population dynamics of mutualistic systems. In *The Biology of Mutualisms* (ed. D.H. Boucher), pp. 248–269. Oxford University Press, Oxford.

Woodward, G., Ebenman, B., Emmerson, M.C., *et al.* (2005) Body size in ecological networks. *Trends in Ecology & Evolution*, **20**, 402–9.

Wootton, J.T. (1994) Predicting direct and indirect effects: an integrated approach using experiments and path analysis. *Ecology*, **75**, 151–65.

Yodzis, P. (1984) How rare is omnivory? *Ecology*, **65**, 321–3.

Chapter 2

Bagdassarian, C.K., Dunham, A.E., Brown, C.G. and Rauscher, D. (2007) Biodiversity maintenance in food webs with regulatory environmental feedbacks. *Journal of Theoretical Biology*, **245**, 705–14.

Baird, D., Asmus, H. and Asmus, R. (2007) Trophic dynamics of eight intertidal communities of the Sylt-Romo Bight ecosystem, northern Wadden Sea. *Marine Ecology-Progress Series*, **351**, 25–41.

Bakker, J.P., Olff, H., Willems, J.H. and Zobel, M. (1996) Why do we need permanent plots in the study of long-term vegetation dynamics? *Journal of Vegetation Science*, **7**, 147–55.

Bascompte, J. and Melian, C.J. (2005) Simple trophic modules for complex food webs. *Ecology*, **86**, 2868–73.

Becks, L., Hilker, F.M., Malchow, H., *et al.* (2005) Experimental demonstration of chaos in a microbial food web. *Nature*, **435**, 1226–9.

Beninca, E., Huisman, J., Heerkloss, R., *et al.* (2008) Chaos in a long-term experiment with a plankton community. *Nature*, **451**, 822–5.

Beukema, J.J., Dekker, R., Essink, K. and Michaelis, H. (2001) Synchronized reproductive success of the main bivalve species in the Wadden Sea: causes and consequences. *Marine Ecology-Progress Series*, **211**, 143–55.

Callaway, R.M. (2007) *Positive Interactions and Interdependence in Plant Communities*. Springer, Dordrecht.

Carpenter, S.R. and Kitchell, J.F., eds. (1993) *The Trophic Cascade in Lakes*. Cambridge University Press, Cambridge.

Carpenter, S.R., Walker, B., Anderies, J.M. and Abel, N. (2001) From metaphor to measurement: resilience of what to what? *Ecosystems*, **4**, 765–81.

Carpenter, S.R., Brock, W.A., Cole, J.J., *et al.* (2008) Leading indicators of trophic cascades. *Ecology Letters*, **11**, 128–38.

Connell, J.H. and Sousa, W.P. (1983) On the evidence needed to judge ecological stability or persistence. *The American Naturalist*, **121**, 789–824.

Costantino, R.F., Desharnais, R.A., Cushing, J.M. and Dennis, B. (1997) Chaotic dynamics in an insect population. *Science*, **275**, 389–91.

DeAngelis, D.L. (1992) *Dynamics of nutrient cycling and food webs*. Chapman and Hall, New York, NY.

DeAngelis, D.L., Post, W.M. and Travis, C.C. (1986) *Positive Feedback in Natural Systems*. Springer-Verlag, Berlin.

de Ruiter, P.C., Neutel, A.M. and Moore, J.C. (1995) Energetics, patterns of interactions strengths, and stability in real ecosystems. *Science*, **269**, 1257–60.

Ellner, S.P. and Turchin, P. (2005) When can noise induce chaos and why does it matter: a critique. *Oikos*, **111**, 620–31.

Elton, C.S. (1927) *Animal Ecology*. Sidgwick and Jackson, London.

Emmons, L.H. (1987) Comparative feeding ecology of felids in a neotropical rainforest. *Behavioural Ecology and Sociobiology*, **20**, 271–83.

Folke, C., Carpenter, S., Walker, B., *et al.* (2004) Regime shifts, resilience, and biodiversity in ecosystem management. *Annual Review of Ecology, Evolution, and Systematics*. **35**, 557–81.

Fretwell, S.D. (1977) The regulation of plant communities by food chains exploiting them. *Perspectives in Biology and Medicine*, **20**, 169–85.

Graham, D.W., Knapp, C.W., Van Vleck, E.S., *et al.* (2007) Experimental demonstration of chaotic instability in biological nitrification. *ISME Journal*, **1**, 385–93.

Hairston, N.G., Smith, F.E. and Slobodkin, L.B. (1960) Community structure, population control and competition. *The American Naturalist*, **44**, 421–5.

Hastings, A., and Powell, T. (1991) Chaos in a three-species food chain. *Ecology*, **72**, 896–903.

Holling, C.S. (1965) The functional response of invertebrate predators to prey density. *Memoires Entomological Society of Canada*, **45**, 3–60.

Holt, R.D. (1977) Predation, apparent competition and the structure of prey communities. *Theoretical Population Biology*, **12**, 197–229.

Holt, R.D. (1997) Community modules. In *Multitrophic Interactions in Terrestrial Ecosystems* (eds A.C. Gange and V.K. Brown), pp. 333–349. Blackwell Scientific, Oxford.

Huisman, J. and Weissing, F.J. (1999) Biodiversity of plankton by species oscillations and chaos. *Nature*, **402**, 407–10.

Huisman, J. and Weissing, F.J. (2001) Fundamental unpredictability in multispecies competition. *The American Naturalist*, **157**, 488–94.

Jefferies, R.L. (1999) Herbivores, nutrients and trophic cascades in terrestrial ecosystems. In *Herbivores: Between Plants and Predators* (eds H. Olff, V.K. Brown and R.H. Drent). Oxford University Press, Oxford.

Karasov, W.H. and Martinez del Rio, C. (2007) *Physiological Ecology. How Animals Process Energy, Nutrients and Toxins*. Princeton University Press, Princeton, NJ.

Kefi S., Rietkerk, M., Alados, C.L., *et al.* (2007) Spatial vegetation patterns and imminent desertification in Mediterranean arid ecosystems. *Nature*, **449**, 213-U5.

Krebs, C.J., Sinclair, A.R.E., Boonstra, R., *et al.* (1999) Community dynamics of vertebrate herbivores: how can we untangle the web? In *Herbivores: Between Plants and Predators* (eds H. Olff, V.K. Brown and R.H. Drent), pp. 447–67. Oxford University Press, Oxford.

Lawton, J.H. and Gaston, K.J. (1989) Temporal patterns in the herbivorous insects of Bracken: a test of community predictability. *Journal of Animal Ecology*, **58**, 1021–34.

Leibold, M.A., Hall, S.R. and Bjornstad, O.N. (2005) Food web architecture and its effects on consumer resource oscillations in experimental pond ecosystems. In *Dynamic Food Webs: Multispecies Assemblages, Ecosystem Development, and Environmental Change* (eds P.C. de Ruiter, V. Wolters and J.C. Moore), pp. 37–47. Academic Press, Burlington, MA.

Loreau, M. and de Mazancourt, C. (2008) Species synchrony and its drivers: neutral and non-neutral community dynamics in fluctuating environments. *The American Naturalist.*, **172**, E48–E66.

Lotka, A.J. (1926) *Elements of Physical Biology*. Williams and Wilkins Co., Baltimore, MD.

Marquis, R.J. and Whelan, C.J. (1994) Insectivorous birds increase growth of white oak through consumption of leaf-chewing insects. *Ecology*, 75, 2007–14.

May, R.M. (1973) *Stability and Complexity in Model Ecosystems*. Princeton University Press, Princeton, NJ.

May, R.M. and McLean, A., eds. (2007) *Theoretical Ecology*, 3rd edn. Oxford University Press, Oxford.

McCann, K., Hastings, A. and Huxel, G.R. (1998) Weak trophic interactions and the balance of nature. *Nature*, **395**, 794–8.

McCauley, E., Nisbet, R.M., Murdoch, W.W., *et al.* (1999) Large-amplitude cycles of *Daphnia* and its algal prey in enriched environments. *Nature*, **402**, 653–6.

Menge, B.A. (1995) Indirect effects in marine rocky intertidal interaction webs: patterns and importance. *Ecological Monographs*, **65**, 21–74.

Morin, P.J. (1999) *Community Ecology*. Wiley-Blackwell, Oxford.

Neutel, A.M., Heesterbeek, J.A.P. and de Ruiter P.C. (2002) Stability in real food webs: weak links in long loops. *Science*, **296**, 1120–3.

Neutel, A.M., Heesterbeek, J.A.P., van de Koppel, J., *et al.* (2007) Reconciling complexity with stability in naturally assembling food webs. *Nature*, **449**, 599-U11.

Nicholson, A.J. and Bailey, V.A. (1935) The balance of animal populations. Part I. *Proceedings of the Zoological Society of London*, **3**, 551–98.

Oksanen, L. (1988) Ecosystem organization: mutualism and cybernetics or plain Darwinian struggle for existence? *The American Naturalist*, **131**, 424–44.

Oksanen, L., Fretwell, S.D., Arruda, J. and Niemela, P. (1981) Exploitation ecosystems in gradients of primary productivity. *The American Naturalist*, **118**, 240–61.

Olff, H., Alonso, D., Berg, M.P., *et al.* (2009) Parallel ecological networks in ecosystems. *Philosophical Transactions of the Royal Society B*. **364**, 1755–79.

Paine, R.T. (1980) Food webs: linkage, interaction strength and community infrastructure. *Journal of Animal Ecology*, **49**, 667–85.

Peterson, G., Allen, C.R. and Holling, C.S. (1998) Ecological resilience, biodiversity, and scale. *Ecosystems*, **1**, 6–18.

Pimm, S.L. (1982) *Food Webs*. Chapman and Hall, London.

Pimm, S.L. (1991) *The Balance of Nature. Ecological Issues in the Conservation of Species and Communities*. University of Chicago Press, Chicago, IL.

Polis, G.A., Myers, C.A. and Holt, R.D. (1989) The ecology and evolution of intraguild predation: potential competitors that eat each other. *Annual Review of Ecology Evolution and Systematics*, **20**, 297–330.

Power, M.E., Matthews, W.J. and Stewart, A.J. (1985) Grazing minnows, piscivorous bass, and stream algae: dynamics of a strong interaction. *Ecology*, **66**, 1448–56.

Prins, H.H.T. and Douglas-Hamilton, I. (1990) Stability in a multispecies assemblage of large herbivores in East-Africa. *Oecologia*, **83**, 392–400.

Rooney, N., McCann, K., Gellner, G. and Moore J.C. (2006) Structural asymmetry and the stability of diverse food webs. *Nature*, **442**, 265–9.

Rosenzweig, M.L. (1971) Paradox of enrichment: destabilisation of exploitation ecosystems in ecological time. *Science*, **171**, 385–7.

Scheffer, M. and Carpenter, S.R. (2003) Catastrophic regime shifts in ecosystems: linking theory to observation. *Trends in Ecology & Evolution*, **18**, 648–56.

Scheffer, M., Hosper, S.H., Meijer, M.-L., *et al.* (1993) Alternative equilibria in shallow lakes. *Trends in Ecology & Evolution*, **8**, 275–9.

Schoener, T.W. (1974) Resource partitioning in ecological communities. *Science*, **185**, 27–39.

Schoener, T.W. (1988) Leaf damage in island buttonwood, *Conocarpus erectus*: correlations with pubescens, island area, isolation and the distribution of major carnivores. *Oikos*, **53**, 253–66.

Schröder, A., Persson, L. and de Roos, A.M. (2005) Direct experimental evidence for alternative stable states: a review. *Oikos*, **110**, 3–19.

Spiller, D.A. and Schoener, T.W. (1989) An experimental study of the effect of lizards on web-spider communities. *Ecological Monographs*, **58**, 57–77.

Tanner, J.T. (1975) Stability and intrinsic growth-rates of prey and predator populations. *Ecology*, **56**, 855–67.

Terborgh, J., Feeley, K., Silman, M., *et al.* (2006) Vegetation dynamics of predator-free land-bridge islands. *Journal of Ecology*, **94**, 253–63.

Tilman, D. (1982) *Resource Competition and Community Structure*. Princeton University Press, Princeton, NJ.

Ulanowicz, R.E. (1997) *Ecology, the Ascendant Perspective*. Columbia University Press, New York, NY.

van der Heide, T., van Nes, E.H., Geerling, G.W., *et al.* (2007) Positive feedbacks in seagrass ecosystems: implications for success in conservation and restoration. *Ecosystems*, **10**, 1311–22.

Vandermeer, J. (1980) Indirect mutualism: variations on a theme by Stephen Levine. *The American Naturalist*, **116**, 441–8.

Vandermeer, J. (1994) The qualitative behavior of coupled predator-prey oscillations as deduced from simple circle maps. *Ecological Modelling*, **73**, 135–48.

Vandermeer, J. (2004) Coupled oscillations in food webs: balancing competition and mutualism in simple ecological models. *The American Naturalist*, **163**, 857–67.

van Nes, E.H. and Scheffer, M. (2007) Slow recovery from perturbations as a generic indicator of a nearby catastrophic shift. *The American Naturalist*, **169**, 738–47.

Vasseur, D.A. and Fox, J.W. (2007) Environmental fluctuations can stabilize food web dynamics by increasing synchrony. *Ecology Letters*, **10**, 1066–74.

Volterra, V. (1926) Variations and fluctuations of the number individuals of animals living together [in Italian]. *Memoires Academia dei Lincei*, **2**, 31–113.

Chapter 3

Abrams, P.A., Menge, B.A., Mittelbach, G.G., *et al.* (1995) The role of indirect effects in food webs. In *Food Webs: Integration of Patterns and Dynamics* (eds G. Polis and K. Winemiller), pp. 371–96. Chapman and Hall, New York, NY.

Berlow, E.L. (1999) Strong effects of weak interactions in ecological communities. *Nature*, **398**, 330–4.

Brose, U., Berlow, E.L. and Martinez, N.D. (2005) Scaling up keystone effects from simple to complex ecological networks. *Ecology Letters*, **8**, 1317–25.

Brose, U., Jonsson, T., Berlow, E.L., *et al.* (2006a) Consumer-resource body-size relationships in natural food webs. *Ecology*, **87**, 2411–17.

Brose, U., Williams, R.J. and Martinez, N.D. (2006b) Allometric scaling enhances stability in complex food webs. *Ecology Letters*, **9**, 1228–36.

Brown, J.H., Gillooly, J.F., Allen, A.P., *et al.* (2004) Toward a metabolic theory of ecology. *Ecology*, **85**, 1771–89.

Camacho, J., Guimera, R. and Amaral, L.A.N. (2002) Robust patterns in food web structure. *Physical Review Letters*, **88**, 228102.

Cattin, M.F., Bersier, L.F., Banasek-Richter, C., *et al.* (2004) Phylogenetic constraints and adaptation explain food-web structure. *Nature*, **427**, 835–9.

Cohen, J.E., and Newman, C.M. (1984) The stability of large random matrices and their products. *Annals of Probability*, **12**, 283–310.

Cohen, J.E., Briand, E.F. and Newman, C.M. (1990) *Community Food Webs: Data and Theory*. Springer-Verlag, New York, NY.

Daily, G.C. (1997) *Nature's Services*. Island Press, Washington, DC.

DeAngelis, D.L. (1975) Stability and connectance in food web models. *Ecology*, **56**, 238–43.

de Ruiter, P., Neutel, A.-M. and Moore, J.C. (1995) Energetics, patterns of interaction strengths, and stability in real ecosystems. *Science*, **269**, 1257–60.

Diehl, S. and Feissel, M. (2001) Intraguild prey suffer from enrichment of their resources: a microcosm experiment with ciliates. *Ecology*, **82**, 2977–83.

Dunne, J.A. (2006) The network structure of food webs. In *Ecological Networks: Linking Structure to Dynamics in Food Webs* (eds M. Pascual and J.A. Dunne), pp. 27–86. Oxford University Press, Oxford.

Dunne, J.A., Williams, R.J. and Martinez, N.D. (2002) Food-web structure and network theory: the role of connectance and size. *Proceedings of the National Academy of Sciences of the United States of America*, **99**, 12917–22.

Dunne, J.A., Williams, R.J. and Martinez, N.D. (2004) Network structure and robustness of marine food webs. *Marine Ecology-Progress Series*, **273**, 291–302.

Egerton, F.N. (2007) Understanding food chains and food webs 1700–1970. *Bulletin of the Ecological Society of America*, 50–69.

Elton, C. (1933) *The Ecology of Animals*. Methuen, London.

Elton, C.S. (1958) *Ecology of Invasions by Animals and Plants*. Chapman & Hall, London.

Fussmann, G.F. and Heber, G. (2002) Food web complexity and chaotic population dynamics. *Ecology Letters*, **5**, 394–401.

Hutchinson, G.E. (1959) Homage to Santa Rosalia, or why are there so many kinds of animals? *The American Naturalist*, **93**, 145–59.

Ives, A.R. and Cardinale, B.J. (2004) Food-web interactions govern the resistance of communities after non-random extinctions. *Nature*, **429**, 174–77.

Kondoh, M. (2003) Foraging adaptation and the relationship between food-web complexity and stability. *Science*, **299**, 1388–91.

Kondoh, M. (2006) Does foraging adaptation create the positive complexity-stability relationship in realistic food-web structure? *Journal of Theoretical Biology*, **238**, 646–51.

Lindeman, R.L. (1942) The trophic-dynamic aspect of ecology. *Ecology*, **23**, 399–418.

Lotka, L. (1925) *Elements of Physical Biology*. Williams & Wilkins, Baltimore.

MacArthur, R.H. (1955) Fluctuations of animal populations, and a measure of community stability. *Ecology*, **36**, 533–6.

Martinez, N.D., Williams, R.J. and Dunne, J.A. (2006) Diversity, complexity, and persistence in large model ecosystems. In *Ecological Networks: Linking Structure to Dynamics in Food Webs* (eds M. Pascual and J.A. Dunne), pp. 163–85. Oxford University Press, Oxford.

May, R.M. (1972) Will a large complex system be stable? *Nature*, **238**, 413–14.

May, R.M. (1973) *Stability and Complexity in Model Ecosystems*. Princeton University Press, Princeton, NJ.

May, R.M. (2001) *Stability and Complexity in Model Ecosystems* (with new Introduction). Princeton University Press, Princeton, NJ.

McCann, K.S. (2000) The diversity-stability debate. *Nature*, **405**, 228–33.

McCann, K.S. and Hastings, A. (1997) Re-evaluating the omnivory-stability relationship in food webs. *Proceedings of the Royal Society of London Series B-Biological Sciences*, **264**, 1249–54.

McCann, K.S. and Yodzis, P. (1994) Biological conditions for chaos in a three-species food chain. *Ecology*, **75**, 561–4.

McCann, K.S., Hastings, A. and Huxel, G.R. (1998) Weak trophic interactions and the balance of nature. *Nature*, **395**, 794–8.

Menge, B.A. (1997) Detection of direct versus indirect effects: were experiments long enough? *The American Naturalist*, **149**, 801–23.

Menge, B.A., Berlow, E.L., Blanchette, C., *et al.* (1994) The keystone species concept: variation in interaction strength in a rocky intertidal habitat. *Ecological Monographs*, **64**, 249–86.

Milo, R., Shen-Orr, S., Itzkovitz, S., *et al.* (2002) Network motifs: simple building blocks of complex networks. *Science*, **298**, 824–7.

Neutel, A.-M., Heesterbeek, J.A.P. and De Ruiter, P.C. (2002) Stability in real food webs: weak links in long loops. *Science*, **296**, 1120–3.

Oaten, A. and Murdoch, W.M. (1975) Functional response and stability in predator-prey systems. *The American Naturalist*, **109**, 289–98.

Odum, E. (1953) *Fundamentals of Ecology*. Saunders, Philadelphia.

Paine, R.T. (1966) Food web complexity and species diversity. *The American Naturalist*, **100**, 65–75.

Paine, R.T. (1974) Intertidal community structure. Experimental studies on the relationship between a dominant competitor and its principal predator. *Oecologia*, **15**, 93–120.

Paine, R.T. (1980) Food webs, linkage interaction strength, and community infrastructure. *Journal of Animal Ecology*, **49**, 667–85.

Pimm, S.L., Lawton, J.H. and Cohen, J.E. (1991) Food web patterns and their consequences. *Nature*, **350**, 669–74.

Power, M.E., Tilman, D., Estes, J., *et al.* (1996) Challenges in the quest for keystones. *BioScience*, **46**, 609–20.

Rall, B.C., Guill, C. and Brose, U. (2008) Food-web connectance and predator interference dampen the paradox of enrichment. *Oikos*, **117**, 202–13.

Real, L.A. (1977) Kinetics of functional response. *The American Naturalist*, **111**, 289–300.

Stouffer, D.B., Camacho, J., Guimera, R., *et al.* (2005) Quantitative patterns in the structure of model and empirical food webs. *Ecology*, **86**, 1301–11.

Stouffer, D.B., Camacho, J. and Amaral, L.A.N. (2006) A robust measure of food web intervality. *Proceedings of the National Academy of Sciences of the United States of America*, **103**, 19015–20.

Stouffer, D.B., Camacho, J., Jiang, W. and Amaral, L.A.N. (2007) Evidence for the existence of a robust pattern of prey selection in food webs. *Proceedings of the Royal Society B-Biological Sciences*, **274**, 1931–40.

Strogatz, S.H. (2001) Exploring complex networks. *Nature*, **410**, 268–76.

Vandermeer, J. (2006) Omnivory and the stability of food webs. *Journal of Theoretical Biology*, **238**, 497–504.

Volterra, V. (1926) Fluctuations in the abundance of a species considered mathematically. *Nature*, **118**, 558–60.

Weitz, J.S. and Levin, S.A. (2006) Size and scaling of predator-prey dynamics. *Ecology Letters*, **9**, 548–57.

Williams, R.J. and Martinez, N.D. (2000) Simple rules yield complex food webs. *Nature*, **404**, 180–3.

Williams, R.J. and Martinez, N.D. (2004) Stabilization of chaotic and non-permanent food web dynamics. *European Physical Journal B*, **38**, 297–303.

Williams, R.J., Martinez, N.D., Berlow, E.L., *et al.* (2002) Two degrees of separation in complex food webs. *Proceedings of the National Academy of Sciences of the United States of America*, **99**, 12913–16.

Yodzis, P. (1981) The stability of real ecosystems. *Nature*, **289**, 674–6.

Yodzis, P. (2000) Diffuse effects in food webs. *Ecology*, **81**, 261–6.

Yodzis, P. and Innes, S. (1992) Body size and consumer-resource dynamics. *The American Naturalist*, **139**, 1151–75.

Chapter 4

Ackerly, D.D. and Cornwell, W.K. (2007) A trait-based approach to community assembly: partitioning of species trait values into within- and among-community components. *Ecology Letters*, **10**, 135–45.

Almany, G.R. (2003) Priority effects in coral reef fish communities. *Ecology*, **84**, 1920–35.

Barkai, A. and McQuaid, C. (1988) Predator-prey role reversal in a marine benthic ecosystem. *Science*, **242**, 62–4.

Belyea, L.R. and Lancaster, J. (1999) Assembly rules within a contingent ecology. *Oikos*, **86**, 402–16.

Bertness, M.D., Trussell, G.C., Ewanchuk, P.J. and Silliman, B.R. (2004) Do alternate stable community states exist in the Gulf of Maine rocky intertidal zone? Reply. *Ecology* **85**, 1165–7.

Cadotte, M.W. (2006) Metacommunity influences on community richness at multiple spatial scales: a microcosm experiment. *Ecology*, **87**, 1008–16.

Cadotte, M.W. (2007) Competition-colonization trade-offs and disturbance effects at multiple scales. *Ecology*, **88**, 823–9.

Cadotte, M.W. and Fukami, T. (2005) Dispersal, spatial scale and species diversity in a hierarchically structured experimental landscape. *Ecology Letters*, **8**, 548–57.

Chase, J.M. (2003) Community assembly: when should history matter? *Oecologia*, **136**, 489–98.

Chase, J.M. (2007) Drought mediates the importance of stochastic community assembly. *Proceedings of the National Academy of Sciences of the United States of America*, **104**, 17430–4.

Clements, F.E. (1916) *Plant Succession: Analysis of the Development of Vegetation*. Publication no. 242. Carnegie Institution of Washington, Washington, DC.

Connor, E.F. and Simberloff, D. (1979) The assembly of species communities: chance or competition? *Ecology*, **60**, 1132–40.

Denslow, J.S. (1980) Patterns of plant species diversity during succession under different disturbance regimes. *Oecologia*, **46**, 18–21.

Diamond, J.M. (1975) Assembly of species communities. In *Ecology and Evolution of Communities* (eds M.L. Cody and J.M. Diamond), pp. 342–444, Belknap, Cambridge, MA.

Drake, J.A. (1991) Community-assembly mechanics and the structure of an experimental species ensemble. *The American Naturalist*, **137**, 1–26.

Fukami, T. (2004a) Assembly history interacts with ecosystem size to influence species diversity. *Ecology*, **85**, 3234–42.

Fukami, T. (2004b) Community assembly along a species pool gradient: implications for multiple-scale patterns of species diversity. *Population Ecology*, **46**, 137–47.

Fukami, T. (2005) Integrating internal and external dispersal in metacommunity assembly: preliminary theoretical analyses. *Ecological Research*, **20**, 623–31.

Fukami, T. (2008) Stochasticity in community assembly, and spatial scale [in Japanese]. In *Community Ecology* [*Gunshuu seitaigaku*], vol. 5 (eds T. Ohgushi, M. Kondoh and T. Noda). Kyoto University Press, Kyoto, Japan.

Fukami, T., Bezemer, T.M., Mortimer, S.R. and van der Putten, W.H. (2005) Species divergence and trait convergence in experimental plant community assembly. *Ecology Letters*, **8**, 1283–90.

Fukami, T., Beaumont, H.J.E., Zhang, X.-X. and Rainey, P.B. (2007) Immigration history controls diversification in experimental adaptive radiation. *Nature*, **446**, 436–9.

Gillespie, R.G. (2004) Community assembly through adaptive radiation in Hawaiian spiders. *Science*, **303**, 356–9.

Gotelli, N.J. (2001) Research frontiers in null model analysis. *Global Ecology and Biogeography*, **10**, 337–43.

Holt, R.D. and Polis, G.A. (1997) A theoretical framework for intraguild predation. *The American Naturalist*, **149**, 745–64.

Hubbell, S.P. (2001) *The Unified Neutral Theory of Biodiversity and Biogeography*. Princeton University Press, Princeton, NJ.

Knowlton, N. (2004) Multiple "stable" states and the conservation of marine ecosystems. *Progress in Oceanography*, **60**, 387–96.

Leibold, M.A., Holyoak, M., Mouquet, N., *et al.* (2004) The metacommunity concept: a framework for multi-scale community ecology. *Ecology Letters*, **7**, 601–13.

Lewontin, R.C. (1969) The meanings of stability. *Brookhaven Symposium on Biology*, **22**, 13–24.

Lockwood, J.L., Powell, R.D., Nott, M.P. and Pimm, S.L. (1997) Assembling ecological communities in space and time. *Oikos*, **80**, 549–53.

Lomolino, M.V. (1990) The target area hypothesis: the influence of island area on immigration rates of non-volant mammals. *Oikos*, **57**, 297–300.

Long, Z.T. and Karel, I. (2002) Resource specialization determines whether history influences community structure. *Oikos*, **96**, 62–9.

Losos, J.B., Jackman, T.R., Larson, A., *et al.* (1998) Contingency and determinism in replicated adaptive radiations of island lizards. *Science*, **279**, 2115–18.

MacArthur, R.H. (1972) *Geographical Ecology: Patterns in the Distribution of Species*. Princeton University Press, Princeton, NJ.

MacArthur, R.H. and Wilson, E.O. (1967) *Theory of Island Biogeography*. Princeton University Press, Princeton, NJ.

McGill, B., Enquist, B.J., Westoby, M. and Weiher, E. (2006) Rebuilding community ecology from functional traits. *Trends in Ecology & Evolution*, **21**, 178–84.

Morin, P.J. (1999) *Community Ecology*. Blackwell, Malden, MA.

Morton, R.D. and Law, R. (1997) Regional species pools and the assembly of local ecological communities. *Journal of Theoretical Biology*, **187**, 321–31.

Mouquet, N., Munguia, P., Kneitel, J.M. and Miller, T.E. (2003) Community assembly time and the relationship between local and regional species richness. *Oikos*, **103**, 618–26.

Olito, C. and Fukami, T. (2009) Long-term effects of predator arrival timing on prey community succession. *The American Naturalist*, **173**, 354–62.

Orrock, J.L. and Fletcher Jr, R.J. (2005) Changes in community size affect the outcome of competition. *The American Naturalist*, **166**, 107–11.

Peterson, C.H. (1984) Does a rigorous criterion for environmental identity preclude the existence of multiple stable points? *The American Naturalist*, **124**, 127–33.

Petraitis, P.S. and Latham, R.E. (1999) The importance of scale in testing the origins of alternative community states. *Ecology*, **80**, 429–42.

Petraitis, P.S., Latham, R.E. and Nesenbaum, R.A. (1989) The maintenance of species diversity by disturbance. *Quarterly Review of Biology*, **64**, 393–418.

Robinson, J.V. and Edgemon, M.A. (1988) An experimental evaluation of the effect of invasion history on community structure. *Ecology*, **69**, 1410–17.

Sale, P.F. (1977) Maintenance of high diversity in coral reef fish communities. *The American Naturalist*, **111**, 337–59.

Samuels, C.L. and Drake, J.A. (1997) Divergent perspectives on community convergence. *Trends in Ecology & Evolution*, **12**, 427–32.

Schreiber, S.J. and Rittenhouse, S. (2004) From simple rules to cycling in community assembly. *Oikos*, **105**, 349–58.

Schröder, A., Persson, L. and de Roos, A.M. (2005) Direct experimental evidence for alternative stable states: a review. *Oikos*, **110**, 3–19.

Shurin, J.B., Amarasekare, P., Chase, J.M., *et al.* (2004) Alternative stable states and regional community structure. *Journal of Theoretical Biology*, **227**, 359–68.

Steiner, C.F. and Leibold, M.A. (2004) Cyclic assembly trajectories and scale-dependent productivity-diversity relationships. *Ecology*, **85**, 107–13.

Thornton, I. (1996) *Krakatau: the Destruction and Reassembly of an Island Ecosystem*. Harvard University Press, Cambridge, MA.

Tilman, D. (1988) *Plant Strategies and the Dynamics and Structure of Plant Communities*. Princeton University Press, Princeton, NJ.

van Geest, G.J., Coops, H., Scheffer, M. and Van Nes, E.H. (2007) Long transients near the ghost of a stable state in eutrophic shallow lakes with fluctuating water levels. *Ecosystems*, **10**, 36–46.

van Nes, E.H., Rip, W.J. and Scheffer, M. (2007) A theory for cyclic shifts between alternative states in shallow lakes. *Ecosystems*, **10**, 17–27.

Walker, L.R., Bellingham, P.J. and Peltzer, D.A. (2006) Plant characteristics are poor predictors of microsite colonization during the first two years of primary succession. *Journal of Vegetation Science*, **17**, 397–406.

Warren, P.H., Law, R. and Weatherby, A.J. (2003) Mapping the assembly of protist communities in microcosms. *Ecology*, **84**, 1001–11.

Weiher, E. (2007) On the status of restoration science: obstacles and opportunities. *Restoration Ecology*, **15**, 340–43.

Weiher, E. and Keddy, P.A. (1995) Assembly rules, null models, and trait dispersion: new questions from old patterns. *Oikos*, **74**, 159–64.

Wilbur, H.M. and Alford, R.A. (1985) Priority effects in experimental pond communities: responses of *Hyla* to *Bufo* and *Rana*. *Ecology*, **66**, 1106–14.

Wilson, D.S. (1992) Complex interactions in metacommunities, with implications for biodiversity and higher levels of selection. *Ecology*, **73**, 1984–2000.

Chapter 5

Adler, P.B., Hille Ris Lambers, J. and Levine J. (2007) A niche for neutrality. *Ecology Letters*, **10**, 95–104.

Allendorf, F., Bayles, D. Bottom, D.L., *et al.* (1997) Prioritizing pacific salmon stocks for conservation. *Conservation Biology*, **11**, 140–52.

Alonso, D., Etienne, R.S. and McKane, A.J. (2006) The merits of neutral theory. *Trends in Ecology & Evolution*, **21**, 451–7.

Amarasekare, P. (2000) The geometry of coexistence. *Biological Journal of the Linnean Society*, **71**, 1–31.

Amarasekare, P., Hoopes, M., Mouquet, N. and Holyoak, M. (2004) Mechanisms of coexistence in competitive metacommunities. *The American Naturalist*, **164**, 310–26.

Andrewartha, H.G. and Birch, L.C. (1954) *The Distribution and Abundance of Animals*. University of Chicago Press, Chicago, IL.

Bengtsson, J. (1989) Interspecific competition increases local extinction rate in a metapopulation system. *Nature*, **340**, 713–15.

Bengtsson, J. (1991) Interspecific competition in metapopulations. *Biological Journal of the Linnean Society*, **42**, 219–37.

Booth, B.D. and Larson, D.W. (1999) Impact of language, history, and choice of system on the study of assembly rules. In *Ecological Assembly Rules: Perspectives, Advances, Retreats* (eds E. Weiher and P.A. Keddy), pp. 206–29. Cambridge University Press, Cambridge.

Cadotte, M.W. (2006) Dispersal and species diversity: a meta-analysis. *The American Naturalist*, **167**, 913–24.

Cadotte, M.W. and Fukami, T. (2005) Dispersal, spatial scale and species diversity in a hierarchically structured experimental landscape. *Ecology Letters*, **8**, 548–57.

Calcagno, V., Mouquet, N., Jarne, P. and David, P. (2006) Coexistence in a metacommunity: the competition-colonization trade-off is NOT dead. *Ecology Letters*, **9**, 897–907.

Chase, J.M. (2003) Community assembly: when does history matter? *Oecologia*, **136**, 489–98.

Chase, J.M. (2005) Towards a really unified theory for metacommunities. *Functional Ecology*, **19**, 182–6.

Chase, J.M. (2007) Drought mediates the importance of stochastic community assembly. *Proceedings of the National Academy of Sciences of the United States of America*, **104**, 17430–4.

Chase, J.M. and Leibold, M.A. (2002) Spatial scale dictates the productivity-diversity relationship. *Nature*, **415**, 427–30.

Chase, J.M and Leibold, M.A. (2003) *Ecological Niches: Linking Classical and Contemporary Approaches*. University of Chicago Press, Chicago, IL.

Chase, J.M. and Ryberg, W.A. (2004) Connectivity, scale dependence, and the productivity-diversity relationship. *Ecology Letters*, **7**, 676–83.

Chase, J.M., Abrams, P.A. Grover, J.P., *et al.* (2002) The interaction between predation and competition: a review and synthesis. *Ecology Letters*, **5**, 302–15.

Chase, J.M., Amarasekare, P., Cottenie, K., *et al.* (2005) Competing theories for competitive metacommunities. In *Metacommunities: Spatial Dynamics and Ecological Communities* (eds M. Holyoak, M. Leibold and R. Holt), pp. 335–54. University of Chicago Press, Chicago, IL.

Chave, J. (2004) Neutral theory and community ecology. *Ecology Letters*, **7**, 241–53.

Chave, J. and Leigh, E.G. (2002) A spatially explicit neutral model of beta-diversity in tropical forests. *Theoretical Population Biology*, **62**, 153–68.

Chesson, P. (2000) Mechanisms of maintenance of species diversity. *Annual Review of Ecology and Systematics*, **31**, 343–66.

Clark, J.S., Dietze, M., Chakraborty, S., *et al.* (2007) Resolving the biodiversity paradox. *Ecology Letters*, **10**, 647–62.

Condit, R., Pitman, N., Leigh Jr, E.G., *et al.* (2002) Beta diversity in tropical forest trees. *Science*, **295**, 666–9.

Connor, E.F. and McCoy, E.D. (1979) The statistics and biology of the species-area relationship. *The American Naturalist*, **113**, 791–833.

Cornell, H.V. (1993) Unsaturated patterns in species assemblages: the role of regional processes in setting local species richness. In *Species Diversity in Ecological Communities: Historical and Geographical Perspectives* (eds R.E. Ricklefs and D. Schluter), pp. 243–52. University of Chicago Press, Chicago, IL.

Damschen, E.I., Haddad, N.M., Orrock, J.L., *et al.* (2006) Corridors increase plant species richness at large scales. *Science*, **313**, 1284–6.

Diamond, J.M. (1975) Assembly of species communities. In *Ecology and Evolution of Communities* (eds M.L. Cody and J.M. Diamond), pp. 342–444. Harvard University Press, Cambridge, MA.

Drakare, S., Lennon, J.J. and Hillebrand, H. (2006) The imprint of the geographical, evolutionary and ecological context on species-area relationships. *Ecology Letters*, **9**, 215–27.

Etienne, R.S., Alonso, D. and McKane, A.J. (2007) The zero-sum assumption in neutral biodiversity theory. *Journal of Theoretical Biology*, **248**, 522–36.

Forbes, A.E. and Chase, J.M. (2002) The role of habitat connectivity and landscape geometry in experimental zooplankton metacommunities. *Oikos*, **96**, 433–40.

Fukami, T. (2004) Community assembly along a species pool gradient: implications for multiple-scale patterns of species diversity. *Population Ecology*, **46**, 137–47.

Gause, G.F. (1934) *The Struggle for Existence*. Williams & Wilkins, Baltimore, MD.

Gilbert, F., Gonzalez, A. and Evans-Freke, I. (1998) Corridors maintain species richness in the fragmented landscapes of a microecosystem. *Proceedings of the Royal Society of London B*, **265**, 577–82.

Gotelli N.J. and Ellison, A.M. (2006) Food-web models predict species abundances in response to habitat change. *PLoS Biology*, **4**, e324.

Gravel, D., Canham, C.D., Beaudet, M. and Messier, C. (2006) Reconciling niche and neutrality: the continuum hypothesis. *Ecology Letters*, **9**, 399–409.

Hanski, I. and Gyllenberg, M. (1997) Uniting two general patterns in the distribution of species. *Science*, **275**, 397–400.

Harrison, S. (1997) How natural habitat patchiness affects the distribution of diversity in Californian serpentine chaparral. *Ecology*, **78**, 1898–906.

Harrison, S. (1999) Local and regional diversity in a patchy landscape: native, alien and endemic herbs on serpentine soils. *Ecology*, **80**, 70–80.

Harrison, S., Davies, K.F., Safford, H.D. and Viers, J.H. (2006a). Beta diversity and the scale-dependence of the productivity-diversity relationship: a test in the Californian serpentine flora. *Journal of Ecology*, **94**, 110–17.

Harrison, S., Safford, H.D., Grace, J.B., *et al.* (2006b). Regional and local species richness in an insular environment: serpentine plants in California. *Ecological Monographs*, **76**, 41–56.

Hastings, A. (1980) Disturbance, coexistence, history, and competition for space. *Theoretical Population Biology*, **18**, 363–73.

Holt, R.D. (1993) Ecology at the mesoscale: the influence of regional processes on local communities. In *Species Diversity in Ecological Communities* (eds R.E. Ricklefs and D. Schluter), pp. 77–88. University of Chicago Press, Chicago, IL.

Holt, R.D. and Hoopes, M.F. (2005) Food web dynamics in a metacommunity context: modules and beyond. In *Metacommunities: Spatial Dynamics and Ecological Communities* (eds M. Holyoak, M. Leibold and R. Holt), pp. 68–93. University of Chicago Press, Chicago, IL.

Holt, R.D., Lawton, J.H., Polis, G.A. and Martinez, N.D. (1999) Trophic rank and the species-area relationship. *Ecology*, **80**, 1495–505.

Holyoak, M. and Loreau, M. (2006) Reconciling empirical ecology with neutral community models. *Ecology*, **87**, 1370–7.

Holyoak, M., Leibold, M.A. and Holt, R.D., eds. (2005) *Metacommunities: Spatial Dynamics and Ecological Communities*. University of Chicago Press, Chicago, IL.

Horn, H. and MacArthur, R.H. (1972) Competition among fugitive species in a harlequin environment. *Ecology*, **53**, 749–52.

Hoyle, M. and Gilbert, F. (2004) Species richness of moss landscapes unaffected by short-term fragmentation. *Oikos*, **105**, 359–67.

Hubbell, S.P. (2001) *The Unified Neutral Theory of Biodiversity and Biogeography*. Princeton University Press, Princeton, NJ.

Huffaker, C.B. (1958) Experimental studies on predation: dispersion factors and predator-prey oscillations. *Hilgardia*, **27**, 343–83.

Hugueny, B., Cornell, H.V. and Harrison, S. (2007) Metacommunity models predict the local-regional richness relationship in a natural system. *Ecology*, **88**, 1696–706.

Kneitel, J.M. and Miller, T.E. (2003) Dispersal rates affect species composition in metacommunities of *Sarracenia purpurea* inquilines. *The American Naturalist*, **162**, 165–71.

Knight, T.M., McCoy, M.W., Chase, J.M., *et al.* (2005) Trophic cascades across landscapes. *Nature*, **430**, 880–3.

Kraft, N.J.B., Cornwell, W.K., Webb, C.O. and Ackerly, D.D. (2007) Trait evolution, community assembly, and the phylogenetic structure of ecological communities. *The American Naturalist*, **170**, 271–83.

Kruess, A and Tscharntke, T. (2000) Species richness and parasitism in a fragmented landscape: experiments and field studies with insects on *Vicia sepium*. *Oecologia*, **122**, 129–37.

Lande, R. (1996) Statistics and partitioning of species diversity, and similarity among multiple communities. *Oikos*, **76**, 5–13.

Leibold, M.A. and McPeek, M.A. (2006) Coexistence of the niche and neutral perspectives in community ecology. *Ecology*, **87**, 1399–410.

Leibold, M.A., Holyoak, M., Mouquet, N., *et al.* (2004) The metacommunity concept: a framework for multi-scale community ecology. *Ecology Letters*, **7**, 601–13.

Levin, S.A. (1974) Dispersion and population interactions. *The American Naturalist*, **108**, 960.

Levins, R. (1969) Some demographic and genetic consequences of environmental heterogeneity for biological control. *Bulletin of the Entomological Society of America*, **15**, 237–40.

Levins, R. and Culver, D. (1971) Regional coexistence of species and competition between rare species. *Proceedings of the National Academy of Sciences of the United States of America*, **68**, 1246–8.

Lomolino, M.V. (2000) Ecology's most general, yet protean pattern: the species-area relationship. *Journal of Biogeography*, **27**, 17–26.

Loreau, M. (2000) Are communities saturated? On the role of α, β, and γ diversity. *Ecology Letters*, **3**, 73–6.

MacArthur, R.H. (1972) *Geographical Ecology*. Princeton University Press, Princeton, NJ.

MacArthur, R.H. and Levins, R. (1964) Competition, habitat selection, and character displacement in a patchy environment. *Proceedings of the National Academy of Sciences of the United States of America*, **51**, 1207–10.

MacArthur, R.H. and Wilson, E.O. (1967) *Theory of Island Biogeography*. Princeton University Press, Princeton, NJ.

McGill, B.J., Maurer, B.A. and Weiser, M.D. (2006) Empirical evaluation of neutral theory. *Ecology*, **87**,1411–23.

McGill, B.J., Etienne, R.S., Gray, J.S., *et al.* (2007) Species abundance distributions: moving beyond single prediction theories to integration within an ecological framework. *Ecology Letters*, **10**, 995–1015.

Mittelbach, G.G., Steiner, C.F., Scheiner, S.M., *et al.* (2001) What is the observed relationship between species richness and productivity? *Ecology*, **82**, 2381–96.

Mouquet, N. and Loreau, M. (2003) Community patterns in source-sink metacommunities. *The American Naturalist*, **162**, 544–57.

Oksanen, T. (1990) Exploitation ecosystems in heterogeneous habitat complexes. *Evolutionary Ecology*, **4**, 220–34.

Ostman, O., Kneitel, J.M. and Chase, J.M. (2006) Disturbance alters habitat isolation's effect on biodiversity in aquatic microcosms. *Oikos*, **114**, 360–6.

Ostman, O., Griffin, N.W., Strasburg, J.L., *et al.* (2007) Habitat area affects arthropod communities directly and indirectly through top predators. *Ecography*, **30**, 359–66.

Park, T. (1948) Experimental studies of interspecies competition. I. Competition between populations of the flour beetles, *Tribolium confusum* Duval and *Tribolium castaneum* Herbst. *Ecological Monographs*, **18**, 265–308.

Park, T. (1954) Experimental studies of interspecies competition. II. Temperature, humidity, and competition in two species of *Tribolium*. *Physiological Zoology*, **27**, 177–238.

Polis, G.A., Power, M.E. and Huxel, G.R., eds. (2004) *Food webs dynamics at the landscape level*. University of Chicago Press, Chicago, IL.

Ricklefs, R.E. (1987) Community diversity: relative roles of local and regional processes. *Science*, **235**, 167–71.

Ricklefs, R.E. (2003) A comment on Hubbell's zero-sum ecological drift model. *Oikos*, **100**, 185–92.

Ricklefs, R.E. (2004) A comprehensive framework for global patterns in biodiversity. *Ecology Letters*, **7**, 1–15.

Ricklefs, R.E., and Schluter, D., eds. (1993) *Species Diversity in Ecological Communities: Historical and Geographical Perspectives*. University of Chicago Press, Chicago, IL.

Rosenzweig, M.L. (1995) *Species Diversity in Space and Time*. Cambridge University Press, Cambridge.

Rosenzweig, M.L. and Ziv, Y. (1999) The echo pattern of species diversity: pattern and processes. *Ecography*, **22**, 614–28.

Ryall, K.L. and Fahrig, L. (2006) Response of predators to loss and fragmentation of prey habitat: a review of theory. *Ecology*, **87**, 1086–93.

Ryberg, W.A. and Chase, J.M. (2007) Predator-dependant species-area curves. *The American Naturalist*, **170**, 636–42.

Scheiner, S.M. (2003) Six types of species-area curves. *Global Ecology and Biogeography*, **12**, 441–7.

Schoener, T.W. (1989) Food webs from the small to the large. *Ecology*, **70**, 1559–89.

Schoener, T.W., Spiller, D.A. and Losos, J.B. (2002) Predation on a common *Anolis* lizard: can the food-web effects of a devastating predator be reversed? *Ecological Monographs*, **72**, 383–407.

Semlitsch, R.D. (1998) Biological determination of terrestrial buffer zones for pond-breeding salamanders. *Conservation Biology*, **12**, 1113–19.

Shulman, R.S. and Chase, J.M. (2007) Increasing isolation reduces predator: prey species richness ratios in aquatic food webs. *Oikos*, **116**, 1581–7.

Shurin, J.B. and Srivastava, D.S. (2005) New perspectives on local and regional diversity: beyond saturation. In *Metacommunities: Spatial Dynamics and Ecological Communities* (eds M. Holyoak, M. Leibold and R. Holt), pp. 399–417. University of Chicago Press, Chicago, IL.

Simberloff, D.S. (1974) Equilibrium theory of island biogeography and ecology. *Annual Review of Ecology and Systematics*, **5**, 161–82.

Simberloff, D.S. and Wilson, E.O. (1969) Experimental zoogeography of islands: the colonization of empty islands. *Ecology*, **50**, 278–96.

Slatkin, M. (1974) Competition and regional coexistence. *Ecology*, **55**, 128–34.

Strong, D.R., Simberloff, D.S., Abele, L.G. and Thistle, A.B., eds. (1984) *Ecological Communities*. Princeton University Press, Princeton, NJ.

Terborgh, J.W. and Faaborg, J. (1980) Saturation of bird communities in the West Indies. *The American Naturalist*, **116**, 178–95.

Terborgh, J., Lopez, L., Nunez, P., *et al.* (2001) Ecological meltdown in predator-free forest fragments. *Science*, **294**, 1923–6.

Tilman, D. (1982) *Resource Competition and Community Structure*. Monographs in Population Biology. Princeton University Press, Princeton, NJ.

Tilman, D. (1994) Competition and biodiversity in spatially structured habitats. *Ecology*, **75**, 2–16.

Tilman, D. (1997) Community invasibility, recruitment limitation, and grassland biodiversity. *Ecology*, **78**, 81–92.

Tilman, D. (2004) Niche tradeoffs, neutrality, and community structure: a stochastic theory of resource competition, invasion, and community assembly. *Proceedings of the National Academy of Sciences of the United States of America*, **101**, 10854–61.

Urban, M.C. and Skelly, D.K. (2006) Evolving metacommunities: toward an evolutionary perspective on metacommunities. *Ecology*, **87**, 1616–26.

van de Koppel, J., van der Wal, D., Bakker, J.P. and Herman, P.J.M. (2005) Self-organization and vegetation collapse in salt-marsh ecosystems. *The American Naturalist*, **165**, E1–E12.

Watts, C.H. and Didham, R.K. (2006) Influences of habitat isolation on invertebrate colonisation of *Sporodanthus ferrugineus* in a mined peat bog. *Restoration Ecology*, **14**, 412–19.

Werner, E.E. and Gilliam, J.F. (1984) The ontogenetic niche and species interactions in size-structured populations. *Annual Review of Ecology and Systematics*, **15**, 393–425.

Whittaker, R.H. (1972) Evolution and measurement of species diversity. *Taxon*, **21**, 213–51.

Whittaker, R.J. and Fernández-Palacios, J.M. (2007) *Island Biogeography: Ecology, Evolution, and Conservation*, 2nd edn. Oxford University Press, Oxford.

Whittaker, R.J., Willis, K.J. and Field, R. (2001) Scale and species richness: towards a general, hierarchical theory of species diversity. *Journal of Biogeography*, **28**, 453–70.

Yu, D.W. and Wilson, H.B. (2001) The competition-colonization trade-off is dead: long live the competition-colonization trade-off. *The American Naturalist*, **158**, 49–63.

Zabel, J. and Tscharntke, T. (1998) Does fragmentation of *Urtica* habitats affect phytophagous and predatory insects differentially? *Oecologia*, **116**, 419–25.

Chapter 6

Bardgett, R.D., Yeates, G.W. and Anderson, J.M. (2005) Patterns and determinants of soil biological diversity. In *Biological Diversity and Function in Soil* (eds R.D. Bardgett, M.B. Usher and D.W. Hopkins), pp. 100–18. Cambridge University Press, Cambridge.

Bengtsson, J. (1994) Temporal predictability in forest soil communities. *Journal of Animal Ecology*, **63**, 653–65.

Bengtsson, J. and Berg, M.P. (2005) Variability in soil food web structure across time and space. In *Dynamic Food Webs* (eds P.C. de Ruiter, V. Wolters and J.C. Moore), pp. 201–10. Elsevier, Amsterdam.

Berg, B. and Matzner, E. (1997) The effect of N deposition on the mineralization of C from plant litter and humus. *Environmental Review*, **5**, 1–25.

Berg, B., Laskowski, R., Caswell, H., eds. (2005) *Litter Decomposition: a Guide to Carbon and Nutrient Turnover*. Advances in Ecological Research, vol. 38. Academic Press, Oxford.

Berg, M.P. and Bengtsson, J. (2007) Temporal and spatial variability in soil food web structure. *Oikos*, **116**, 1789–804.

Berg, M.P. and Verhoef, H.A. (1998) Ecological characteristics of a nitrogen-saturated coniferous forest in the Netherlands. *Biology and Fertility of Soils*, **26**, 258–67.

Berg, M.P., Kniese, J.P., Bedaux, J.J.M. and Verhoef, H.A. (1998a) Dynamics and stratification of functional groups of micro- and mesoarthropods in the organic layer of a Scots pine forest. *Biology and Fertility of Soils*, **26**, 268–84.

Berg, M.P., Kniese, J.P., Zoomer, R. and Verhoef, H.A. (1998b) Long-term decomposition of successive organic strata in a nitrogen saturated Scots pine forest soil. *Forest Ecology and Management*, **107**, 159–72.

Berg, M.P., de Ruiter, P.C., Didden, W., *et al.* (2001) Community food web, decomposition and nitrogen mineralisation in a stratified Scots pine forest soil. *Oikos*, **94**, 130–42.

Bosatta, E. and Ågren, G.I. (1991) Dynamics of carbon and nitrogen in the organic matter of the soil: a generic theory. *The American Naturalist*, **138**, 227–45.

Briones, M.J.I. and Ineson, P. (2002) Use of ^{14}C carbon dating to determine feeding behaviour of enchytraeids. *Soil Biology and Biochemistry*, **34**, 881–4.

Briones, M.J.I., Ineson, P. and Piearce, T.G. (1997) Effects of climate change on soil fauna; responses of enchytraeids, Diptera larvae and tardigrades in a transplant experiment. *Applied Soil Ecology*, **6**, 117–34.

Brose, U., Pavao-Zuckerman, M., Eklöf, A., *et al.* (2005) Spatial aspects of food webs. In *Dynamic Food Webs* (eds P.C. de Ruiter, V. Wolters and J.C. Moore), pp. 463–9. Elsevier, Amsterdam.

Chesson, A. (1997) Plant degradation by ruminants: parallels with litter decomposition in soils. In *Driven by Nature. Plant Litter Quality and Decomposition* (eds G. Gadisch and K.E. Giller), pp. 47–66. CAP International, Wallingford.

Closs, G.P. and Lake, P.S. (1994) Spatial and temporal variation in the structure of an intermittent-stream food web. *Ecological Monograph*, **64**, 1–21.

Cohen, J.E., Jonsson, T. and Carpenter, S.R. (2003) Ecological community description using food web, species abundance, and body-size. *Proceedings of the National Academy of Sciences of the United States of America*, **100**, 1781–6.

DeAngelis, D.L. (1992) *Dynamics of Nutrient Cycling and Food Webs*. Chapman & Hall, London.

de Ruiter, P.C., Moore, J.C., Zwart, K.B., *et al.* (1993) Simulation of nitrogen mineralization in the below-ground food webs of two winter wheat fields. *Journal of Applied Ecology*, **30**, 95–106.

de Ruiter, P.C., Neutel, A.M. and Moore, J.C. (2005) Energetics, patterns of interaction strengths, and stability in real ecosystems. *Science*, **269**, 1257–60.

Dilly, O. and Irmler, U. (1998) Succession in the food web during the decomposition of leaf litter in a black alder (*Alnus glutinosa* (Gaertn.) L.) forest. *Pedobiologia*, **42**, 109–23.

Ettema, C.H. and Wardle, D.A. (2002) Spatial soil ecology. *Trends in Ecology & Evolution*, **17**, 177–83.

Faber, J.H. (1991) Functional classification of soil fauna: a new approach. *Oikos*, **62**, 110–17.

Findlay, S., Pace, M. and Fisher, D. (1996) Spatial and temporal variability in the lower food web of the tidal freshwater Hudson River. *Estuaries*, **19**, 866–73.

Gadisch, G. and Giller, K.E. (1997) *Driven by Nature. Plant Litter Quality and Decomposition*. CAB International, Wallingford.

Hall, S.J. and Raffaelli, D.G. (1997) Food web patterns: what do we really know? In *Multitrophic Interactions in Terrestrial Systems* (eds A. Gange and A.C. Brown), pp. 395–416. Blackwell University Press, Cambridge.

Hättenschwiler, S. and Vitousek, P.M. (2000) The role of polyfenols in terrestrial nutrient cycling. *Trends in Ecology & Evolution*, **15**, 238–43.

Hättenschwiler, S., Tiunov, A.V. and Scheu, S. (2005) Biodiversity and litter decomposition in terrestrial ecosystems. *Annual Review of Ecology, Evolution and Systematics*, **36**, 191–218.

Hedlund, K., Griffiths, B., Christensen, S., *et al.* (2004) Trophic interactions in changing landscapes: responses of soil food webs. *Basic and Applied Ecology*, **5**, 495–503.

Heemsbergen, D.A., Berg, M.P., Loreau, M., *et al.* (2004) Biodiversity effects on soil processes explained by interspecific functional dissimilarity. *Science*, **306**, 1019–20.

Holt, R.D. (1996) Food webs in space: an island biogeographic perspective. In *Food Webs: Integration of Patterns and Dynamics* (eds G.A. Polis and K.O. Winemiller), pp. 313–23. Chapman & Hall, London.

Hooper, D.U., Chapin, F.S., Ewel, J.J., *et al.* (2005) Effect of biodiversity on ecosystem functioning: a consensus of current knowledge. *Ecological Monographs*, **75**, 3–35.

Hunt, H.W., Coleman, D.C., Ingham, E.R., *et al.* (1987) The detrital food web in a shortgrass prairie. *Biology and Fertility of Soils*, **3**, 57–68.

Jennings, S. and Mackinson, S. (2003) Abundance-body mass relationships in size-structured food webs. *Ecology Letters*, **6**, 971–4.

Jones, C.G. and Lawton, J.H. (1995) *Linking Species and Ecosystems*. Chapman & Hall, London.

Kendrick, W.B. and Burges, A. (1962) Biological aspects of the decay of *Pinus sylvestris* leaf litter. *Nova Hedwigia*, **4**, 313–44.

Klironomos, J.N., Rillig, M.C. and Allen, M.F. (1999) Designing belowground field experiments with help of semi-variance and power analyses. *Applied Soil Ecology*, **12**, 227–38.

Kondoh, M. (2003) Foraging adaptation and the relationship between food-web complexity and stability. *Science*, **299**, 1388–91.

Kondoh, M. (2005) Linking flexible food web structure to population stability: a theoretical consideration on adaptive food webs. In *Dynamic Food Webs* (eds P.C. de Ruiter, V. Wolters and J.C. Moore), pp. 101–13. Elsevier, Amsterdam.

Lavelle, P. and Spain, A.V. (2001) *Soil Ecology*. Kluwer Academic Publishers, Amsterdam.

Laverman, A.M., Borgers, P. and Verhoef, H.A. (2002) Spatial variation in net nitrate production in a N-saturated coniferous forest soil. *Forest Ecology and Management*, **161**, 123–32.

Legendre, P. and Legendre, L. (1998) *Numerical Ecology. Development in Environmental Modeling*, vol. 20. Elsevier, Amsterdam.

Loreau, M., Naeem, S. and Inchausti, P. (2002) *Biodiversity and Ecosystem Functioning: Synthesis and Perspectives*. Oxford University Press, Oxford.

McCann, K., Rasmussen, J., Umbanhowar, J. and Humphries, M. (2005) The role of space, time, and variability in food web dynamics. In *Dynamic Food Webs* (eds P.C. de Ruiter, V. Wolters and J.C. Moore), pp. 56–70. Elsevier, Amsterdam.

Moore, J.C., Walter, D.E. and Hunt, W.J. (1988) Arthropod regulation of micro- and mesobiota in below-ground detrital food webs. *Annual Review of Entomology*, **33**, 419–39.

Morin, P.J. (1999) *Community Ecology*. Blackwell Publishers, London.

Orwin, K.H., Wardle, D.A. and Greenfield, L.G. (2006) Context-dependent changes in the resistance and resilience of soil microbes to an experimental disturbance for three primary plant chronosequences. *Oikos*, **112**, 196–208.

Ostfeld, R.S. and Keesing, F. (2000) Pulsed resources and community dynamics of consumers in terrestrial ecosystems. *Trends in Ecology & Evolution*, **15**, 232–7.

Pimm, S.L. (1982) *Food Webs*. Chapman and Hall, London.

Pokarzhevskii, A.D., Van Straalen, N.M., Zaboev, D.P. and Zaitsev, A.S. (2003) Microbial links and element flows in nested detrital food-webs. *Pedobiologia*, **47**, 213–24.

Ponge, J.F. (1991) Succession of fungi and fauna during decomposition of needles in a small area of Scots pine litter. *Plant and Soil*, **138**, 99–113.

Rooney, N., McCann, K., Gellner, G. and Moore, J.C. (2006) Structural asymmetry and the stability of diverse food webs. *Nature*, **442**, 265–9.

Saetre, P. and Bååth, E. (2000) Spatial variation and patterns of the soil microbial community structure in a mixed spruce-birch stand. *Soil Biology and Biochemistry*, **32**, 909–17.

Schoenly, K. and Cohen, J.E. (1991) Temporal variation in food web structure: sixteen empirical cases. *Ecological Monographs*, **61**, 267–98.

Schröter, D., Wolters, V. and de Ruiter P.C. (2003) C and N mineralisation in the decomposer food webs of a European forest transect. *Oikos*, **102**, 294–308.

Setälä, H. and Aarnio, T. (2002) Vertical stratification and trophic interactions among organisms of a soil decomposer food web: a field experiment using ^{15}N as a tool. *European Journal of Soil Biology*, **38**, 29–34.

Shah, V. and Nerud, F. (2002) Lignin degrading system of white-rot fungi and its exploitation for dye decolorization. *Canadian Journal of Microbiology*, **48**, 857–70.

Teng, J. and McCann, K. (2004) The dynamics of compartmented and reticulate food webs in relation to energetic flows. *The American Naturalist*, **164**, 86–100.

Vanni, M.J. (2002) Nutrient cycling by animals in freshwater ecosystem. *Annual Review of Ecology and Systematics*, **33**, 341–70.

Wardle, D.A. (2002) *Communities and Ecosystems. Linking the Aboveground and Belowground Components*. Princeton University Press, Princeton, NJ.

Wardle, D.A. and Lavelle, P. (1997) Linkages between soil biota, plant litter quality and decomposition. In *Driven by Nature. Plant Litter Quality and Decomposition* (eds G. Gadisch and K.E. Giller), pp. 107–24. CAP International, Wallingford.

Warren, P.H. (1989) Spatial and temporal variation in the structure of a freshwater food web. *Oikos*, **55**, 299–311.

Chapter 7

Abraham, K.F., Jefferies, R.L. and Alisauskas, R.T. (2005) The dynamics of landscape change and snow geese in mid-continent North America. *Global Change Biology*, **11**, 841–55.

Al Mufti, M.M., Sydes, C.L., Furness, S.B., *et al.* (1977) Quantitative-analysis of shoot phenology and dominance in herbaceous vegetation. *Journal of Ecology*, **65**, 759–91.

Bakker, E.S., Ritchie, M.E., Olff, H., *et al.* (2006) Herbivore impact on grassland plant diversity depends on habitat productivity and herbivore size. *Ecology Letters*, **9**, 780–8.

Bakker, J.P. and Berendse, F. (1999) Constraints in the restoration of ecological diversity in grassland and heathland communities. *Trends in Ecology & Evolution*, **14**, 63–8.

Bardgett, R.D., Wardle, D.A. and Yeates, G.W. (1998) Linking above-ground and below-ground interactions: how plant responses to foliar herbivory influence soil organisms. *Soil Biology and Biochemistry*, **30**, 1867–78.

Bardgett, R.D., Bowman, W.D., Kaufmann, R. and Schmidt, S.K. (2005) A temporal approach to linking aboveground and belowground ecology. *Trends in Ecology & Evolution*, **20**, 634–41.

Bell, J.R., Traugott, M., Sunderland, K.D., *et al.* (2008) Beneficial links for the control of aphids: the effects of compost applications on predators and prey. *Journal of Applied Ecology*, **45**, 1266–73.

Bever, J.D. (2003) Soil community feedback and the coexistence of competitors: conceptual frameworks and empirical tests. *New Phytologist*, **157**, 465–73.

Bezemer, T.M. and van der Putten, W.H. (2007) Diversity and stability in plant communities. *Nature*, **446**, E6–E7.

Bezemer, T.M., De Deyn, G.B., Bossinga, T.M., *et al.* (2005) Soil community composition drives aboveground plant-herbivore-parasitoid interactions. *Ecology Letters*, **8**, 652–61.

Blomqvist, M.M., Olff, H., Blaauw, M.B., *et al.* (2000) Interactions between above- and belowground biota: importance for small-scale vegetation mosaics in a grassland ecosystem. *Oikos*, **90**, 582–98.

Blossey, B. and Nötzold, R. (1995) Evolution of increased competitive ability in invasive nonindigenous plants: a hypothesis. *Journal of Ecology*, **83**, 887–9.

Both, C. and Visser, M.E. (2001) Adjustment to climate change is constrained by arrival date in a long-distance migrant bird. *Nature*, **411**, 296–8.

Brown, V.K. and Gange, A.C. (1992) Secondary succession: how is it mediated by insect herbivory. *Vegetatio*, **101**, 3–13.

Burdon, J.J. and Marshall, D.R. (1981) Biological-control and the reproductive mode of weeds. *Journal of Applied Ecology*, **18**, 649–58.

Callaway, R.M. and Ridenour, W.M. (2004) Novel weapons: invasive success and the evolution of increased competitive ability. *Frontiers in Ecology and the Environment*, **2**, 436–43.

Cappuccino, N. and Arnason, J.T. (2006) Novel chemistry of invasive exotic plants. *Biology Letters*, **2**, 189–93.

Cardinale, B.J., Srivastava, D.S., Duffy, J.E., *et al.* (2006) Effects of biodiversity on the functioning of trophic groups and ecosystems. *Nature*, **443**, 989–92.

Davidson, D.W. (1993) The effects of herbivory and granivory on terrestrial plant succession. *Oikos*, **68**, 23–35.

De Deyn, G.B., Raaijmakers, C.E., Zoomer, H.R., *et al.* (2003) Soil invertebrate fauna enhances grassland succession and diversity. *Nature*, **422**, 711–13.

De Deyn, G.B., Raaijmakers, C.E. and van der Putten, W.H. (2004) Plant community development is affected by nutrients and soil biota. *Journal of Ecology*, **92**, 824–34.

Ehrlich, P.R. and Raven, P.H. (1964) Butterflies and plants: a study in coevolution. *Evolution*, **18**, 586–608.

Elton, C.S. (1958) *The Ecology of Invasions by Animals and Plants*. Methuen, London.

Engelkes, T., Morriën, E., Verhoeven, K.J.F., *et al.* (2008) Successful range expanding plants experience less above-ground and below-ground enemy impacts. *Nature*, **456**, 946–8.

Eppinga, M.B., Rietkerk, M., Dekker, S.C., *et al.* (2006) Accumulation of local pathogens: a new hypothesis to explain exotic plant invasions. *Oikos*, **114**, 168–76.

Fukami, T., Wardle, D.A., Bellingham, P.J., *et al.* (2006) Above- and below-ground impacts of introduced predators in seabird-dominated island ecosystems. *Ecology Letters*, **9**, 1299–307.

Gange, A.C., Brown, V.K. and Aplin, D.M. (2003) Multitrophic links between arbuscular mycorrhizal fungi and insect parasitoids. *Ecology Letters*, **6**, 1051–5.

Hairston, N.G., Smith, F.E. and Slobodkin, L.B. (1960) Community structure, population control, and competition. *The American Naturalist*, **94**, 421–5.

Hierro, J.L., Maron, J.L. and Callaway, R.M. (2005) A biogeographical approach to plant invasions: the importance of studying exotics in their introduced and native range. *Journal of Ecology*, **93**, 5–15.

Holtkamp, R., Kardol, P., Van der Wal, A., *et al.* (2008) Soil food web structure during ecosystem development after land abandonment. *Applied Soil Ecology*, **39**, 23–34.

Ineson, P., Levin, L.A., Kneib, R.T., *et al.* (2004) Cascading effects of deforestation on ecosystem services across soils and freshwater and marine sediments. In *Sustaining Biodiversity and Ecosystem Services in Soils and Sediments* (ed. D.H. Wall), pp. 225–48. Island Press, Washington, DC.

Jobin, A., Schaffner, U. and Nentwig, W. (1996) The structure of the phytophagous insect fauna on the introduced weed *Solidago altissima* in Switzerland. *Entomologia Experimentalis et Applicata*, **79**, 33–42.

Jonsson, T., Cohen, J.E. and Carpenter, S.R. (2005) Food webs, body size, and species abundance in ecological community description. *Advances in Ecological Research*, **36**, 1–84.

Karban, R. and Baldwin, I.T. (1997) *Induced Responses to Herbivory*. The University of Chicago Press, Chicago, IL.

Kardol, P., Bezemer, T.M. and van der Putten, W.H. (2006) Temporal variation in plant-soil feedback controls succession. *Ecology Letters*, **9**, 1080–8.

Kardol, P., Cornips, N.J., Van Kempen, M.M.L., *et al.* (2007) Microbe-mediated plant-soil feedback causes historical contingency effects in plant community assembly. *Ecological Monographs*, **77**, 147–62.

Kardol, P., Van der Wal, A., Bezemer, T.M., *et al.* (2008) Restoration of species-rich grasslands on ex-arable land: seed addition outweighs soil fertility reduction. *Biological Conservation*, **141**, 2208–17.

Keane, R.M. and Crawley, M.J. (2002) Exotic plant invasions and the enemy release hypothesis. *Trends in Ecology & Evolution*, **17**, 164–70.

Klironomos, J.N. (2002) Feedback with soil biota contributes to plant rarity and invasiveness in communities. *Nature*, **417**, 67–70.

Levine, J.M., Vila, M., D'antonio, C.M., *et al.* (2003) Mechanisms underlying the impacts of exotic plant invasions. *Proceedings of the Royal Society of London Series B-Biological Sciences*, **270**, 775–81.

Loreau, M., Naeem, S., Inchausti, P., *et al.* (2001) Biodiversity and ecosystem functioning: current knowledge and future challenges. *Science*, **294**, 804–8.

Malmstrom, C.M., McCullough, A.J., Johnson, H.A., *et al.* (2005) Invasive annual grasses indirectly increase virus incidence in California native perennial bunchgrasses. *Oecologia*, **145**, 153–64.

Mangla, S., Inderjit and Callaway, R.M. (2008) Exotic invasive plant accumulates native soil pathogens which inhibit native plants. *Journal of Ecology*, **96**, 58–67.

Marrs, R.H. (1993) Soil fertility and nature conservation in Europe: theoretical considerations and practical management solutions. *Advances in Ecological Research*, **24**, 241–300.

Memmott, J., Fowler, S.V., Paynter, Q., *et al.* (2000) The invertebrate fauna on broom, *Cytisus scoparius*, in two native and two exotic habitats. *Acta Oecologica-International Journal of Ecology*, **21**, 213–22.

Menendez, R., Gonzalez-Megias, A., Lewis, O.T., *et al.* (2008) Escape from natural enemies during climate-driven range expansion: a case study. *Ecological Entomology*, **33**, 413–21.

Mitchell, C.E. and Power, A.G. (2003) Release of invasive plants from fungal and viral pathogens. *Nature*, **421**, 625–7.

Moore, J.C., McCann, K., Setälä, H. and de Ruiter, P.C. (2003) Top-down is bottom-up: does predation in the rhizosphere regulate aboveground dynamics? *Ecology*, **84**, 846–57.

Neutel, A.M., Heesterbeek, J.A.P., van de Koppel, J., *et al.* (2007) Reconciling complexity with stability in naturally assembling food webs. *Nature*, **449**, 599–U511.

Olff, H. and Ritchie, M.E. (1998) Effects of herbivores on grassland plant diversity. *Trends in Ecology & Evolution*, **13**, 261–5.

Reinhart, K.O., Packer, A., van der Putten, W.H. and Clay, K. (2003) Plant-soil biota interactions and spatial distribution of black cherry in its native and invasive ranges. *Ecology Letters*, **6**, 1046–50.

Richardson, D.M., Allsopp, N., D'Antonio, C.M., *et al.* (2000) Plant invasions: the role of mutualisms. *Biological Reviews*, **75**, 65–93.

Sanchez-Pinero, F. and Polis, G.A. (2000) Bottom-up dynamics of allochthonous input: direct and indirect effects of seabirds on islands. *Ecology*, **81**, 3117–32.

Schädler, M., Jung, G., Brandl, R. and Auge, H. (2004) Secondary succession is influenced by belowground insect herbivory on a productive site. *Oecologia*, **138**, 242–52.

Scheffer, M. and van Nes, E.H. (2006) Self-organized similarity, the evolutionary emergence of groups of similar species. *Proceedings of the National Academy of Sciences of the United States of America*, **103**, 6230–5.

Schmitz, O.J., Kalies, E.L. and Booth, M.G. (2006) Alternative dynamic regimes and trophic control of plant succession. *Ecosystems*, **9**, 659–72.

Soler, R., Bezemer, T.M., Cortesero, A.M., *et al.* (2007) Impact of foliar herbivory on the development of a root-feeding insect and its parasitoid. *Oecologia*, **152**, 257–64.

Stinson, K.A., Campbell, S.A., Powell, J.R., *et al.* (2006) Invasive plant suppresses the growth of native tree seedlings by disrupting belowground mutualisms. *PLoS Biology*, **4**, 727–31.

Stohlgren, T.J., Binkley, D., Chong, G.W., *et al.* (1999) Exotic plant species invade hot spots of native plant diversity. *Ecological Monographs*, **69**, 25–46.

Suding, K.N., Gross, K.L. and Houseman, G.R. (2004) Alternative states and positive feedbacks in restoration ecology. *Trends in Ecology & Evolution*, **19**, 46–53.

Tilman, D. (1982) *Resource Competition and Community Structure*. Princeton University Press, Princeton, NJ.

Tscharntke, T. (1997) Vertebrate effects on plant-invertebrate food webs. In *Multitrophic Interactions in Terrestrial Systems* (eds A.C. Gange and V.K. Brown), pp. 277–97. Blackwell Science Ltd, Oxford.

Tscharntke, T. and Hawkins, B.A. (2002) *Multitrophic Level Interactions*. Cambridge University Press, Cambridge.

Tscharntke, T., Klein, A.M., Kruess, A., *et al.* (2005) Landscape perspectives on agricultural intensification and biodiversity: ecosystem service management. *Ecology Letters*, **8**, 857–74.

van de Koppel, J., Bardgett, R.D., Bengtsson, J., *et al.* (2005) The effects of spatial scale on trophic interactions. *Ecosystems*, **8**, 801–7.

van der Putten, W.H., Vet, L.E.M., Harvey, J.A. and Wackers, F.L. (2001) Linking above- and belowground multitrophic interactions of plants, herbivores, pathogens, and their antagonists. *Trends in Ecology & Evolution*, **16**, 547–54.

van der Putten, W.H., Kowalchuk, G.A., Brinkman, E.P., *et al.* (2007) Soil feedback of exotic savanna grass relates to pathogen absence and mycorrhizal selectivity. *Ecology*, **88**, 978–88.

van der Wal, A., van Veen, J.A., Smant, W., *et al.* (2006) Fungal biomass development in a chronosequence of land abandonment. *Soil Biology and Biochemistry*, **38**, 51–60.

van Grunsven, R.H.A., van der Putten, W.H., Bezemer, T.M., *et al.* (2007) Reduced plant-soil feedback of plant species expanding their range as compared to natives. *Journal of Ecology*, **95**, 1050–7.

van Ruijven, J., De Deyn, G.B., Raaijmakers, C.E., *et al.* (2005) Interactions between spatially separated herbivores indirectly alter plant diversity. *Ecology Letters*, **8**, 30–7.

Visser, M.E. and Holleman, L.J.M. (2001) Warmer springs disrupt the synchrony of oak and winter moth phenology. *Proceedings of the Royal Society of London Series B-Biological Sciences*, **268**, 289–94.

Vitousek, P.M., Walker, L.R., Whiteaker, L.D., *et al.* (1987) Biological invasion by *Myrica faya* alters ecosystem development in Hawaii. *Science*, **238**, 802–4.

Wardle, D.A., Bardgett, R.D., Klironomos, J.N., *et al.* (2004) Ecological linkages between aboveground and belowground biota. *Science*, **304**, 1629–33.

Williamson, M. (1996) *Biological Invasions*. Chapman and Hall, London.

Wolfe, L.M., Elzinga, J.A. and Biere, A. (2004) Increased susceptibility to enemies following introduction in the invasive plant *Silene latifolia*. *Ecology Letters*, **7**, 813–20.

Chapter 8

Bakun, A.(2006) Wasp-waist populations and marine ecosystem dynamics: Navigating the 'predator pit' topographies. *Progress in Oceanography*, **68**, 271–88.

Barkai, A. and McQuaid, C. (1988) Predator-prey role reversal in a marine benthic ecosystem. *Science*, **242**, 62–4.

Barrett, J.H., Locker, A.M. and Roberts, C.M. (2004) The origins of intensive marine fishing in medieval Europe: the English evidence. *Proceedings of the Royal Society of London Series B-Biological Sciences*, **271**, 2417–21.

Bascompte, J., Melian, C.J. and Sala, E. (2005) Interaction strength combinations and the overfishing of a marine food web. *Proceedings of the National Academy of Sciences of the United States of America*, **102**, 5443–7.

Baum, J.K., Myers, R.A., Kehler, D.G., *et al.* (2003) Collapse and conservation of shark populations in the Northwest Atlantic. *Science*, **299**, 389–92.

Berlow, E.L., Neutel, A.M., Cohen, J.E., *et al.* (2004) Interaction strengths in food webs: issues and opportunities. *Journal of Animal Ecology*, **73**, 585–98.

Borer, E.T., Seabloom, E.W., Shurin, J.B., *et al.* (2005) What determines the strength of a trophic cascade? *Ecology*, **86**, 528–37.

Borer, E.T., Halpern, B.S. and Seabloom, E.W. (2006) Asymmetry in community regulation: effects of predators and productivity. *Ecology*, **87**, 2813–20.

Botsford, L.W. (1981) The effects of increased individual growth rates on depressed population size. *The American Naturalist*, **117**, 38–63.

Botsford, L.W., Castilla, J.C. and Peterson, C.H. (1997) The management of fisheries and marine ecosystems. *Science*, **277**, 509–15.

Brett, M.T. and Goldman, C.R. (1996) A meta-analysis of the freshwater trophic cascade. *Proceedings of the National Academy of Sciences of the United States of America*, **93**, 7723–6.

Brose, U., Jonsson, T., Berlow, E.L., *et al.* (2006) Consumer-resource body-size relationships in natural food webs. *Ecology*, **87**, 2411–17.

Byrnes, J.E., Reynolds, P.L. and Stachowicz, J.J. (2007) Invasions and extinctions reshape coastal marine food webs. *PLoS ONE*, **2**, e295.

Cardillo, M., Mace, G.M., Jones, K.E., *et al.* (2005) Multiple causes of high extinction risk in large mammal species. *Science*, **309**, 1239–41.

Carpenter, R.C. (1990) Mass mortality of *Diadema antillarum*. I. Long-term effects on sea urchin population-dynamics and coral reef algal communities. *Marine Biology*, **104**, 67–77.

Carpenter, S.R. (1996) Microcosm experiments have limited relevance for community and ecosystem ecology. *Ecology*, **77**, 677–80.

Carpenter, S.R. and Kitchell, J.F. (1993) *The Trophic Cascade in Lakes*. Cambridge University Press, Cambridge.

Carpenter, S.R., Kitchell, J.F. and Hodgson, J.R. (1985) Cascading trophic interactions and lake productivity. *BioScience*, **35**, 634–9.

Carr, M.H., Anderson, T.W. and Hixon, M.A. (2002) Biodiversity, population regulation, and the stability of coral-reef fish communities. *Proceedings of the National Academy of Sciences of the United States of America*, **99**, 11241–5.

Cebrian, J. (1999) Patterns in the fate of production in plant communities. *The American Naturalist*, **154**, 449–68.

Christensen, V. and Walters, C.J. (2004) Ecopath with Ecosim: methods, capabilities and limitations. *Ecological Modelling*, **172**, 109–39.

Clark, J.S., Carpenter, S.R., Barber, M., *et al.* (2001) Ecological forecasts: an emerging imperative. *Science*, **293**, 657–60.

Cloern, J.E. (2001) Our evolving conceptual model of the coastal eutrophication problem. *Marine Ecology-Progress Series*, **210**, 223–53.

Cohen, J.E., Jonsson, T. and Carpenter, S.R. (2003) Ecological community description using the food web, species abundance, and body size. *Proceedings of the National Academy of Sciences of the United States of America*, **100**, 1781–6.

Collie, J.S., Richardson, K. and Steele, J.H. (2004) Regime shifts: can ecological theory illuminate the mechanisms? *Progress in Oceanography*, **60**, 281–302.

Conover, D.O. and Munch, S.B. (2002) Sustaining fisheries yields over evolutionary time scales. *Science*, **297**, 94–6.

Cottingham, K.L., Brown, B.L. and Lennon, J.T. (2001) Biodiversity may regulate the temporal variability of ecological systems. *Ecology Letters*, **4**, 72–85.

Crooks, K.R. and Soule, M.E. (1999) Mesopredator release and avifaunal extinctions in a fragmented system. *Nature*, **400**, 563–6.

Cury, P., Bakun, A., Crawford, R.J.M., *et al.* (2000) Small pelagics in upwelling systems: patterns of interaction and structural changes in 'wasp-waist' ecosystems. *ICES Journal of Marine Science*, **57**, 603–18.

Daskalov, G.M. (2002) Overfishing drives atrophic cascade in the Black Sea. *Marine Ecology-Progress Series*, **225**, 53–63.

Daskalov, G.M., Grishin, A.N., Rodionov, S. and Mihneva, V. (2007) Trophic cascades triggered by overfishing reveal possible mechanisms of ecosystem regime shifts. *Proceedings of the National Academy of Sciences of the United States of America*, **104**, 10518–23.

Davenport, A.C. and Anderson, T.W. (2007) Positive indirect effects of reef fishes on kelp performance: the importance of mesograzers. *Ecology*, **88**, 1548–61.

Dayton, P.K., Tegner, M.J., Edwards, P.B. and Riser, K.L. (1998) Sliding baselines, ghosts, and reduced expectations in kelp forest communities. *Ecological Applications*, **8**, 309–22.

de Roos, A.M., Boukal, D.S. and Persson, L. (2006) Evolutionary regime shifts in age and size at maturation of exploited fish stocks. *Proceedings of the Royal Society B-Biological Sciences*, **273**, 1873–80.

de Ruiter, P.C., Wolters, V. and Moore, J.C. (2005) *Dynamic Food Webs. Multispecies Assemblages, Ecosystem Development and Environmental Change*. Academic Press, New York, NY.

Deason, E.E. and Smayda, T.J. (1982) Ctenophore-zooplankton-phytoplankton interactions in Narragansett Bay, Rhode Island, USA, during 1972–1977. *Journal of Plankton Research*, **4**, 203–17.

Del Giorgio, P.A. and Gasol, J.M. (1995) Biomass distribution in fresh-water plankton communities. *The American Naturalist*, **146**, 135–52.

Dirzo, R. and Raven, P.H. (2003) Global state of biodiversity and loss. *Annual Review of Environment and Resources*, **28**, 137–67.

Doak, D.F., Bigger, D., Harding, E.K., *et al.* (1998) The statistical inevitability of stability-diversity relationships in community ecology. *The American Naturalist*, **151**, 264–76.

Dobson, A., Lodge, D., Alder, J., *et al.* (2006) Habitat loss, trophic collapse, and the decline of ecosystem services. *Ecology*, **87**, 1915–24.

Duffy, J.E. (2002) Biodiversity and ecosystem function: the consumer connection. *Oikos*, **99**, 201–19.

Duffy, J.E. (2003) Biodiversity loss, trophic skew and ecosystem functioning. *Ecology Letters*, **6**, 680–7.

Duffy, J.E. and Hay, M.E. (2000) Strong impacts of grazing amphipods on the organization of a benthic community. *Ecological Monographs*, **70**, 237–63.

Duffy, J.E. and Stachowicz, J.J. (2006) Why biodiversity is important to oceanography: potential roles of genetic, species, and trophic diversity in pelagic ecosystem processes. *Marine Ecology-Progress Series*, **311**, 179–89.

Duffy, J.E., Richardson, J.P. and France, K.E. (2005) Ecosystem consequences of diversity depend on food chain length in estuarine vegetation. *Ecology Letters*, **8**, 301–9.

Duffy, J.E., Cardinale, B.J., France, K.E., *et al.* (2007) The functional role of biodiversity in ecosystems: incorporating trophic complexity. *Ecology Letters*, **10**, 522–38.

Dulvy, N.K., Sadovy, Y. and Reynolds, J.D. (2003) Extinction vulnerability in marine populations. *Fish and Fisheries*, **4**, 25–64.

Dulvy, N.K., Freckleton, R.P. and Polunin, N.V.C. (2004) Coral reef cascades and the indirect effects of predator removal by exploitation. *Ecology Letters*, **7**, 410–16.

Dunne, J.A., Williams, R.J. and Martinez, N.D. (2004) Network structure and robustness of marine food webs. *Marine Ecology-Progress Series*, **273**, 291–302.

Elton, C.S. (1958) *The Ecology of Invasions by Animals and Plants*. Methuen and Co., London.

Emmerson, M.C. and Huxham, M. (2002) How can marine ecology contribute to the biodiversity-ecosystem functioning debate? In *Biodiversity and Ecosystem Functioning: Synthesis and Perspectives* (eds M. Loreau, S. Naeem and P. Inchausti), pp. 139–46. Oxford University Press, New York, NY.

Emmerson, M.C. and Raffaelli, D. (2004) Predator-prey body size, interaction strength and the stability of a real food web. *Journal of Animal Ecology*, **73**, 399–409.

Emmerson, M.C. and Yearsley, J.M. (2004) Weak interactions, omnivory and emergent food-web properties. *Proceedings of the Royal Society of London Series B-Biological Sciences*, **271**, 397–405.

Essington, T.E., Beaudreau, A.H. and Wiedenmann, J. (2006) Fishing through marine food webs. *Proceedings of the National Academy of Sciences of the United States of America*, **103**, 3171–5.

Estes, J.A. and Duggins, D.O. (1995) Sea otters and kelp forests in Alaska: generality and variation in a community ecological paradigm. *Ecological Monographs*, **65**, 75–100.

Estes, J.A. and Palmisano, J.F. (1974) Sea otters: their role in structuring nearshore communities. *Science*, **185**, 1058–60.

Estes, J.A., Tinker, M.T., Williams, T.M. and Doak, D.F. (1998) Killer whale predation on sea otters linking oceanic and nearshore ecosystems. *Science*, **282**, 473–6.

FAO. (2007) *The State of World Fisheries and Aquaculture 2006*. Food and Agriculture Organization of the United Nations, Rome.

Feng, J.F., Wang, H.L., Huang, D.W. and Li, S.P. (2006) Alternative attractors in marine ecosystems: a comparative analysis of fishing effects. *Ecological Modelling*, **195**, 377–84.

Folke, C., Carpenter, S., Walker, B., *et al.* (2004) Regime shifts, resilience, and biodiversity in ecosystem management. *Annual Review of Ecology Evolution and Systematics*, **35**, 557–81.

France, K.E. and Duffy, J.E. (2006) Consumer diversity mediates invasion dynamics at multiple trophic levels. *Oikos*, **113**, 515–29.

Frank, K.T., Petrie, B., Choi, J.S. and Leggett, W.C. (2005) Trophic cascades in a formerly cod-dominated ecosystem. *Science*, **308**, 1621–3.

Frank, K.T., Petrie, B., Shackell, N.L. and Choi, J.S. (2006) Reconciling differences in trophic control in mid-latitude marine ecosystems. *Ecology Letters*, **9**, 1096–105.

Frank, K.T., Petrie, B. and Shackell, N.L. (2007) The ups and downs of trophic control in continental shelf ecosystems. *Trends in Ecology & Evolution*, **22**, 236–42.

Friedlander, A.M. and DeMartini, E.E. (2002) Contrasts in density, size, and biomass of reef fishes between the northwestern and the main Hawaiian islands: the effects of fishing down apex predators. *Marine Ecology-Progress Series*, **230**, 253–64.

Guidetti, P. (2006) Marine reserves reestablish lost predatory interactions and cause community changes in rocky reefs. *Ecological Applications*, **16**, 963–76.

Haahtela, I. (1984) A hypothesis of the decline of the bladder wrack (*Fucus vesiculosus* L.) in SW Finland in 1975–1981. *Limnologica*, **15**, 345–50.

Hairston, N.G., Smith, F.E. and Slobodkin, L.G. (1960) Community structure, population control, and competition. *The American Naturalist*, **94**, 421–5.

Hairston, N.G., Ellner, S.P., Geber, M.A., *et al.* (2005) Rapid evolution and the convergence of ecological and evolutionary time. *Ecology Letters*, **8**, 1114–27.

Halpern, B.S. and Warner, R.R. (2002) Marine reserves have rapid and lasting effects. *Ecology Letters*, **5**, 361–6.

Halpern, B.S., Borer, E.T., Seabloom, E.W. and Shurin, J.B. (2005) Predator effects on herbivore and plant stability. *Ecology Letters*, **8**, 189–94.

Hay, M.E. (1984) Patterns of fish and urchin grazing on Caribbean coral reefs: are previous results typical? *Ecology Letters*, **65**, 446–54.

Hay, M.E. and Taylor, P.R. (1985) Competition between herbivorous fishes and urchins on Caribbean reefs. *Oecologia*, **65**, 591–8.

Hays, G.C., Richardson, A.J. and Robinson, C. (2005) Climate change and marine plankton. *Trends in Ecology & Evolution*, **20**, 337–44.

Hector, A. and Hooper, R. (2002) Ecology: Darwin and the first ecological experiment. *Science*, **295**, 639–40.

Hilborn, R., Branch, T.A., Ernst, B., *et al.* (2003a) State of the world's fisheries. *Annual Review of Environment and Resources*, **28**, 359–99.

Hilborn, R., Quinn, T.P., Schindler, D.E. and Rogers, D.E. (2003b) Biocomplexity and fisheries sustainability. *Proceedings of the National Academy of Sciences of the United States of America*, **100**, 6564–8.

Hillebrand, H. and Cardinale, B.J. (2004) Consumer effects decline with prey diversity. *Ecology Letters*, **7**, 192–201.

Hixon, M.A. and Carr, M.H. (1997) Synergistic predation, density dependence, and population regulation in marine fish. *Science*, **277**, 946–9.

Hobbs, R.J., Arico, S., Aronson, J., *et al.* (2006) Novel ecosystems: theoretical and management aspects of the new ecological world order. *Global Ecology and Biogeography*, **15**, 1–7.

Hsieh, C.H., Reiss, C.S., Hunter, J.R., *et al.* (2006) Fishing elevates variability in the abundance of exploited species. *Nature*, **443**, 859–62.

Hughes, T.P. (1994) Catastrophes, phase shifts, and large-scale degradation of a Caribbean coral reef. *Science*, **265**, 1547–51.

Hunt, G.L. and McKinnell, S. (2006) Interplay between top-down, bottom-up, and wasp-waist control in marine ecosystems. *Progress in Oceanography*, **68**, 115–24.

Hutchings, J.A. and Baum, J.K. (2005) Measuring marine fish biodiversity: temporal changes in abundance, life history and demography. *Philosophical Transactions of the Royal Society B-Biological Sciences*, **360**, 315–38.

Hutchings, J.A. and Reynolds, J.D. (2004) Marine fish population collapses: consequences for recovery and extinction risk. *BioScience*, **54**, 297–309.

Jackson, J.B.C., Kirby, M.X., Berger, W.H., *et al.* (2001) Historical overfishing and the recent collapse of coastal ecosystems. *Science*, **293**, 629–38.

Jennings, S. and Kaiser, M.J. (1998) The effects of fishing on marine ecosystems. *Advances in Marine Biology*, **34**, 201–352.

Jennings, S., Reynolds, J.D. and Mills, S.C. (1998) Life history correlates of responses to fisheries exploitation. *Proceedings of the Royal Society of London Series B-Biological Sciences*, **265**, 333–9.

Jennings, S., Greenstreet, S.P.R. and Reynolds, J.D. (1999a) Structural change in an exploited fish community: a consequence of differential fishing effects on species with contrasting life histories. *Journal of Animal Ecology*, **68**, 617–27.

Jennings, S., Reynolds, J.D. and Polunin, N.V.C. (1999b) Predicting the vulnerability of tropical reef fishes to exploitation with phylogenies and life histories. *Conservation Biology*, **13**, 1466–75.

Jennings, S., Pinnegar, J.K., Polunin, N.V.C. and Boon, T.W. (2001) Weak cross-species relationships between body size and trophic level belie powerful size-based trophic structuring in fish communities. *Journal of Animal Ecology*, **70**, 934–44.

Johnson, C.N., Isaac, J.L. and Fisher, D.O. (2007) Rarity of a top predator triggers continent-wide collapse of mammal prey: dingoes and marsupials in Australia. *Proceedings of the Royal Society B-Biological Sciences*, **274**, 341–6.

Kangas, P., Autio, H., Haellfors, G., *et al.* (1982) A general model of the decline of *Fucus vesiculosus* at Tvaerminne, South Coast of Finland in 1977–81. *Acta Botanica Fennica*, **118**, 1–27.

Knowlton, N. (1992) Thresholds and multiple stable states in coral reef community dynamics. *American Zoologist*, **32**, 674–82.

Knowlton, N. (2004) Multiple 'stable' states and the conservation of marine ecosystems. *Progress in Oceanography*, **60**, 387–96.

Law, R. (2000) Fishing, selection, and phenotypic evolution. *ICES Journal of Marine Science*, **57**, 659–68.

Leibold, M.A. (1989) Resource edibility and the effects of predators and productivity on the outcome of trophic interactions. *The American Naturalist*, **134**, 922–49.

Leibold, M.A. (1996) A graphical model of keystone predators in food webs: trophic regulation of abundance, incidence, and diversity patterns in communities. *The American Naturalist*, **147**, 784–812.

Leibold, M.A., Holyoak, M., Mouquet, N., *et al.* (2004) The metacommunity concept: a framework for multi-scale community ecology. *Ecology Letters*, **7**, 601–13.

Lewontin, R.C. (1969) The meaning of stability. *Brookhaven Symposia in Biology* **22**, 13–23.

Loreau, M., Downing, A., Emmerson, M., *et al.* (2002) A new look at the relationship between diversity and stability. In *Biodiversity and Ecosystem Functioning: Synthesis and Perspectives* (eds M. Loreau, S. Naeem and P. Inchausti), pp. 79–91. Oxford University Press, New York, NY.

Lotze, H.K., Reise, K., Worm, B., *et al.* (2005) Human transformations of the Wadden Sea ecosystem through time: a synthesis. *Helgoland Marine Research*, **59**, 84–95.

MacArthur, R. (1955) Fluctuations of animal populations and a measure of community stability. *Ecology Letters*, **36**, 533–6.

McCann, K.S. (2000) The diversity-stability debate. *Nature*, **405**, 228–33.

McCann, K.S., Hastings, A. and Huxel, G.R. (1998) Weak trophic interactions and the balance of nature. *Nature*, **395**, 794–8.

McClanahan, T.R., Kamukuru, A.T., Muthiga, N.A., *et al.* (1996) Effect of sea urchin reductions on algae, coral, and fish populations. *Conservation Biology*, **10**, 136–54.

Menge, B.A. (1995) Indirect effects in marine rocky intertidal interaction webs: patterns and importance. *Ecological Monographs*, **65**, 21–74.

Micheli, F. (1999) Eutrophication, fisheries, and consumer-resource dynamics in marine pelagic ecosystems. *Science*, **285**, 1396–8.

Micheli, F., Cottingham, K.L., Bascompte, J., *et al.* (1999) The dual nature of community variability. *Oikos*, **85**, 161–9.

Micheli, F., Halpern, B.S., Botsford, L.W. and Warner, R.R. (2004) Trajectories and correlates of community change in no-take marine reserves. *Ecological Applications*, **14**, 1709–23.

Mork, M. (1996) The effect of kelp in wave damping. *Sarsia*, **80**, 323–7.

Myers, R.A. and Worm, B. (2003) Rapid worldwide depletion of predatory fish communities. *Nature*, **423**, 280–3.

Myers, R.A., Baum, J.K., Shepherd, T.D., *et al.* (2007) Cascading effects of the loss of apex predatory sharks from a coastal ocean. *Science*, **315**, 1846–50.

Odum, E.P. (1971) *Fundamentals of Ecology*, 3rd edn. W.B. Saunders Company, Philadelphia, PA.

Oksanen, L. (2001) Logic of experiments in ecology: is pseudoreplication a pseudoissue? *Oikos*, **94**, 27–38.

Olsen, E.M., Heino, M., Lilly, G.R., *et al.* (2004) Maturation trends indicative of rapid evolution preceded the collapse of northern cod. *Nature*, **428**, 932–5.

Orr, J.C., Fabry, V.J., Aumont, O., *et al.* (2005) Anthropogenic ocean acidification over the twenty-first century and its impact on calcifying organisms. *Nature*, **437**, 681–6.

Overland, J.E., Percival, D.B. and Mofjeld H.O. (2006) Regime shifts and red noise in the North Pacific. *Deep-Sea Research Part I-Oceanographic Research Papers*, **53**, 582–8.

Pace, M.L., Cole, J.J., Carpenter, S.R. and Kitchell, J.F. (1999) Trophic cascades revealed in diverse ecosystems. *Trends in Ecology & Evolution*, **14**, 483–8.

Paine, R.T. (1980) Food webs: linkage, interaction strength and community infrastructure. *Journal of Animal Ecology*, **49**, 667–85.

Parker, J.D., Burkepile, D.E. and Hay, M.E. (2006) Opposing effects of native and exotic herbivores on plant invasions. *Science*, **311**, 1459–61.

Parmesan, C. (2006) Ecological and evolutionary responses to recent climate change. *Annual Review of Ecology Evolution and Systematics*, **37**, 637–69.

Pauly, D. and Christensen, V. (1995) Primary production required to sustain global fisheries. *Nature*, **374**, 255–7.

Pauly, D., Christensen, V., Dalsgaard, J., *et al.* (1998) Fishing down marine food webs. *Science*, **279**, 860–3.

Pederson, H.G. and Johnson, C.R. (2006) Predation of the sea urchin *Heliocidaris erythrogramma* by rock lobsters (*Jasus edwardsii*) in no-take marine reserves. *Journal of Experimental Marine Biology and Ecology*, **336**, 120–34.

Perry, A.L., Low, P.J., Ellis, J.R. and Reynolds, J.D. (2005) Climate change and distribution shifts in marine fishes. *Science*, **308**, 1912–15.

Petchey, O.L., Downing, A.L., Mittelbach, G.G., *et al.* (2004) Species loss and the structure and functioning of multitrophic aquatic systems. *Oikos*, **104**, 467–78.

Peterson, C.H. (1984) Does a rigorous criterion for environmental identity preclude the existence of multiple stable points? *The American Naturalist*, **124**, 127–33.

Polis, G.A. (1999) Why are parts of the world green? Multiple factors control productivity and the distribution of biomass. *Oikos*, **86**, 3–15.

Polis, G.A., Anderson, W.B. and Holt, R.D. (1997) Toward an integration of landscape and food web ecology: the dynamics of spatially subsidized food webs. *Annual Review of Ecology and Systematics*, **28**, 289–316.

Power, M.E. (1990) Effects of fish in river food webs. *Science*, **250**, 811–14.

Raffaelli, D.G. and Hall, S.J. (1996) Assessing the relative importance of trophic links in food webs. In *Food Webs: Integration of Patterns and Dynamics* (eds G. Polis and K. Winemiller), pp. 185–91. Chapman and Hall, New York, NY.

Reznick, D.N. and Ghalambor, C.K. (2005) Can commercial fishing cause evolution? Answers from guppies (*Poecilia reticulata*). *Canadian Journal of Fisheries and Aquatic Sciences*, **62**, 791–801.

Ruiz, G.M., Fofonoff, P.W., Carlton, J.T., *et al.* (2000) Invasion of coastal marine communities in North America: apparent patterns, processes, and biases. *Annual Review of Ecology and Systematics*, **31**, 481–531.

Sala, E. and Graham, M.H. (2002) Community-wide distribution of predator-prey interaction strength in kelp forests. *Proceedings of the National Academy of Sciences of the United States of America*, **99**, 3678–83.

Sala, E., Boudouresque, C.F. and Harmelin-Vivien, M. (1998) Fishing, trophic cascades, and the structure of algal assemblages: evaluation of an old but untested paradigm. *Oikos*, **82**, 425–39.

Sax, D.F. and Gaines, S.D. (2003) Species diversity: from global decreases to local increases. *Trends in Ecology & Evolution*, **18**, 561–6.

Sax, D.F., Stachowicz, J.J. and Gaines, S.D. (2005) *Species Invasions: Insights into Ecology, Evolution, and Biogeography*. Sinauer Associates, Sunderland, MA.

Scheffer, M. and Carpenter, S.R. (2003) Catastrophic regime shifts in ecosystems: linking theory to observation. *Trends in Ecology & Evolution*, **18**, 648–56.

Scheffer, M. and Jeppesen, E. (2007) Regime shifts in shallow lakes. *Ecosystems*, **10**, 1–3.

Scheffer, M., Carpenter, S.R., Foley, J.A., *et al.* (2001) Catastrophic shifts in ecosystems. *Nature*, **413**, 591–6.

Schiel, D.R., Steinbeck, J.R. and Foster, M.S. (2004) Ten years of induced ocean warming causes comprehensive changes in marine benthic communities. *Ecology*, **85**, 1833–9.

Shackell, N.L. and Frank, K.T. (2007) Compensation in exploited marine fish communities on the Scotian Shelf, Canada. *Marine Ecology-Progress Series*, **336**, 235–47.

Shears, N.T. and Babcock, R.C. (2003) Continuing trophic cascade effects after 25 years of no-take marine reserve protection. *Marine Ecology-Progress Series*, **246**, 1–16.

Shiomoto, A., Tadokoro, K., Nagasawa, K. and Ishida, Y. (1997) Trophic relations in the subarctic North Pacific ecosystem: possible feeding effect from pink salmon. *Marine Ecology-Progress Series*, **150**, 75–85.

Shurin, J.B., Borer, E.T., Seabloom, E.W., *et al.* (2002) A cross-ecosystem comparison of the strength of trophic cascades. *Ecology Letters*, **5**, 785–91.

Shurin, J.B., Gruner, D.S. and Hillebrand, H. (2006) All wet or dried up? Real differences between aquatic and terrestrial food webs. *Proceedings of the Royal Society B-Biological Sciences*, **273**, 1–9.

Silliman, B.R. and Bertness, M.D. (2002) A trophic cascade regulates salt marsh primary production. *Proceedings of the National Academy of Sciences of the United States of America*, **99**, 10500–5.

Soule, M.E., Estes, J.A., Miller, B. and Honnold, D.L. (2005) Strongly interacting species, conservation policy, management, and ethics. *BioScience*, **55**, 168–76.

Srivastava, D.S. and Vellend, M. (2005) Biodiversity-ecosystem function research: is it relevant to conservation? *Annual Review of Ecology Evolution and Systematics*, **36**, 267–94.

Stachowicz, J.J., Whitlatch, R.B. and Osman, R.W. (1999) Species diversity and invasion resistance in a marine ecosystem. *Science*, **286**, 1577–9.

Stachowicz, J.J., Fried, H., Osman, R.W. and Whitlatch, R.B. (2002a) Biodiversity, invasion resistance, and marine ecosystem function: reconciling pattern and process. *Ecology*, **83**, 2575–90.

Stachowicz, J.J., Terwin, J.R., Whitlatch, R.B. and Osman, R.W. (2002b) Linking climate change and biological invasions: ocean warming facilitates nonindigenous species invasions. *Proceedings of the National Academy of Sciences of the United States of America*, **99**, 15497–500.

Stachowicz, J.J., Bruno, J.F. and Duffy, J.E. (2007) Understanding the effects of marine biodiversity on community and ecosystem processes. *Annual Review of Ecology, Evolution, and Systematics*, **38**, 739–66.

Steele, J.H. (1985) A comparison of terrestrial and marine ecological systems. *Nature*, **313**, 355–8.

Steele, J.H. (1991) Marine functional diversity. *BioScience*, **41**, 470–74.

Steele, J.H. (2004) Regime shifts in the ocean: reconciling observations and theory. *Progress in Oceanography*, **60**, 135–41.

Steele, J.H. and Schumacher, M. (2000) Ecosystem structure before fishing. *Fisheries Research*, **44**, 201–5.

Steiner, C.F. (2001) The effects of prey heterogeneity and consumer identity on the limitation of trophic-level biomass. *Ecology*, **82**, 2495–506.

Steneck, R.S., Vavrinec, J. and Leland, A.V. (2004) Accelerating trophic-level dysfunction in kelp forest ecosystems of the western North Atlantic. *Ecosystems*, **7**, 323–32.

Stevens, J.D., Bonfil, R., Dulvy, N.K. and Walker, P.A. (2000) The effects of fishing on sharks, rays, and chimaeras (chondrichthyans), and the implications for marine ecosystems. *Ices Journal of Marine Science*, **57**, 476–94.

Stibor, H., Vadstein, O., Diehl, S., *et al.* (2004) Copepods act as a switch between alternative trophic cascades in marine pelagic food webs. *Ecology Letters*, **7**, 321–8.

Strathmann, R.R. (1990) Why life histories evolve differently in the sea. *American Zoologist*, **30**, 197–207.

Strong, D.R. (1992) Are trophic cascades all wet: differentiation and donor-control in speciose ecosystems. *Ecology*, **73**, 747–54.

Sutherland, J.P. (1974) Multiple stable points in natural communities. *The American Naturalist*, **108**, 859–73.

Swain, D.P. and Sinclair, A.F. (2000) Pelagic fishes and the cod recruitment dilemma in the Northwest Atlantic. *Canadian Journal of Fisheries and Aquatic Sciences*, **57**, 1321–5.

Tegner, M.J. and Dayon, P.K. (1987) El Niño effects on southern California kelp forest communities. *Advances in Ecological Research*, **17**, 243–79.

Thompson, J.N. (1998) Rapid evolution as an ecological process. *Trends in Ecology & Evolution*, **13**, 329–32.

Thompson, R.M., Hemberg, M., Starzomski, B.M. and Shurin, J.B. (2007) Trophic levels and trophic tangles: the prevalence of omnivory in real food webs. *Ecology*, **88**, 612–17.

Tilman, D., Lehman, C.L. and Thomson, K.T. (1997) Plant diversity and ecosystem productivity: theoretical considerations. *Proceedings of the National Academy of Sciences of the United States of America*, **94**, 1857–61.

van Nes, E.H., Amaro, T., Scheffer, M. and Duineveld, G.C.A. (2007) Possible mechanisms for a marine benthic regime shift in the North Sea. *Marine Ecology-Progress Series*, **330**, 39–47.

Verity, P.G. and Smetacek, V. (1996) Organism life cycles, predation, and the structure of marine pelagic ecosystems. *Marine Ecology-Progress Series*, **130**, 277–93.

Voigt, W., Perner, J., Davis, A.J., *et al.* (2003) Trophic levels are differentially sensitive to climate. *Ecology*, **84**, 2444–53.

Watling, L. and Norse, E.A. (1998) Disturbance of the seabed by mobile fishing gear: a comparison to forest clearcutting. *Conservation Biology*, **12**, 1180–97.

Wilcove, D.S., Rothstein, D., Dubow, J., *et al.* (1998) Quantifying threats to imperiled species in the United States. *Bioscience*, **48**, 607–15.

Williams, R.J. and Martinez, N.D. (2004) Limits to trophic levels and omnivory in complex food webs: theory and data. *The American Naturalist*, **163**, 458–68.

Winder, M. and Schindler, D.E. (2004) Climate change uncouples trophic interactions in an aquatic system. *Ecology*, **85**, 2100–6.

Wing, S.R. and Wing, E.S. (2001) Prehistoric fisheries in the Caribbean. *Coral Reefs*, **20**, 1–8.

Woodward, G., Ebenman, B., Emmerson, M.C., *et al.* (2005) Body size in ecological networks. *Trends in Ecology & Evolution*, **20**, 402–9.

Wootton, J.T. (1997) Estimates and tests of per capita interaction strength: diet, abundance, and impact of intertidally foraging birds. *Ecological Monographs* **67**, 45–64.

Wootton, J.T. and Emmerson, M.C. (2005) Measurement of interaction strength in nature. *Annual Review of Ecology Evolution and Systematics*, **36**, 419–44.

Worm, B. and Myers, R.A. (2003) Meta-analysis of cod-shrimp interactions reveals top-down control in oceanic food webs. *Ecology*, **84**, 162–73.

Worm, B., Barbier, E.B., Beaumont, N., *et al.* (2006) Impacts of biodiversity loss on ocean ecosystem services. *Science*, **314**, 787–90.

Yachi, S. and Loreau, M. (1999) Biodiversity and ecosystem productivity in a fluctuating environment: the insurance hypothesis. *Proceedings of the National Academy of Sciences of the United States of America*, **96**, 1463–8.

Chapter 9

Andersson, E., Barthel, S. and Ahrné, K. (2007) Measuring social-ecological dynamics behind the generation of ecosystem services. *Ecological Applications*, **17**, 1267–78.

Andrewartha, H.G. and Birch, L.C. (1954) *The Distribution and Abundance of Animals*. University of Chicago Press, Chicago, IL.

Barbosa, P., ed. (1998) *Conservation Biological Control*. Academic Press, San Diego, CA.

Bell, G. (2001) Ecology: neutral macroecology. *Science*, **293**, 2413–18.

Bengtsson, J. (1998) Which species? What kind of diversity? Which ecosystem function? Some problems in studies of relations between biodiversity and ecosystem function. *Applied Soil Ecology*, **10**, 191–9.

Bengtsson, J., Angelstam, P., Elmqvist, T., *et al.* (2003) Reserves, resilience and dynamic landscapes. *Ambio*, **32**, 389–96.

Bengtsson, J., Ahnström, J. and Weibull, A.-C. (2005) The effects of organic farming on biodiversity and abundance: a meta-analysis. *Journal of Applied Ecology*, **42**, 261–9.

Benton, T.G., Vickery, J.A. and Wilson, J.D. (2003) Farmland biodiversity: is habitat heterogeneity the key? *Trends in Ecology & Evolution*, **18**, 182–8.

Biesmeijer, J.C., Roberts, S.P.M., Reemer, M., *et al.* (2006) Parallel declines in pollinators and insect-pollinated plants in Britain and the Netherlands. *Science*, **313**, 351–4.

Chamberlain, D.E., Fuller, R.J., Bunce, R.G.H., *et al.* (2000) Changes in the abundance of farmland birds in relation to the timing of agricultural intensification in England and Wales. *Journal of Applied Ecology*, **37**, 771–88.

Chase, J.M. and Leibold, M.A. (2002) *Ecological Niches: Linking Classical and Contemporary Approaches*. University of Chicago Press, Chicago, IL.

Daily, G.C., ed. (1997) *Nature's Services: Societal Dependence on Natural Ecosystems*. Island Press, Washington, DC.

Donald, P.F., Green, R.E. and Heath, M.F. (2001) Agricultural intensification and the collapse of Europe's farmland bird populations. *Proceedings of the Royal Society of London Series B*, **268**, 25–9.

Ekbom, B.S., Wiktelius, S. and Chiverton, P.A. (1992) Can polyphagous predators control the bird cherry-oat aphid (*Rhopalosiphum padi*) in spring cereals. *Entomologia Experimentalis et Applicata*, **65**, 215–23.

Elmqvist, T., Folke, C., Nyström, M., *et al.* (2003) Response diversity and ecosystem resilience. *Frontiers in Ecology and the Environment*, **1**, 488–94.

Eriksson, O. (1996) Regional dynamics of plants: a review of evidence for remnant, source-sink and metapopulations. *Oikos*, **77**, 248–58.

Folke, C., Jansson, A., Larsson, J. and Costanza, R. (1997) Ecosystem appropriation by cities. *Ambio*, **26**, 167–72.

Fuller, R.J., Norton, L.R., Feber, R.E., *et al.* (2005) Benefits of organic farming to biodiversity vary among taxa. *Biology Letters*, **1**, 431–4.

Hanski, I. and Ranta, E. (1983) Coexistence in patchy environment: three species of *Daphnia* in rock pools. *Journal of Animal Ecology*, **52**, 263–79.

Harrison, S.P. (1991) Local extinction in a metapopulation context: an empirical evaluation. *Biological Journal of the Linnean Society*, **42**, 73–88.

Holt, R.D. (2002) Food webs in space: on the interplay of dynamic instability and spatial processes. *Ecological Research*, **17**, 261–73.

Holyoak, M., Leibold, M.A. and Holt, R.D., eds. (2005) *Metacommunities: Spatial Dynamics and Ecological Communities*. University of Chicago Press, Chicago, IL.

Holzschuh, A., Steffan-Dewenter, I., Kleijn, D. and Tscharntke, T. (2007) Diversity of flower-visiting bees in cereal fields: effects of farming system, landscape composition and regional context. *Journal of Applied Ecology*, **44**, 41–9.

Hubbell, S.P. (2001) *The Unified Theory of Biodiversity and Biogeography*. Princeton University Press, Princeton, NJ.

Kleijn, D. and van Langevelde, F. (2006) Interacting effects of landscape context and habitat quality on flower visiting insects in agricultural landscapes. *Basic and Applied Ecology*, **7**, 201–14.

Lawton, J.H. (2007) Ecology, politics and policy. *Journal of Applied Ecology*, **44**, 465–74.

Leibold, M.A., Holyoak, M., Mouquet, N., *et al.* (2004) The metacommunity concept: a framework for multi-scale community ecology. *Ecology Letters*, **7**, 601–13.

Levins, R. (1969) Some genetic and demographic consequences of environmental heterogeneity for biological control. *Bulletin of the Entomological Society of America*, **15**, 237–40.

Lindenmayer, D., Hobbs, R.J., Montague-Drake, R., *et al.* (2008) A checklist for ecological management of landscapes for conservation. *Ecology Letters*, **11**, 78–91.

Loreau, M., Naeem, S., Inchausti, P., *et al.* (2001) Biodiversity and ecosystem functioning: current knowledge and future challenges. *Science*, **294**, 804–8.

Loreau, M., Mouquest, N. and Gonzalez, A. (2003) Biodiversity as spatial insurance in heterogenous landscapes. *Proceedings of the National Academy of Sciences of the United States of America*, **100**, 12765–70.

Ma, M. (2006) Plant species diversity of buffer zones in agricultural landscapes: in search of determinants from the local to regional scale. PhD thesis, Helsinki University, Finland.

MacArthur, R.H. and Wilson, E.O. (1967) *Theory of Island Biogeography*. Princeton University Press, Princeton, NJ.

Mouillot, D. (2007) Niche-assembly vs. dispersal-assembly rules in coastal fish metacommunities: implications for management of biodiversity in brackish lagoons. *Journal of Applied Ecology*, **44**, 760–7.

Mouquet, N. and Loreau, M. (2003) Community patterns in source-sink metacommunities. *The American Naturalist*, **162**, 544–57.

Nee, S. and May, R.M. (1992) Dynamics of metapopulations: habitat destruction and competitive coexistence. *Journal of Animal Ecology*, **61**, 37–40.

Norberg, J., Swaney, D.P., Dushoff, J., *et al.* (2001) Phenotypic diversity and ecosystem functioning in changing environments: a theoretical framework. *Proceedings of the National Academy of Sciences of the United States of America*, **98**, 11376–81.

Öberg, S., Ekbom, B. and Bommarco, R. (2007) Influence of habitat type and surrounding landscape on spider diversity in Swedish agroecosystems. *Agriculture Ecosystems and Environment*, **122**, 211–19.

Odling-Smee, F.J., Laland, K.N. and Feldman, M.W. (2003) *Niche Construction: The Neglected Process in Evolution*. Princeton University Press, Princeton, NJ.

Oksanen, T. (1990) Exploitation ecosystems in heterogenous habitat complexes. *Evolutionary Ecology*, **4**, 220–34.

Östman, Ö., Ekbom, B. and Bengtsson, J. (2001) Landscape heterogeneity and farming practice influence biological control. *Basic and Applied Ecology*, **2**, 365–71.

Östman, Ö., Ekbom, B. and Bengtsson, J. (2003) Yield increase attributable to aphid predation by ground-living natural enemies in spring barley in Sweden. *Ecological Economics*, **45**, 149–58.

Palumbi, S.R. (2001) Humans as the world's greatest evolutionary force. *Science*, **293**, 1786–90.

Peterson, D.L. and Parker, V.T., eds. (1998) *Ecological Scale: Theory and Applications*. Columbia University Press, New York, NY.

Polis, G.A., Anderson, W.B. and Holt, R.D. (1997) Toward an integration of landscape and food web ecology: the dynamics of spatially subsidized food webs. *Annual Review of Ecology and Systematics*, **28**, 289–316.

Ricklefs, R.E. and Schluter, D., eds. (1993) *Species Diversity in Ecological Communities: Historical and Geographical Perspectives*. University of Chicago Press, Chicago, IL.

Roschewitz, I., Gabriel, D., Tscharntke, T. and Thies, C. (2005) The effects of landscape complexity on arable weed species diversity in organic and conventional farming. *Journal of Applied Ecology*, **42**, 873–82.

Rundlöf, M. (2007) Biodiversity in agricultural landscapes: landscape and scale-dependent effects of organic farming. PhD thesis, Lund University, Sweden.

Rundlöf, M. and Smith, H.G. (2006) The effect of organic farming on butterfly diversity depends on landscape context. *Journal of Applied Ecology*, **43**, 1121–7.

Rundlöf, M., Bengtsson, J. and Smith, H.G. (2008) Local and landscape effects of organic farming on butterfly species richness and abundance. *Journal of Applied Ecology*, **45**, 813–20.

Schmidt, M.H. and Tscharntke, T. (2005) The role of perennial habitats for Central European farmland spiders. *Agricultture Ecosystems and Environment*, **105**, 235–42.

Schoener, T.W. (1976) The species-area relation within archipelagos: models and evidence from island land birds. In *Proceedings of the 16th International Ornithological Congress Canberra, 1974* (eds H.J. Firth and J.H. Calaby), pp. 629–42. Australian Academy of Science, Canberra, ACT.

Steffan-Dewenter, I., Münzenberg, U., Bürger, C., *et al.* (2002) Scale-dependent effects of landscape context on three pollinator guilds. *Ecology*, **83**, 1421–32.

Thies, C. and Tscharntke, T. (1999) Landscape structure and biological control in agroecosystems. *Science*, **285**, 893–5.

Tremlova, K. and Münzbergova, Z. (2007) Importance of species traits for species distribution in fragmented landscapes. *Ecology*, **88**, 965–77.

Tscharntke, T., Klein, A.M., Kruess, A., *et al.* (2005) Landscape perspectives on agricultural intensification and biodiversity ecosystem service management. *Ecology Letters*, **8**, 857–74.

van de Koppel, J., Bardgett, R., Bengtsson, J., *et al.* (2005) Effects of spatial scale on trophic interactions. *Ecosystems*, **8**, 801–7.

Vandermeer, J. and Perfecto, I. (2007) The agricultural matrix and a future paradigm for conservation. *Conservation Biology*, **21**, 274–7.

Vitousek, P.M., Mooney, H.A., Lubchenco, J. and Melillo, J.M. (1997) Human domination of earth's ecosystems. *Science*, **277**, 494–9.

Weibull, A.C. and Östman, Ö. (2003) Species composition in agroecosystems: the effect of landscape, habitat and farm management. *Basic and Applied Ecology*, **4**, 349–61.

Weibull, A.C., Bengtsson, J. and Nohlgren, E. (2000) Diversity of butterflies in the agricultural landscape: the role of farming system and landscape heterogeneity. *Ecography*, **23**, 743–50.

Weibull, A.C., Östman, Ö. and Granqvist, Å. (2003) Species richness in agroecosystems: the effect of landscape, habitat and farm management. *Biodiversity and Conservation*, **12**, 1335–55.

Worster, D. (1994) *Nature's Economy: A History of Ecological Ideas*, 2nd edn. Cambridge University Press, Cambridge.

Chapter 10

Andresen, H., Bakker, J.P., Brongers, M., *et al.* (1990) Long-term changes of salt marsh communities by cattle grazing. *Vegetatio*, **89**, 137–48.

Bakker, J.P., Esselink, P., Dijkema, K.S., *et al.* (2002) Restoration of salt marshes. *Hydrobiologia*, **478**, 29–51.

Bakker, J.P., Bos, D. and de Vries, Y. (2003a) To graze or not to graze: that is the question. In *Challenges to the Wadden Sea Area* (eds W. Wolff, K.M. Essink, A. Kellermann and M.A. van Leeuwe), pp. 67–87. Proceedings of the 10th International Scientific Wadden Sea Symposium. Ministry of Agriculture, Nature Management and Fisheries and Department of Marine Biology, University of Groningen.

Bakker, J.P., Bos, D., Stahl, J., *et al.* (2003b) Biodiversität und Landnutzung in Salzwiesen. *Nova Acta Leopoldina*, **87**, 163–94.

Bakker, J.P., Bunje, J., Dijkema, K.S., *et al.* (2005a) Salt Marshes. In *Wadden Sea Quality Status Report 2004* (eds K. Essink, C. Dettmann, H. Farke, *et al.*), pp. 163–79. Wadden Sea Ecosystem No 19. Trilateral Monitoring and Assessment Group. Common Wadden Sea Secretariat, Wilhelmshaven, Germany.

Bakker, J.P., Bouma, T.J. and van Wijnen, H.J. (2005b) Interactions between microorganisms and intertidal plant communities. In *Interactions Between Macro- and Microorganisms in Marine Sediments* (eds K. Kristensen, J.E. Kostka and R.R. Haese), pp. 179–98. Coastal and Estuarine Studies 60. American Geophysical Union, Washington, DC.

Berg, G., Groeneweg, M., Esselink, P. and Kiehl, K. (1997) Micropatterns in a *Festuca rubra*-dominated salt-marsh vegetation induced by sheep grazing. *Plant Ecology*, **132**, 1–14.

Bertness, M.D. and Shumway, S.W. (1992) Changes in the composition and standing crop of salt-marsh communities in response to the removal of a grazer. *Journal of Ecology*, **74**, 693–706.

Bortolous, A. and Iribarne, O. (1999) Effects of the SW Atlantic burrowing crab *Chasmagnathus granulata* on a *Spartina* salt marsh. *Marine Ecology-Progress Series*, **178**, 79–88.

Bos, D., Bakker, J.P., de Vries, Y. and van Lieshout, S. (2002) Effects of changes in grazing management over 25 years on plant species richness and important plants for geese in back-barrier salt marshes in the Wadden Sea. *Applied Vegetation Science*, **5**, 45–54.

Bos, D., Loonen, M., Stock, M., *et al.* (2005) Utilisation of Wadden Sea salt marshes by geese in relation to livestock grazing. *Journal for Nature Conservation*, **15**, 1–15.

Cramp, S. and Simmons, K.E.L., eds. (1983) *Handbook of the Birds of Europe, The Middle East and North Africa: the Birds of the Western Palaearctic*. Vol. III. *Waders to Gulls*. Oxford University Press, Oxford.

Davy, A.J., Bakker, J.P. and Figueroa, M.E. (2009) Human impact on European salt marshes. In *Anthropogenic Modification of North American Salt Marshes* (eds B.R. Silliman, M.D. Bertness and D. Strong). University of California Press, Berkeley, CA.

de Leeuw, J., de Munck, W., Olff, H. and Bakker, J.P. (1993) Does zonation reflect the succession of salt marsh vegetation? A comparison of an estuarine and a coastal bar island marsh in the Netherlands. *Acta Botanica Neerlandica*, **42**, 435–45.

Dijkema, K.S. (2007) *Ecologische Onderbouwing van het Kwelderherstelplan in Groningen*. Imares, Texel.

Eskildsen, K., Fiedler, U. and Hälterlein, B. (2000) Die Entwicklung der Brutvogelbestände auf der Hamburger Hallig. In *Die Salzwiesen der Hamburger Hallig*, vol. 11 (eds M. Stock and K. Kiehl), pp. 61–5. Nationalpark Schleswig-Holsteinisches Wattenmeer, Toenning.

Esselink, P. (2000) Nature management of coastal salt marshes: interactions between anthropogenic influences and natural dynamics. PhD Thesis, University of Groningen, The Netherlands.

Esselink, P., Zijlstra, W., Dijkema, K.S. and Van Diggelen, R. (2000) The effects of decreased management on plant-species distribution in a salt marsh nature reserve in the Wadden Sea. *Biological Conservation*, **93**, 61–76.

Esselink, P., Fresco, L.F.M. and Dijkema, K.S. (2002) Vegetation change in a man-made salt marsh affected by a reduction in both grazing and drainage. *Applied Vegetation Science*, **5**, 17–32.

Hälterlein, B. (1998) *Brutvogelbestände im Schleswig-Holsteinischen Wattenmeer*. UBA-Texte 76–97. Umweltbundesamt, Berlin.

Hälterlein, B., Bunje, J. and Potel, P. (2003) Zum Einfluß der Salzwiesennutzung an der Nordseeküste auf die Vogelwelt: Übersicht über die aktuellen Forschungsergebnisse. *Vogelkundliche Berichte Niedersachsen*, 35: 179–86.

Huisman, J., Grover, J.P., van der Wal, R. and van Andel, J. (1999) Compensation for light, plant species replacement, and herbivore abundance along productivity gradients. In *Herbivores Between Plants and Predators* (eds H. Olff, V.K. Brown and R.H. Drent), pp. 239–70. Blackwell Scientific, Oxford.

Hulbert, I.A.R. and Andersen, R. (2001) Food competition between a large ruminant and a small hindgut fermenter: the case of the roe deer and mountain hare. *Oecologia*, **128**, 499–508.

Irmler, U. and Heydemann, B. (1986) Die Ökologische Problematik der Beweidung von Salzwiesen and der niedersächsischen Küste: am Beispiel der Leybucht. *Naturschutz und Landschaftspflege Niedersachsen*, **11**, 1–115.

Jefferies, R.J., Drent, R.H. and Bakker, J.P. (2006) Connecting arctic and temperate wetlands and agricultural landscapes: the dynamics of goose populations in response to global change. In *Wetlands and Natural Resource Management* (eds J.T.A. Verhoeven, B. Beltman, R. Bobbink and D.F. Whigham), pp. 293–314. Ecological Studies, vol. 190. Springer, Berlin.

Kiehl, K. (1997) *Vegetationsmuster in Vorlandsalzwiesen in Abhängigkeit von Beweidung und abiotischen Standortfaktoren.* Arbeitsgemeinschaft Geobotanik in Schleswig-Holstein und Hamburg e.V., Kiel.

Kiehl, K., Eischeid, I., Gettner, S. and Walter, J. (1996) Impact of different sheep grazing intensities on salt marsh vegetation in northern Germany. *Journal of Vegetation Science*, **7**, 99–106.

Koffijberg, K. (in press) *Overzicht van de relatie tussen kustbroedvogels en beheer van kwelders langs de Noord-Groninger kust.*

Kuijper, D.P.J. (2004) Small herbivores losing control: plant-herbivore interactions along a natural productivity gradient. PhD Thesis, University of Groningen, The Netherlands.

Kuijper, D.P.J. and Bakker, J.P. (2003) Large-scale effects of a small herbivore on salt-marsh vegetation succession, a comparative study. *Journal of Coastal Conservation*, **9**, 179–88.

Kuijper, D.P.J. and Bakker, J.P. (2005) Top-down control of small herbivores on salt-marsh vegetation along a productivity gradient. *Ecology*, **86**, 914–23.

Kuijper, D.P.J. and Bakker, J.P. (2008) Unpreferred plants affect patch choice and spatial distribution of brown hares. *Acta Oecologica*, **34**, 339–44.

Kuijper, D.P.J., Nijhoff, N. and Bakker, J.P. (2004) Herbivory and competition slow down invasion of a tall grass along a productivity gradient. *Oecologia*, **141**, 452–9.

Kuijper, D.P.J., Beek, P., van Wieren, S.E. and Bakker, J.P. (2008) Time-scale effects in the interaction between a large and a small herbivore. *Basic and Applied Ecology*, **9**, 126–34.

Leendertse, P.C., Roozen, A.J.M. and Rozema, J. (1997) Long-term changes (1953–1990) in the salt marsh vegetation at the Boschplaat on Terschelling in relation to sedimentation and flooding. *Plant Ecology*, **132**, 49–58.

Meyer, H., Fock, H., Haase, A., *et al.* (1995) Structure of the invertebrate fauna in salt marshes of the Wadden Sea coast of Schleswig-Holstein influenced by sheep grazing. *Helgoländer Meerseuntersuchungen*, **49**, 563–89.

Norris, K., Cook, B., O'Dowd, B. and Durdin, C. (1997) The density of redshank *Tringa totanus* breeding on the saltmarshes of the Wash in relation to habitat and its grazing management. *Journal of Applied Ecology*, **34**, 999–1013.

Norris, K., Brindy, E., Cook, T., *et al.* (1998) Is the density of redshank *Tringa totanus* nesting on saltmarshes in Great Britain declining due to changes in grazing management? *Journal of Applied Ecology*, **35**, 621–34.

Oksanen, L. and Oksanen, T. (2000) The logic and realism of the hypothesis of exploitation ecosystems. *The American Naturalist*, **155**, 703–23.

Oksanen, L., Fretwell, S.D., Arruda, J. and Niemelä, P. (1981) Exploitation ecosystems in gradients of primary production. *The American Naturalist*, **118**, 240–61.

Olff, H., de Leeuw, J., Bakker, J.P., *et al.* (1997) Vegetation succession and herbivory on a salt marsh: changes induced by sea level rise and silt deposition along an elevational gradient. *Journal of Ecology*, **85**, 799–814.

Oltmanns, B. (2003) Von der Hellerweide zur Salzwiese: Veränderungen der Brutvogelgemeinschaft in der Leybucht durch die Nutzungsaufgabe. *Vogelkundliche Berichte Niedersachsen*, **35**, 157–66.

Pétillon, J., Ysnel, F., Canard, A. and Lefeuvre, J.C. (2005) Impact of an invasive plant (*Elymus athericus*) on the conservation value of tidal salt marshes in western France and implications for management: Responses of spider populations. *Biological Conservation*, **126**, 103–17.

Pétillon, J., Georges, A., Canard, A., *et al.* (2008) Arthropod community structure and indicator value: comparison of spider (Araneae) and ground beetle (Coleoptera, Carabidae) groups in different salt-marsh systems. *Basic and Applied Ecology*, **9**, 743–51.

Prop, J. and Deerenberg, C. (1991) Spring staging in Brent Geese (*Branta bernicla*): feeding constraints and the impact of diet on the accumulation of body reserves. *Oecologia*, **87**, 19–28.

Schrader, S. (2003) Zehn Jahre später: Brutvogelbestände in unterschiedlich beweideten Salzwiesen der schleswig-holsteinischen Festlandsküste. *Vogelkundliche Berichte Niedersachsen*, **35**, 167–72.

Schröder, H.K., Kiehl, K. and Stock, M. (2002) Directional and non-directional vegetation changes in a temperate salt marsh in relation to biotic and abiotic factors. *Applied Vegetation Science*, **5**, 33–44.

Schultz, W. (1987) Einfluss der Beweidung von Salzwiesen auf die Vogelfauna. In *Salzwiesen: geformt von Küstenschutz, Landwirtschaft oder Natur?* (eds N. Kempf, J. Lamp and P. Prokosch), pp. 255–70. WWF-Deutschland, Husum.

Silliman, B.R., van de Koppel, J., Bertness, M.D., *et al.* (2005) Drought, snails and large-scale die-off of Southern U.S. salt marshes. *Science*, **310**, 1803–6.

Smith, R.K., Jenning, N.V., Robinson, A. and Harris, S. (2004) Conservation of European hares (*Lepus europaeus*) in Britain: is increasing heterogeneity in farmland the answer? *Journal of Applied Ecology*, **41**, 1092–102.

Smith, T.J. and Odum, W.E. (1983) The effects of grazing by snow geese on coastal salt marshes. *Ecology*, **62**, 98–106.

Stahl, J. (2001) Limits to the co-occurrence of avian herbivores: how geese share scarce resources. PhD Thesis, University of Groningen, The Netherlands.

Stahl, J., Bos, D. and Loonen, M.J.J.E. (2002) Foraging along a salinity gradient: the effect of tidal inundation on site choice by Dark-bellied Brent Geese *Branta bernicla* and Barnacle Geese *B. leucopsis*. *Ardea*, **90**, 201–12.

Stahl, J., van der Graaf, A.J., Drent, R.H. and Bakker, J.P. (2006) Subtle interplay of competition and facilitation among small herbivores in coastal grasslands. *Functional Ecology*, **20**, 908–15.

Thyen, S. (2005) Reproduction of coastal breeding birds in the Wadden Sea: variation, influencing factors and monitoring. PhD Thesis, University of Oldenburg, Germany.

Thyen, S. and Exo, K.M. (2003) Sukzession der Salzrasen an der niedersächsischen Küste: Chance oder Risiko für Brutvögel der Außengroden. *Vogelkundliche Berichte Niedersachsen*, **35**, 173–8.

Thyen, S. and Exo, K.M. (2005) Interactive effects of time and vegetation on reproduction of redshanks (*Tringa totanus*) breeding in Wadden Sea salt marshes. *Journal of Ornithology*, **146**, 215–25.

van de Koppel, J., Huisman, J., van der Wal, R. and Olff, H. (1996) Patterns of herbivory along a productivity gradient: An empirical and theoretical investigation. *Ecology*, **77**, 736–45.

van der Graaf, A.J., Stahl, J. and Bakker, J.P. (2005) Compensatory growth of Festuca rubra after grazing: can migratory herbivores increase their own harvest during staging? *Functional Ecology*, **19**, 961–70.

van der Graaf, A.J.,Coehoorn, P. and Stahl, J. (2006) Sward height and bite size affect the functional response of Branta leucopsis. *Journal of Ornithology*, **147**, 479–84.

van der Wal, R., Kunst, P. and Drent, R.H. (1998) Interactions between hare and Brent Goose in a salt marsh system: evidence for food competition? *Oecologia*, **117**, 227–34.

van der Wal, R., van Lieshout, S., Bos, D. and Drent, R.H. (2000a) Are spring staging Brent Geese evicted by vegetation succession? *Ecography*, **23**, 60–9.

van der Wal, R., Egas, M., van der Veen, A. and Bakker, J.P. (2000b) Effects of resource competition on plant performance along a natural productivity gradient. *Journal of Ecology*, **88**, 317–30.

van der Wal, R., van Wieren, S., van Wijnen, H., *et al.* (2000c) On facilitation between herbivores: how Brent Geese profit from Brown Hares. *Ecology*, **81**, 969–80.

van Dijk, A.J. and Bakker, J.P. (1980) Beweiding en broedvogels op de Oosterkwelder van Schiermonnikoog. *Waddenbulletin*, **15**, 134–40.

van Wijnen, H.J. and Bakker, J.P. (1997) Nitrogen accumulation and plant species replacement in three salt-marsh systems in the Wadden Sea. *Journal of Coastal Conservation*, **3**, 19–26.

van Wijnen, H.J., Bakker, J.P. and de Vries, Y. (1997) Twenty years of salt marsh succession on a Dutch coastal barrier island. *Journal of Coastal Conservation*, **3**, 9–18.

Chapter 11

Abrams, P.A. (1995) Implications of dynamically variable traits for identifying, classifying, and measuring direct and indirect effects in ecological communities. *The American Naturalist*, **146**, 112–34.

Agrawal, A.A., Ackerly, D.D., Adler, E., *et al.* (2007) Filling key gaps in population and community ecology. *Frontiers in Ecology and the Environment*, **5**, 145–52.

Antonovics, J. (1978) The population genetics of species mixtures. In *Plant Relations in Pastures* (ed. J.R. Wilson), pp. 223–52. CSIRO, East Melbourne.

Baker, H.G. (1965) Characteristics and modes of origin of weeds. In *The Genetics of Colonizing Species* (eds H.G. Baker and G.L. Stebbins), pp. 147–69. Academic Press, New York, NY.

Barrett, R.D.H., MacLean, R.C. and Bell, G. (2005) Experimental evolution of *Pseudomonas fluorescens* in simple and complex environments. *The American Naturalist*, **166**, 470–80.

Bell, G. (1991) The ecology and genetics of fitness in *Chlamydomonas*. 3. Genotype-by-environment interaction within strains. *Evolution*, **45**, 668–79.

Bolker, B., Holyoak, M., Krivan, V., *et al.* (2003) Connecting theoretical and empirical studies of trait-mediated interactions. *Ecology*, **84**, 1101–14.

Booth, R.E. and Grime, J.P. (2003) Effects of genetic impoverishment on plant community diversity. *Journal of Ecology*, **91**, 721–30.

Callaway, R.M. and Aschehoug, E.T. (2000) Invasive plants versus their new and old neighbors: a mechanism for exotic invasion. *Science*, **290**, 521–3.

Chown, S.L., Slabber, S., McGeoch, M.A., *et al.* (2007) Phenotypic plasticity mediates climate change responses among invasive and indigenous arthropods. *Proceedings of the Royal Society B-Biological Sciences*, **274**, 2531–7.

Crawford, K.M., Crutsinger, G.M. and Sanders, N.J. (2007) Host-plant genotypic diversity mediates the distribution of an ecosystem engineer. *Ecology*, **88**, 2114–20.

Crutsinger, G.M., Collins, M.D., Fordyce, J.A., *et al.* (2006) Plant genotypic diversity predicts community structure and governs an ecosystem process. *Science*, **313**, 966–8.

Davis, A.J., Lawton, J.H., Shorrocks, B. and Jenkinson, L.S. (1998) Individualistic species responses invalidate simple physiological models of community dynamics under global environmental change. *Journal of Animal Ecology*, **67**, 600–12.

Dawkins, R. (1982) *The Extended Phenotype*. Oxford University Press, Oxford.

De Meester, L., Louette, G., Duvivier, C., *et al.* (2007) Genetic composition of resident populations influences establishment success of immigrant species. *Oecologia*, **153**, 431–40.

Dungey, H.S., Potts, B.M., Whitham, T.G. and Li, H.F. (2000) Plant genetics affects arthropod community richness and composition: evidence from a synthetic eucalypt hybrid population. *Evolution*, **54**, 1938–46.

Fabricius, K.E., Mieog, J.C., Colin, P.L., *et al.* (2004) Identity and diversity of coral endosymbionts (zooxanthellae) from three Palauan reefs with contrasting bleaching, temperature and shading histories. *Molecular Ecology*, **13**, 2445–58.

Fridley, J.D., Grime, J.P. and Bilton, M. (2007) Genetic identity of interspecific neighbours mediates plant responses to competition and environmental variation in a species-rich grassland. *Journal of Ecology*, **95**, 908–15.

Gamfeldt, L., Wallen, J., Jonsson, P.R., *et al.* (2005) Increasing intraspecific diversity enhances settling success in a marine invertebrate. *Ecology*, **86**, 3219–24.

Harvey, J.A., van Dam, N.M. and Gols, R. (2003) Interactions over four trophic levels: foodplant quality affects development of a hyperparasitoid as mediated through a herbivore and its primary parasitoid. *Journal of Animal Ecology*, **72**, 520–31.

Havill, N.P. and Raffa, K.F. (2000) Compound effects of induced plant responses on insect herbivores and parasitoids: implications for tritrophic interactions. *Ecological Entomology*, **25**, 171–9.

Hector, A., Schmid, B., Beierkuhnlein, C. *et al.* (1999) Plant diversity and productivity experiments in European grasslands. *Science*, **286**, 1123–7.

Heemsbergen, D.A., Berg, M.P., Loreau, M., *et al.* (2004) Biodiversity effects on soil processes explained by interspecific functional dissimilarity. *Science*, **306**, 1019–20.

Hooper, D.U., Chapin, F.S., Ewel, J.J., *et al.* (2005) Effects of biodiversity on ecosystem functioning: a consensus of current knowledge. *Ecological Monographs*, **75**, 3–35.

Hughes, A.R. and Stachowicz, J.J. (2004) Genetic diversity enhances the resistance of a seagrass ecosystem to disturbance. *Proceedings of the National Academy of Sciences of the United States of America*, **101**, 8998–9002.

Huston, M.A. (1997) Hidden treatments in ecological experiments: re-evaluating the ecosystem function of biodiversity. *Oecologia*, **110**, 449–60.

Jaenike, J. (1978) A hypothesis to account for the maintenance of sex within populations. *Evolutionary Theory* **3**, 191–4.

Jiang, L. and Morin, P.J. (2004) Temperature-dependent interactions explain unexpected responses to environmental warming in communities of competitors. *Journal of Animal Ecology*, **73**, 569–76.

Johnson, M.T.J. and Agrawal, A.A. (2005) Plant genotype and environment interact to shape a diverse arthropod community on evening primrose (*Oenothera biennis*). *Ecology*, **86**, 874–85.

Johnson, M.T.J. and Stinchcombe, J.R. (2007) An emerging synthesis between community ecology and evolutionary biology. *Trends in Ecology & Evolution*, **22**, 250–7.

Johnson, M.T.J., Lajeunesse, M.J. and Agrawal, A.A. (2006) Additive and interactive effects of plant genotypic diversity on arthropod communities and plant fitness. *Ecology Letters*, **9**, 24–34.

Koch, A.M., Croll, D. and Sanders, I.R. (2006) Genetic variability in a population of arbuscular mycorrhizal fungi causes variation in plant growth. *Ecology Letters*, **9**, 103–10.

Krebs, R.A. and Holbrook, S.H. (2001) Reduced enzyme activity following Hsp70 overexpression in *Drosophila melanogaster*. *Biochemical Genetics*, **39**, 73–82.

Lawton, J.H. (1999) Are there general laws in ecology? *Oikos*, **84**, 177–92.

Liefting, M. and Ellers, J. (2008) Habitat-specific differences in thermal plasticity in natural populations of a soil arthropod. *Biological Journal of the Linnean Society* **94**, 265–71.

Loeschcke, V., Bundgaard, J. and Barker, J.S.F. (1999) Reaction norms across and genetic parameters at different temperatures for thorax and wing size traits in *Drosophila aldrichi* and *D. buzzatii*. *Journal of Evolutionary Biology*, **12**, 605–23.

Long, Z.T. and Morin, P.J. (2005) Effects of organism size and community composition on ecosystem functioning. *Ecology Letters*, **8**, 1271–82.

Loreau, M. and Hector, A. (2001) Partitioning selection and complementarity in biodiversity experiments. *Nature*, **412**, 72–6.

Mack, R.N., Simberloff, D., Lonsdale, W.M., *et al.* (2000) Biotic invasions: causes, epidemiology, global consequences, and control. *Ecological Applications*, **10**, 689–710.

Madritch, M., Donaldson, J.R. and Lindroth, R.L. (2006) Genetic identity of *Populus tremuloides* litter influences decomposition and nutrient release in a mixed forest stand. *Ecosystems*, **9**, 528–37.

Mattila, H.R. and Seeley, T.D. (2007) Genetic diversity in honey bee colonies enhances productivity and fitness. *Science*, **317**, 362–4.

McKay, J.K. and Latta, R.G. (2002) Adaptive population divergence: markers, QTL and traits. *Trends in Ecology & Evolution*, **17**, 285–91.

Mealor, B.A. and Hild, A.L. (2006) Potential selection in native grass populations by exotic invasion. *Molecular Ecology* **15**, 2291–300.

Mealor B.A., Hild A.L. and Shaw N.L. (2004) Native plant community composition and genetic diversity associated with long-term weed invasions. *Western North American Naturalist* **64**, 503–13.

Merila, J. and Crnokrak, P. (2001) Comparison of genetic differentiation at marker loci and quantitative traits. *Journal of Evolutionary Biology*, **14**, 892–903.

Mulder, C.P.H., Uliassi, D.D. and Doak, D.F. (2001) Physical stress and diversity-productivity relationships: the role of positive interactions. *Proceedings of the National Academy of Sciences of the United States of America*, **98**, 6704–8.

Munkvold, L., Kjoller, R., Vestberg, M., *et al.* (2004) High functional diversity within species of arbuscular mycorrhizal fungi. *New Phytologist*, **164**, 357–64.

Muth, N.Z. and Pigliucci, M. (2007) Implementation of a novel framework for assessing species plasticity in biological invasions: responses of *Centaurea* and *Crepis* to phosphorus and water availability. *Journal of Ecology*, **95**, 1001–13.

Peacor, S.D. and Werner, E.E. (2000) Predator effects on an assemblage of consumers through induced changes in consumer foraging behavior. *Ecology*, **81**, 1998–2010.

Prasad, R.P. and Snyder, W.E. (2006) Diverse trait-mediated interactions in a multi-predator, multi-prey community. *Ecology*, **87**, 1131–7.

Reed, D.H. and Frankham, R. (2001) How closely correlated are molecular and quantitative measures of genetic variation? A meta-analysis. *Evolution*, **55**, 1095–103.

Relyea, R.A. and Yurewicz, K.L. (2002) Predicting community outcomes from pairwise interactions: integrating density- and trait-mediated effects. *Oecologia*, **131**, 569–79.

Reusch, T.B.H., Ehlers, A., Hammerli, A. and Worm, B. (2005) Ecosystem recovery after climatic extremes enhanced by genotypic diversity. *Proceedings of the National Academy of Sciences of the United States of America*, **102**, 2826–31.

Richards, C.L., Bossdorf, O., Muth, N.Z., *et al.* (2006) Jack of all trades, master of some? On the role of phenotypic plasticity in plant invasions. *Ecology Letters*, **9**, 981–93.

Roscher, C., Schumacher, J., Foitzik, O. and Schulze, E.D. (2007) Resistance to rust fungi in *Lolium perenne* depends on within-species variation and performance of the host species in grasslands of different plant diversity. *Oecologia*, **153**, 173–83.

Scheiner, S.M. and Lyman, R.F. (1989) The genetics of phenotypic plasticity. 1. Heritability. *Journal of Evolutionary Biology*, **2**, 95–107.

Schlichting, C.D. and Pigliucci, M. (1998) *Phenotypic Evolution: a Reaction Norm Perspective*. Sinauer Associates, Inc., Sunderland, MA.

Schmid, B. (1994) Effects of genetic diversity in experimental stands of *Solidago altissima*: evidence for the potential role of pathogens as selective agents in plant populations. *Journal of Ecology*, **82**, 165–75.

Semlitsch, R.D., Hotz, H. and Guex, G.D. (1997) Competition among tadpoles of coexisting hemiclones of hybridogenetic *Rana esculenta*: support for the frozen niche variation model. *Evolution*, **51**, 1249–61.

Sih, A., Englund, G. and Wooster, D. (1998) Emergent impacts of multiple predators on prey. *Trends in Ecology & Evolution*, **13**, 350–5.

Srivastava, D.S. and Lawton, J.H. (1998) Why more productive sites have more species: an experimental test of theory using tree-hole communities. *The American Naturalist*, **152**, 510–29.

Stiling, P. and Rossi, A.M. (1996) Complex effects of genotype and environment on insect herbivores and their enemies. *Ecology*, **77**, 2212–18.

Strauss, S.Y., Lau, J.A. and Carroll, S.P. (2006) Evolutionary responses of natives to introduced species: what do introductions tell us about natural communities? *Ecology Letters* **9**, 357–74.

Strayer, D.L., Eviner, V.T., Jeschke, J.M. and Pace, M.L. (2006) Understanding the long-term effects of species invasions. *Trends in Ecology & Evolution*, **21**, 645–51.

Sznajder, B. and Harvey, J.A. (2003) Second and third trophic level effects of differences in plant species reflect dietary specialisation of herbivores and their endoparasitoids. *Entomologia Experimentalis et Applicata*, **109**, 73–82.

Tagg, N., Innes, D.J. and Doncaster, C.P. (2005) Outcomes of reciprocal invasions between genetically diverse and genetically uniform populations of *Daphnia obtusa* (Kurz). *Oecologia*, **143**, 527–36.

Thompson, J.N. (2005) *The Geographic Mosaic of Coevolution*. University of Chicago Press, Chicago, IL.

Vavrek, M.C. (1998) Within-population genetic diversity of *Taraxacum officinale* (Asteraceae): differential genotype response and effect on interspecific competition. *American Journal of Botany*, **85**, 947–54.

Vellend, M. and Geber, M.A. (2005) Connections between species diversity and genetic diversity. *Ecology Letters*, **8**, 767–81.

Verschoor, A.M., Vos, M. and van der Stap, I. (2004) Inducible defences prevent strong population fluctuations in bi- and tritrophic food chains. *Ecology Letters*, **7**, 1143–8.

Via, S., Gomulkiewicz, R., De Jong, G., *et al.* (1995) Adaptive phenotypic plasticity: consensus and controversy. *Trends in Ecology & Evolution*, **10**, 212–17.

Werner, E.E. (1992) Individual behavior and higher order species interactions. *The American Naturalist*, **140**, S5–S32.

Whitham, T.G., Young, W.P., Martinsen, G.D., *et al.* (2003) Community and ecosystem genetics: a consequence of the extended phenotype. *Ecology*, **84**, 559–73.

Whitlock, R., Grime, J.P., Booth, R. and Burke, T. (2007) The role of genotypic diversity in determining grassland community structure under constant environmental conditions. *Journal of Ecology*, **95**, 895–907.

Whittaker, R.H. (1975) *Communities and Ecosystems*. MacMillan Publishing Company, New York, NY.

Wimp, G.M., Martinsen, G.D., Floate, K.D., *et al.* (2005) Plant genetic determinants of arthropod community structure and diversity. *Evolution*, **59**, 61–9.

Wojdak, J.M. and Mittelbach, G.G. (2007) Consequences of niche overlap for ecosystem functioning: an experimental test with pond grazers. *Ecology*, **88**, 2072–83.

Chapter 12

Abrams, P.A. (1991) The effects of interacting species on predator-prey coevolution. *Theoretical Population Biology*, **39**, 241–62.

Abrams, P.A. (1993) Effect of increased productivity on the abundances of trophic levels. *The American Naturalist*, **141**, 351–71.

Abrams, P.A. and Chen, X. (2002) The evolution of traits affecting resource acquisition and predator vulnerability: character displacement under real and apparent competition. *The American Naturalist*, **160**, 692–704.

Abrams, P.A. and Matsuda, H. (1997) Prey adaptation as a cause of predator-prey cycles. *Evolution*, **51**, 1742–50.

Amaral, L.A.N. and Ottino, J.M. (2004) Complex networks, augmenting the framework for the study of complex systems. *The European Physical Journal B*, **38**, 147–62.

Anderson, P.E. and Jensen, H.J. (2005) Network properties, species abundance and evolution in a model of evolutionary ecology. *Journal of Theoretical Biology*, **232**, 551–8.

Baird, D. and Ulanowicz, R.E. (1989) The seasonal dynamics of the Chesapeake Bay ecosystem. *Ecological Monographs*, **59**, 329–64.

Balanya, J., Oller, J.M., Huey, R.B., *et al.* (2006) Global genetic change tracks global climate warming in *Drosophila subobscura*. *Science*, **313**, 1773–5.

Barabasi, A.L. and Albert, R. (1999) Emergence of scaling in random networks. *Science*, **286**, 509–12.

Bascompte, J., Jordano, P. and Olesen, J.M. (2006) Asymmetric coevolutionary networks facilitate biodiversity maintenance. *Science*, **312**, 431–3.

Berlow, E.L., Neutel, A.M., Cohen, J.E., *et al.* (2004) Interaction strengths in food webs: issues and opportunities. *Journal of Animal Ecology*, **73**, 585–98.

Brown, J.H. (2004) Toward a metabolic theory of ecology. *Ecology*, **85**, 1771–89.

Brown, J.H. and Gillooly, J.F. (2003) Ecological unification: high quality data facilitate theoretical unification. *Proceedings of the National Academy of Sciences of the United States of America*, **100**, 1467–8.

Brown, J.H. and Maurer, B.A. (1986) Body size, ecological dominance and cope's rule. *Nature*, **324**, 248–50.

Byström, P., Andersson, J., Persson, L. and de Roos, A.M. (2004) Size-dependent resource limitation and foraging-predation risk trade-offs: growth and habitat use in young arctic char. *Oikos*, **104**, 109–21.

Caldarelli, G., Higgs, P.G. and McKane, A.J. (1998) Modelling coevolution in multispecies communities. *Journal of Theoretical Biology*, **193**, 345–58.

Callaway, R.M., Brooker, R.W., Choler, P., *et al.* (2002) Positive interactions among alpine plants increase with stress. *Nature*, **417**, 844–8.

Cattin, M.F., Bersier, L.F., Banasek-Richter, C., Baltensperger, R. and Gabriel, J.P. (2004) Phylogenetic constraints and adaptation explain food-web structure. *Nature*, **427**, 835–9.

Christensen, K., di Collobiano, S.A., Hall, M. and Jensen, H.J. (2002) Tangled nature: a model of evolutionary ecology. *Journal of Theoretical Biology*, **216**, 73–84.

Christian, R.R. and Luczkovich, J.J. (1999) Organizing and understanding a winter's seagrass foodweb network through effective trophic levels. *Ecological Modelling*, **117**, 99–124.

Cohen, J.E. (1989) Food-webs and community structure. In *Perspectives in Ecological Theory* (ed. J.E. Cohen), pp. 181–202. Princeton University Press, Princeton, NJ.

Cohen, J.E., Briand, F. and Newman, C.M., eds. (1990) *Community Food Webs: Data and Theory*, vol. 20. Springer, Berlin.

Cohen, J.E., Beaver, R.A., Cousins, S.H., *et al.* (1993a) Improving food webs. *Ecology*, **74**, 252–8.

Cohen, J.E., Pimm, S.L., Yodzis, P. and Saldana, J. (1993b) Body sizes of animal predators and animal prey in food webs. *Journal of Animal Ecology*, **62**, 67–78.

Cohen, J.E., Jonsson, T. and Carpenter, S.R. (2003) Ecological community description using the food web, species abundance and body size. *Proceedings of the National Academy of Sciences of the United States of America*, **100**, 1781–6.

Coltman, D.W., O'Donoghue, P., Jorgenson, J.T., Hogg, J.T., Strobeck, C. and Festa-Blanchet, M. (2003) Undesirable evolutionary consequences of trophy hunting. *Nature*, **426**, 655–8.

Crumrine, P.W. (2005) Size structure and sustainability in an odonate intraguild predation system. *Oecologia*, **145**, 132–9.

Cyr, H. (2000) Individual energy use and the allometry of population density. In *Scaling in Biology*, pp. 267–95. Oxford University Press, New York, NY.

Damuth, J. (1981) Population density and body size in mammals. *Nature*, **290**, 699–700.

Damuth, J. (1991) Of size and abundance. *Nature*, **351**, 268–9.

Damuth, J. (1993) Cope's rule, the island rule and the scaling of mammalian population density. *Nature*, **365**, 748–50.

de Mazancourt, C. and Dieckmann, U. (2004) Trade-off geometries and frequency-dependent selection. *The American Naturalist*, **164**, 765–78.

Dercole, F., Ferrière, R., Gragnani, A., and Rinaldi, S. (2006) Coevolution of slow-fast populations: evolutionary sliding, evolutionary pseudo equilibria and complex red queen dynamics. *Proceedings of the Royal Society, Biological Sciences*, **273**, 983–90.

de Ruiter, P.C., Neutel, A.M. and Moore, J.C. (1995) Energetics, patterns of interaction strengths and stability in real ecosystems. *Science*, **269**, 1257–60.

Drossel, B., Higgs, P.G. and McKane, A.J. (2001) The influence of predator-prey population dynamics on the long-term evolution of food web structure. *Journal of Theoretical Biology*, **208**, 91–107.

Dunne, J.A., Williams, R.J. and Martinez, N.D. (2002) Network structure and biodiversity loss in food webs: robustness increases with connectance. *Ecology Letters*, **5**, 558–67.

Emmerson, M.C. and Raffaelli, D. (2004) Predator-prey body size, interaction strength and the stability of a real food web. *Journal of Animal Ecology*, **73**, 399–409.

Franks, S.J., Sim, S. and Weis, A.E. (2007) Rapid evolution of flowering time by an annual plant in response to a climate fluctuation. *Proceedings of the National Academy of Sciences of the United States of America*, **104**, 1278–82.

Frost, P.C., Benstead, J.P., Cross, W.F., Hillebrand, H., Larson, J.H., Xenopoulos, M.A. and Yoshida, T. (2006) Threshold elemental ratios of carbon and phosphorus in aquatic consumers. *Ecology Letters*, **9**, 774–9.

Fussmann, G.F., Loreau, M. and Abrams, P.A. (2007) Eco-evolutionary dynamics of communities and ecosystems. *Functional Ecology*, **21**, 465–77.

Goldwasser L. and Roughgarden, J. (1993) Construction and analysis of a large carribbean food web. *Ecology*, **74**, 1216–33.

Goudard A. and Loreau, M. (2008) Non-trophic interactions, biodiversity and ecosystem functioning: an interaction web model. *The American Naturalist*, **171**, 91–106.

Greenwood, J.J.D., Gregory, R.D., Harris, S., Morris, P.A. and Yalden, D.W. (1996) Relations between abundance, body-size and species number in british birds and mammals. *Philosophical Transactions of the Royal Society of London, Biological Sciences*, **351**, 265–78.

Grimm, V., Revilla, E., Berger, U., *et al.* (2005) Pattern-oriented modeling of agent-based complex systems: lessons from ecology. *Science*, **310**, 987–91.

Gross, T., Ebenhöh, W. and Feudel, U. (2004) Enrichment and foodchain stability: the impact of different forms of predator-prey interaction. *Journal of Theoretical Biology*, **227**, 349–58.

Grover, J.P. (2003) The impact of variable stoichiometry on predator-prey interactions: a multinutrient approach. *The American Naturalist*, **162**, 29–43.

Gyllenberg, M. and Metz, J.A.J. (2001) On fitness in structured metapopulations. *Journal of Mathematical Biology*, **43**, 545–60.

Hairston, N.G. Jr., Ellner, S.P., Geber, M.A., Yoshida, T. and Fox, J.A. (2005) Rapid evolution and the convergence of ecological and evolutionary time. *Ecology Letters*, **8**, 1114–27.

Hall, S.J. and Raffaelli, D. (1991) Food web patterns: lessons from a species rich web. *Journal of Animal Ecology*, **60**, 823–42.

Havens, K. (1992) Scale and structure in natural food webs. *Science*, **257**, 1107–9.

Heath, D.D., Heath, J.W., Bryden, C.A., Johnson, R.M. and Fox, C.W. (2003) Rapid evolution of egg size in captive salmon. *Science*, **299**, 1738–40.

Hendry, A.P., Wenburg, J.K., Bentzen, P., Volk, E.C. and Quinn, T.P. (2000) Rapid evolution of reproductive isolation in the wild: evidence from introduced salmon. *Science*, **290**, 516–8.

Huey, R.B., Gilchrist, G.W., Carlson, M.L., Berrigan, D. and Serra, L. (2000) Rapid evolution of a geographic cline in size in an introduced fly. *Science*, **287**, 308–9.

Ito, H.C. and Dieckmann, U. (2007) A new mechanism for recurrent adaptive radiations. *The American Naturalist*, **170**, E96–E111.

Ito, H.C. and Ikegami, T. (2006) Food web formation with recursive evolutionary branching. *Journal of Theoretical Biology*, **238**, 1–10.

Jennings, S., Pinnegar, J.K., Polunin, N.V.C. and Warr, K.J. (2002a) Linking size-based and trophic analyses of benthic community structure. *Marine Ecology-Progress Series*, **226**, 77–85.

Jennings, S., Warr, K.J. and Mackinson, S. (2002b) Use of size-based production and stable isotope analyses to predict transfer efficiencies and predator-prey body mass ratios in food webs. *Marine Ecology-Progress Series*, **240**, 11–20.

Jetz, W., Carbone, C., Fulford, J. and Brown, J.H. (2004) The scaling of animal space use. *Science*, **306**, 266–68.

Jordano, P., Bascompte, J. and Olesen, J.M. (2003) Invariant properties in coevolutionary networks of plant-animal interactions. *Ecology Letters*, **6**, 69–81.

Justic, D., Rabalais, N.N. and Turner, R.E. (1995) Stoichiometric nutrient balance and origin of coastal eutrophication. *Marine Pollution Bulletin*, **30**, 41–6.

King, R.B. (2002) Predicted and observed maximum prey size-snake size allometry. *Functional Ecology*, **16**, 766–72.

Kisdi, E. (2002) Dispersal: risk spreading versus local adaptation. *The American Naturalist*, **159**, 579–96.

Klausmeier, C.A., Litchman, E. Daufresne, T. and Levin, S. A. (2004) Optimal nitrogen-to-phosphorus stoichiometry of phytoplankton. *Nature*, **429**, 171–4.

Kleiber, M., ed. (1961) *The Fire of Life: an Introduction to Animal Energetics*. Wiley, New York, NY.

Kodric-Brown, A., Sibly, R.M. and Brown, J.H. (2006) The allometry of ornaments and weapons. *Proceedings of the National Academy of Sciences of the United States of America*, **103**, 8733–8.

Kokkoris, GD, Troumbis, AY and Lawton, JH. (1999) Patterns of species interaction strength in assembled theoretical competition communities. *Ecology Letters*, **2**, 70–4.

Kokkoris, GD, Jansen, V.A.A., Loreau, M. and Troumbis, AY. (2002) Variability in interactions strength and implications for biodiversity. *Journal of Animal Ecology*, **71**, 362–71.

Kooijman, S.A.L.M. (1998) The synthesizing unit as model for the stoichiometric fusion and branching of metabolic fluxes. *Biophysical Chemistry*, **73**, 179–88.

Krause, A.E. Frank, K.A. Mason, D.M. Ulanowicz, R.E. and Taylor, W.W. (2003) Compartments revealed in food-web structure. *Nature*, **426**, 282–5.

Krivan, V. and Eisner, J. (2003) Optimal foraging and predator-prey dynamics III. *Theoretical Population Biology*, **63**, 269–79.

Lafferty, K.D., Dobson, A.P. and Kuris, A.M. (2006) Parasites dominate food web links. *Proceedings of the National Academy of Sciences of the United States of America*, **103**, 11211–6.

Leibold, M.A., Holyoak, M., Mouquet, N., *et al.* (2004) The metacommunity concept: a framework for multi-scale community ecology. *Ecology Letters*, **7**, 601–13.

Levin, S.A. and Udovic, J.D. (1977) A mathematical model of coevolving populations. *The American Naturalist*, **111**, 657–75.

Lewis, H.M. and Law, R. (2007) Effects of dynamics on ecological networks. *Journal of Theoretical Biology*, **247**, 64–76.

Loeuille, N. and Loreau, M. (2004) Nutrient enrichment and food chains: can evolution buffer top-down control? *Theoretical Population Biology*, **65**, 285–98.

Loeuille, N. and Loreau, M. (2005) Evolutionary emergence of size-structured food webs. *Proceedings of the National Academy of Sciences of the United States of America*, **102**, 5761–6.

Loeuille, N. and Loreau, M. (2006) Evolution of body size in food webs: does the energetic equivalence rule hold? *Ecology Letters*, **9**, 171–8.

Loeuille, N. and Loreau, M. (2008) Evolution in metacommunities: On the relative importance of species sorting and monopolization in structuring communities. *The American Naturalist*, **171**, 788–99.

Loeuille, N. Loreau, M. and Ferrière, R. (2002) Consequences of plant-herbivore coevolution on the dynamics and functioning of ecosystems. *Journal of Theoretical Biology*, **217**, 369–81.

Loladze, I. and Kuang, Y. (2000) Stoichiometry in producer-grazer systems: linking energy flow with element cycling. *Bulletin of Mathematical Biology*, **62**, 1137–62.

Long, Z.T. and Morin, P.J. (2005) Effects of organism size and community composition on ecosystem functioning. *Ecology Letters*, **8**, 1271–82.

Loreau, M., Mouquet, N. and Gonzalez, A. (2003) Biodiversity as spatial insurance in heterogeneous landscapes. *Proceedings of the National Academy of Sciences of the United States of America*, **100**, 12765–70.

Maiorana, V.C. and Van Valen, L.M. (1990) *The unit of evolutionary ecology. Proceedings of the fourth international congress of systematic and evolutionary biology*, Energy and community evolution, pp. 655–65. Discorides Press, Portland, OR.

Makino, W., Cotner, J.B., Sterner, R.W. and Elser, J.J. (2003) Are bacteria more like plants or animals? Growth rate and resource dependence of bacterial c:n:p stoichiometry. *Functional Ecology*, **17**, 121–30.

Marquet, P.A., Navarrete, S.A. and Castilla, J.C. (1990) Scaling population density to body size in rocky intertidal communities. *Science*, **250**, 1125–7.

Marquet, P.A., Navarrete, S.A. and Castilla, J.C. (1995) Body size, population density and the energetic equivalence rule. *Journal of Animal Ecology*, **66**, 325–32.

Martinez, N.D. (1991) Artifacts or attributes? Effects of resolution on the little rock lake food web. *Ecological Monographs*, **61**, 367–92.

Martinez, N.D., Hawkins, B.A., Dawah, H.A. and Feifarek, B.P. (1999) Effects of sampling effort on characterization of food web structure. *Ecology*, **80**, 1044–55.

Matsumura, M., Trafelet-Smith, G.M., Gratton, C., Finke, D.L., Fagan, W.F. and Denno, R.F. (2004) Does intraguild predation enhance predator performance? a stoichiometric perspective. *Ecology*, **85**, 2601–15.

Mauricio, R. (1998) Costs of resistance to natural enemies in field populations of the annual plant *Arabidopsis thaliana*. *The American Naturalist*, **151**, 20–8.

May, R.M. (1973) editor. *Stability and complexity in model ecosystems*. Princeton University Press, Princeton, N.J.

McCann, K. (2000) The diversity-stability debate. *Nature*, **405**, 228–33.

Memmott, J., Martinez, N.D. and Cohen, J.E. (2000) Predators, parasitoids and pathogens: species richness, trophic generality and body sizes in a natural food web. *Journal of Animal Ecology*, **69**, 1–15.

Metz, J.A.J. and Gyllenberg, M. (2001) How should we define fitness in structured metapopulation models? including an application to the calculation of evolutionarily stable dispersal strategies. *Proceedings of the Royal Society, Biological Sciences*, **268**, 499–508.

Michalet, R., Brooker, R.W., Cavieres, L.A., *et al.* (2006) Do biotic interactions shape both sides of the humped-back model of species richness in plant communities? *Ecology Letters*, **9**, 767–73.

Milo, R., Itzkovitz, S., Kashtan, N., Levitt, R., Shen-Orr, S., Ayzenshtat, I., Sheffer, M. and Alon, U. (2004) Superfamilies of evolved and designed networks. *Science*, **303**, 1538–42.

Montoya, J.M., Pimm, S.L. and Solé, R.V. (2006) Ecological networks and their fragility. *Nature*, **442**, 259–64.

Morton R.D. and Law, R. (1997) Regional species pool and the assembly of local ecological communities. *Journal of Theoretical Biology*, **187**, 321–31.

Nee, S., Read, A.F. and Harvey, P.H. (1991) The relationship between abundance and body size in british birds. *Nature*, **351**, 312–13.

Neira S. and Arancibia, H. (2004) Trophic interactions and community structure in the upwelling system off Central Chile. *Journal of Experimental Marine Biology and Ecology*, **312**, 349–66.

Neira, S., Arancibia, H. and Cubillos, L. (2004) Comparative analysis of trophic structure of commercial fishery species off Central Chile in 1992 and 1998. *Ecological Modelling*, **172**, 233–48.

Neubert, M.G., Blumenshine, S.C., Duplisea, D.E., Jonsson, T. and Rashleigh, B. (2000) Body size and food web structure: testing the equiprobability assumption of the cascade model. *Oecologia*, **123**, 241–51.

Neutel, A.M., Heesterbeek, J.A.P. and de Ruiter, P.C. (2002) Stability in real food webs: weak links in long loops. *Science*, **296**, 1120–4.

Paine, R.T. (1966) Food web complexity and species diversity. *The American Naturalist*, **100**, 65–75.

Paine, R.T. (1980) Food webs: linkage, interaction strength and community infrastructure. *Journal of Animal Ecology*, **49**, 666–85.

Paine, R.T. (1988) Road maps of interactions or grist for theoretical development. *Ecology*, **69**, 1648–54.

Pauly, D., Christensen, V., Dalsgaard, J., *et al.* (1998) Fishing down marine food webs. *Science*, **279**, 860–3.

Peters, R.H. (1983) editor. *The ecological implications of body size*. Cambridge University Press, Cambridge, U.K.

Pimentel, D. (1961) Animal population regulation by the genetic feed-back mechanism. *The American Naturalist*, **95**, 65–79.

Pimm, S.L. (1979) The structure of food webs. *Theoretical Population Biology*, **16**, 144–58.

Pimm, S.L. editor. (1982) *Food webs*. Chapmann and Hall, New York.

Pimm, S.L. and Rice, J. (1987) The dynamics of multispecies, multi-life-stage models of aquatic food webs. *Theoretical Population Biology*, **32**, 303–25.

Polis, G.A. (1991) Complex trophic interactions in deserts: an empirical critique of food web theory. *The American Naturalist*, **138**, 123–55.

Post, W.M. and Pimm, S.L. (1983) Community assembly and food web stability. *Mathematical Biosciences*, **64**, 169–92.

Price, M.V. (1978) The role of microhabitat in structuring desert rodents communities. *Ecology*, **59**, 910–21.

Proulx, S.R., Promislow, D.E.L. and Phillips, P.C. (2005) Network thinking in ecology and evolution. *Trends in Ecology & Evolution*, **20**, 345–53.

Quince, C., Higgs, P.G. and McKane, A.J. (2005) Topological structures and interaction strength in model food webs. *Ecological Modelling*, **187**, 389–412.

Reale, D., McAdam, A.G., Boutin, S. and Berteaux, D. (2003) Genetic and plastic responses of a northern mammal to climate change. *Proceedings of the Royal Society, Biological Sciences*, **270**, 591–6.

Reich, P.B., Tjoelker, M.G., Machado, J.L. and Oleksyn, J. (2006) Universal scaling of respiratory metabolism, size and nitrogen in plants. *Nature*, **439**, 457–61.

Reznick, D.N., Shaw, F.H., Rodd, F.H. and Shaw, R.G. (1997) Evaluation of the rate of evolution in natural populations of guppies. *Science*, **275**, 1934–7.

Rossberg, A.G., Matsuda, H., Amemiya, T. and Itoh, K. (2006) Food webs: experts consuming families of experts. *Journal of Theoretical Biology*, **241**, 552–63.

Rossberg, A. G., Gishii, R., Amemiya, T. and Itoh, K. (2008) The top-down mechanism for body mass-abundance scaling. *Ecology*, **89**, 567–80.

Russo, S.E., Robinson, S.K. and Terborgh, J. (2003) Size-abundance relationships in an Amazonian bird community: implications for the energetic equivalence rule. *The American Naturalist*, **161**, 267–83.

Saloniemi, I. (1993) A coevolutionary predator-prey model with quantitative characters. *The American Naturalist*, **141**, 880–96.

Sànchez, F. and Olaso, I. (2004) Effects of fisheries on the cantabrian sea shelf ecosystem. *Ecological Modelling*, **172**, 151–74.

Savage, V.M., Gillooly, J.F., Brown, J.H., West, G.B. and Charnov, E.L. (2004) Effects of body size and temperature on population growth. *The American Naturalist*, **163**, 429–41.

Schade, J.D., Kyle, M., Hobbie, S.E., Fagan, W.F. and Elser, J.J. (2003) Stoichiometric tracking of soil nutrients by a desert insect herbivore. *Ecology Letters*, **6**, 96–101.

Sherry, R.A., Zhou, X., Gu, S., *et al.* (2007). Divergence of reproductive phenology under climate warming. *Proceedings of the National Academy of Sciences of the United States of America*, **104**, 198–202.

Solow, A.R. and Beet, A.R. (1998) On lumping species in food webs. *Ecology*, **79**, 2013–8.

Steiner, C.F. and Leibold, M.A. (2004) Cyclic assembly trajectories and scale-dependent productivity-diversity relationships. *Ecology*, **85**, 107–13.

Strauss, S.Y., Rudgers, J.A., Lau, J.A. and Irwin, R.E. (2002) Direct and ecological costs of resistance to herbivory. *Trends in Ecology & Evolution*, **17**, 278–85.

Taylor, PJ. (1988) The construction and turnover of complex community models having generalized Lotka-Volterra dynamics. *Journal of Theoretical Biology*, **135**, 569–88.

Tewfik, A., Rasmussen, J.B. and McCann, K.S. (2005) Anthropogenic enrichment alters a marine benthic food web. *Ecology*, **86**, 2726–36.

Trites, A.W., Livingston, P.A., Mackinson, S., Vasconcellos, M.C., Springer, A.M. and Pauly, D. (1999) Ecosystem change and the decline of marine mammals in the eastern Bering sea: testing the ecosystem shift and commercial whaling hypotheses. *Fisheries Centre Research Reports*, **7**, 1–71.

Turner, R.E., Qureshi, N., Rabalais, N.N., Dortch, Q., Justic, D., Shaw, R.F. and Cope, J. (1998) Fluctuating silicate:nitrate rations and coastal plankton food webs. *Proceedings of the National Academy of Sciences of the United States of America*, **95**, 13048–51.

Urban, M.C. (2006) Maladaptation and mass effects in a metacommunity: consequences for species coexistence. *The American Naturalist*, **168**, 28–40.

van Baalen, M. and Sabelis, M.W. (1993) Coevolution of patch selection strategies of predator and prey and the consequences for ecological stability. *The American Naturalist*, **142**, 646–70.

Vázquez D.P. and Aizen, M.A. (2004) Asymmetric specialization: a pervasive feature of plant-pollinator interactions. *Ecology*, **85**, 1251–7.

Vermeij, G.J. editor. (1987) *Evolution and escalation: an ecological history of life*. Princeton University Press, Princeton, NJ.

Warren, P.H. (1989) Spatial and temporal variation in the structure of freshwater food webs. *Oikos*, **55**, 299–311.

Warren, P.H. and Lawton, J.H. (1987) Invertebrate predator-prey body size relationships: an explanation for upper triangular food-webs and patterns in food-web structure? *Oecologia*, **74**, 231–5.

Williams, R.J. and Martinez, N.D. (2000) Simple rules yield complex food webs. *Nature*, **404**, 180–3.

Williams, T.M., Estes, J.A., Doak, D.F. and Springer, A.M. (2004) Killer appetites: assessing the role of predators in ecological communities. *Ecology*, **85**, 3373–84.

Winemiller, K.O. (1990) Spatial and temporal variation in tropical fish trophic networks. *Ecological Monographs*, **60**, 331–67.

Wing, S.L., Harrington, G.J., Smith, F.A., *et al.* (2005) Transient floral change and rapid global warming at the paleocene-eocene boundary. *Science*, **310**, 993–6.

Woodward, G. and Hildrew, G.A. (2002) Body size determinants of niche overlap and intraguild predation within a complex food web. *Journal of Animal Ecology*, **71**, 1063–74.

Yachi, S. and Loreau, M. (1999) Biodiversity and ecosystem productivity in a fluctuating environment: the insurance hypothesis. *Proceedings of the National Academy of Sciences of the United States of America*, **96**, 1463–8.

Yamauchi, A. and Yamamura, N. (2005) Effects of defense evolution and diet choice on population dynamics in a one-predator-two prey system. *Ecology*, **86**, 2513–24.

Yodzis, P. (2000) Diffuse effects in food webs. *Ecology*, **81**, 261–6.

Yoshida, T, Jones, L.E., Ellner, S.P., et al. (2003) Rapid evolution drives ecological dynamics in a predator-prey system. *Nature*, **424**, 303–6

Chapter 13

Alexander, I.J. and Lee, S.S. (2005) Mycorrhizas and ecosystem processes in tropical rain forest: implications for diversity. In *Biotic Interactions in the Tropics: their Role in the Maintenance of Species Diversity* (eds D.F.R.P. Burslem, M.A. Pinard and S.E. Hartley), pp. 165–203. Cambridge University Press, Cambridge.

Allen, E.B. and Allen, M.F. (1990) The mediation of competition by mycorrhizae in successional and patchy environments. In *Perspectives on Plant Competition* (eds J.B. Grace and D. Tilman), pp. 367–85. Academic Press, San Diego, CA.

Allen, M.F. (1991) *The Ecology of Mycorrhizae*. Cambridge University Press, Cambridge.

Anstett, M.C., Hossaert-McKey, M. and Kjellberg, F. (1997) Figs and fig pollinators: evolutionary conflicts in a coevolved mutualism. *Trends in Ecology & Evolution*, **12**, 94–9.

Axelrod, R. and Hamilton, W.D. (1981) The evolution of cooperation. *Science*, **211**, 1390–5.

Beismeijer, J.C., Roberts, S.P.M., Reemer, M., *et al.* (2006) Parallel declines in pollinators and insect pollinated plants in Britain and the Netherlands. *Science*, **313**, 351–4.

Bever, J.D. and Simms, E.L. (2000) Evolution of nitrogen fixation in spatially structured populations of *Rhizobium*. *Heredity*, **85**, 366–72.

Bidartondo, M.I., Redecker, D., Hijri, I., *et al.* (2002) Epiparasitic plants specialized on arbuscular mycorrhizal fungi. *Nature*, **419**, 389–92.

Björkman, E. (1960) *Monotropa hypopitys* L.: an epiparasite on tree roots. *Physiologia Plantarum*, **13**, 308–27.

Bobbink, R., Hornung, M. and Roelofs, J.G.M. (1998) The effects of air-borne nitrogen pollutants on species diversity in natural and semi-natural European vegetation. *Journal of Ecology*, **86**, 717–38.

Boucher, D.H., James, S., Keeler, K.H. (1982) The ecology of mutualism. *Annual Review of Ecology and Systematics*, **13**, 315–47.

Bronstein, J.L. (2001) The exploitation of mutualisms. *Ecology Letters*, **4**, 277–87.

Bronstein, J.L., Alarcón R. and Geber, M. (2006) The evolution of plant-insect mutualisms. *New Phytologist*, **172**, 412–28.

Callaway, R.M., Thelen, G.C., Rodriguez, A. and Holben, W.E. (2004) Soil biota and exotic plant invasion. *Nature*, **427**, 731–3.

Cameron, D.D., Leake, J.R. and Read, D.J. (2006) Mutualistic mycorrhiza in orchids: evidence from plant-fungus carbon and nitrogen transfers in the green-leaved terrestrial orchid *Goodyera repens*. *New Phytologist*, **171**, 405–16.

Cohen, J.E. (1998) Cooperation and self interest: pareto-inefficiency of nash equilibria in finite random games. *Proceedings of the National Academy of Sciences of the United States of America*, **95**, 9724–31.

Connell, J.H. and Lowman, M.D. (1989) Low-diversity tropical rain forests: some possible mechanisms for their existence. *The American Naturalist*, **134**, 88–119.

de Faria, S.M., Lewis, G.P., Sprent, J.I. and Sutherland, J.M. (1989) Occurrence of nodulation in the Leguminosae. *New Phytologist*, **111**, 607–19.

Denison, R.F. (2000) Legume sanctions and the evolution of symbiotic cooperation by rhizobia. *The American Naturalist*, **156**, 567–76.

Denison, R.F. and Kiers, E.T. (2004a) Lifestyle alternatives for rhizobia: mutualism, parasitism, and forgoing symbiosis. *FEMS Microbiology Letters*, **237**, 187–93.

Denison, R.F. and Kiers, E.T. (2004b) Why are most rhizobia beneficial to their plant hosts, rather than parasitic? *Microbes and Infection*, **6**, 1235–9.

Denison, R.F., Bledsoe, C., Kahn, M., *et al.* (2003) Cooperation in the rhizosphere and the 'free rider' problem. *Ecology*, **84**, 838–45.

Doebeli, M. and Knowlton, N. (1998) The evolution of interspecific mutualisms. *Proceedings of the National Academy of Sciences of the United States of America*, **95**, 8676–80.

Finkes, L.K., Cady, A.B., Mulroy, J.C., *et al.* (2006) Plant–fungus mutualism affects spider composition in successional fields. *Ecology Letters*, **9**, 347–56.

Fitter, A.H. (1977) Influence of mycorrhizal infection on competition for phosphorous and potassium by two grasses. *New Phytologist*, **79**, 119–25.

Fitter, A.H., Graves, J.D., Watkins, N.K., *et al.* (1998) Carbon transfer between plants and its control in networks of arbuscular mycorrhiza. *Functional Ecology*, **12**, 406–12.

Fontaine, C., Dojaz, I., Merguet, J. and Loreau, M. (2006) Functional diversity of plant–pollinator interaction webs enhances the persistence of plant communities. *PLoS Biology*, **4**, e1.

Francis, R. and Read, D.J. (1994) The contributions of mycorrhizal fungi to the determination of plant community structure. *Plant and Soil*, **159**, 11–25.

Gange, A.C., Brown, V.K. and Sinclair, G.S. (1993) Vesicular-arbuscular mycorrhizal fungi: a determinant of plant community structure in early succession. *Functional Ecology*, **7**, 616–22.

Ghazoul, J. (2006) Floral diversity and the facilitation of pollination. *Journal of Ecology*, **94**, 295–304.

Gomez, J.M., Bosch, J., Perfectti, F., *et al.* (2007) Pollinator diversity affects plant reproduction and recruitment: the tradeoffs of generalization. *Oecologia*, **153**, 597–605.

Grime, J.P., Mackey, J.M.L., Hillier, S.H. and Read, D.J. (1987) Floristic diversity in a model system using experimental microcosms. *Nature*, **328**, 420–2.

Gutschick, V.P. (1981) Evolved strategies in nitrogen acquisition by plants. *The American Naturalist*, **118**, 607–37.

Hagen, M.J. and Hamerick, J.L. (1996) Population level processes in *Rhizobium leguminosarum* bv *trifolii*: the role of founder effects. *Molecular Ecology*, **5**, 707–14.

Hardin, G. (1968) Tragedy of commons. *Science*, **162**, 1243–8.

Hartnett, D.C. and Wilson, W.T. (1999) Mycorrhizae influence plant community structure and diversity in tall grass prairie. *Ecology*, **80**, 1187–95.

Hay, M.E., Parker, J.D., Burkepile, D.E., *et al.* (2004) Mutualisms and aquatic community structure: the enemy of my enemy is my friend. *Annual Review of Ecology and Systematics*, **35**, 175–97.

Helgason, T., Daniell, T.J., Husband, R., *et al.* (1998) Ploughing up the wood-wide web. *Nature*, **394**, 431.

Helgason, T., Merryweather, J.W., Young, J.P.W. and Fitter, A.H. (2007) Specificity and resilience in the arbuscular mycorrhizal fungi of a natural woodland community. *Journal of Ecology*, **95**, 623–30.

Hempel, S., Renker, C. and Buscot, F. (2007) Differences in the species composition of arbuscular mycorrhizal fungi in spore, in a grassland ecosystem root and soil communities. *Environmental Microbiology*, **9**, 1930–8.

Herre, E.A., Knowlton, N., Mueller, U.G. and Rehner, S.A. (1999) The evolution of mutualisms exploring the paths between conflict and cooperation. *Trends in Ecology & Evolution*, **14**, 49–53.

Hetrick, B.A.D., Wilson, G.W.T. and Hartnett, D.C. (1989) Relationships between mycorrhizal dependency and competitive ability of two tall grass prairie forbs. *Canadian Journal of Botany*, **70**, 1521–8.

Heywood, V.H., ed. (1993) *Flowering Plants of the World*. Oxford University Press, New York, NY.

Hoeksema, J.D. and Bruna, E.M. (2000) Pursuing the big questions about interspecific mutualism: a review of theoretical approaches. *Oecologia*, **125**, 321–30.

Hoeksema, J.D. and Schwartz, M.W. (2003) Expanding comparative-advantage biological market models: contingency of mutualism on partners' resource requirements and acquisition trade-offs. *Proceedings of the Royal Society of London B*, **270**, 913–19.

Hoeksema, J.D. and Schwartz, M.W. (2006) Modeling interspecific mutualisms as biological markets. In *Economics in Nature* (eds R. Noë, J.A.R.A.M. van Hooff and P. Hammerstein). Cambridge University Press, Cambridge.

Holland, J.N., DeAngelis, D.L. and Schultz, S.T. (2004) Evolutionary stability of mutualism: interspecific population regulation as an evolutionarily stable strategy. *Proceedings of Royal Society of London B*, **271**, 1807–14.

Horton, T.R. and van der Heijden, M.G.A. (2008) The role of symbioses in seedling establishment and survival. In *Seedling Ecology and Evolution* (eds M. Leck, V.T. Parker and B. Simpson). Cambridge University Press, Cambridge.

James, T.Y., Kauff, F., Schoch, C.L., *et al.* (2006) Reconstructing the early evolution of fungi using a six-gene phylogeny. *Nature*, **443**, 818–22.

Johnson, N.C., Graham, J.H. and Smith, F.A. (1997) Functioning of mycorrhizal associations along the mutualism–parasitism continuum. *New Phytologist*, **135**, 575–85.

Johnson, S.D., Peter, C.I., Nilsson, L.A. and Agren, J. (2003) Pollination success in a deceptive orchid is enhanced by co-occurring rewarding magnet plants. *Ecology*, **84**, 2919–27.

Kato, M., Takimura, A. and Kawakita, A. (2003) An obligate pollination mutualism and reciprocal diversification in the tree genus *Glochidion* (Euphorbiaceae). *Proceedings of the National Academy of Sciences of the United States of America*, **100**, 5264–7.

Kawakita, A. and Kato, M. (2004) Evolution of obligate pollination mutualism in New Caledonian *Phyllanthus* (Euphorbiaceae). *American Journal of Botany*, **91**, 410–15.

Kearns, C.A., Inouye, D.W. and Waser, N.M. (1998) Endangered mutualisms: the conservation of plant pollinator interactions. *Annual Review of Ecology and Systematics*, **29**, 83–112.

Kiers, E.T. and Denison, R.F. (2008) Sanctions, cooperation, and the stability of plant-rhizosphere mutualisms. *Annual Review of Ecology, Evolution and Systematics*, **39**, 215–36.

Kiers, E.T. and van der Heijden, M.G.A. (2006) Mutualistic stability in the arbuscular mycorrhizal symbiosis: exploring hypotheses of evolutionary cooperation. *Ecology*, **87**, 1627–36.

Kiers, E.T., West, S.A. and Denison, R.F. (2002) Mediating mutualisms: farm management practices and evolutionary changes in symbiont co-operation. *Journal of Applied Ecology*, **39**, 745–54.

Kiers, E.T., Rousseau, R.A., West, S.A. and Denison, R.F. (2003) Host sanctions and the legume–rhizobium mutualism. *Nature*, **425**, 78–81.

Kiers, E.T., Rousseau, R.A. and Denison, R.F. (2006) Measured sanctions: legume hosts detect quantitative variation in rhizobium cooperation and punish accordingly. *Evolutionary Ecology Research*, **8**, 1077–86.

Klironomos, J.N. (2002) Feedback with soil biota contributes to plant rarity and invasiveness in communities. *Nature*, **417**, 67–70.

Kogel, K.H., Franken, P. and Hückelhoven, R. (2006) Endophyte or parasite: what decides? *Current Opinion in Plant Biology*, **9**, 358–63.

Lach, L. (2003) Invasive ants: unwanted partners in ant-plant interactions? *Annals of the Missouri Botanical Garden*, **90**, 91–108.

Leake, J.R. (1994) The biology of myco-heterotrophic ('saprophytic') plants. *New Phytologist*, **127**, 171–216.

Leake, J.R., Johnson, D., Donnelly, D.P., *et al.* (2004) Networks of power and influence: the role of mycorrhizal mycelium in controlling plant communities and agroecosystem functioning. *Canadian Journal of Botany*, **82**, 1016–45.

Lemons, A., Clay, K. and Rudgers, J.A. (2005) Connecting plant–microbial interactions above and belowground: a fungal endophyte affects decomposition. *Oecologia*, **145**, 595–604.

Leuchtmann, A. (1992) Systematics, distribution, and host specificity of grass endophytes. *Natural Toxins*, **1**, 150–62.

Lipsitch, M., Nowak, M.A., Ebert, D., *et al.* (1995) The population dynamics of vertically and horizontally transmitted parasites. *Proceedings of the Royal Society of London B*, **260**, 321–7.

Machado, C.A., Robbins, N., Gilbert, M.T.P. and Herre, E.A. (2005) Critical review of host specificity and its coevolutionary implications in the fig-fig-wasp mutualism. *Proceedings of the National Academy of Sciences of the United States of America*, **102**, 6558–65.

Maherali, H. and Klironomos, J.N. (2007) Influence of phylogeny on fungal community assembly and ecosystem functioning. *Science*, **316**, 1746–8.

Malloch, D.W., Pirozynski, K.A., Raven, P.H. (1980) Ecological and evolutionary significance of mycorrhizal symbioses in vascular plants (a review). *Proceedings of the National Academy of Sciences of the United States of America*, **77**, 2113–18.

Marler, M.J., Zabinski, C.A. and Callaway, R.M. (1999) Mycorrhizae indirectly enhance competitive effects of an invasive forb on a native bunchgrass. *Ecology*, **80**, 1180–6.

Mehdiabadi, N.J., Hughes, B. and Mueller, U.G. (2006) Cooperation, conflict, and coevolution in the attine ant-fungus symbiosis. *Behavioral Ecology*, **17**, 291–6.

Memmott, J., Craze, P.G., Waser, N.M. and Price, M.V. (2007) Global warming and the disruption of plant-pollinator interactions. *Ecology Letters*, **10**, 710–17.

Moran, N.A. (2006) Symbiosis. *Current Biology*, **16**, R866–R871.

Moran, N.A. (2007) Symbiosis as an adaptive process and source of phenotypic complexity. *Proceedings of the National Academy of Sciences of the United States of America*, **104**, 8627–33.

Morris, S.J. and Blackwood, C.B. (2007) The ecology of soil organisms. In *Soil Microbiology, Ecology and Biochemistry* (ed. E.A. Paul), pp. 195–229. Academic Press, Burlington, VT.

Morton, J.B. and Franke, M. and Bentivenga, S.P. (1994) Developmental foundations for morphological diversity among endomycorrhizal fungi in Glomales (Zygomycetes). In *Mycorrhiza: Structure, Function, Molecular Biology, and Biotechnology* (eds A. Varma and B. Hock), pp. 669–83. Springer-Verlag, Berlin.

Nowak, M.A. and Sigmund, K. (1992) Tit-for-tat in heterogeneous populations. *Nature*, **355**, 250–3.

O'Connor, P.J., Smith, S.E. and Smith, F.A. (2002) Arbuscular mycorrhizas influence plant diversity and community structure in a semiarid herbland. *New Phytologist*, **154**, 209–18.

Olde Venterink, H., Pieterse, N.M., Belgers, J.D.M., *et al.* (2002) N, P and K budgets along nutrient availability and productivity gradients in wetlands. *Ecological Applications*, **12**, 1010–26.

Palmer, T.M. (2003) Spatial habitat heterogeneity influences competition and coexistence in an African acacia ant guild. *Ecology*, **84**, 2843–55.

Palmer, T.M., Stanton, M.L., Young, T.P., *et al.* (2008) Breakdown of ant-plant mutualism follows the loss of large herbivores from an African savanna. *Science*, **319**, 192–5.

Parker, M.A. (2001) Mutualism as a constraint on invasion success for legumes and rhizobia. *Diversity and Distributions*, **7**, 125–36.

Parker, M.A., Malek, W. and Parker, I.M. (2006) Growth of an invasive legume is symbiont limited in newly occupied habitats. *Diversity and Distributions*, **12**, 563–71.

Pellmyr, O. and Leebens-Mack, J. (1999) Forty million years of mutualism: evidence for Eocene origin of the yucca-yucca moth association. *Proceedings of the National Academy of Sciences of the United States of America*, **96**, 9178–83.

Pepper, I.L. (2000) Beneficial and pathogenic microbes in agriculture. In *Environmental Microbiology* (eds R.M. Maier, I.L. Pepper and C.P. Gerba), pp. 425–46. Academic Press, California.

Pfeffer, P.E., Douds Jr, D.D., Bücking, H., *et al.* (2004) The fungus does not transfer carbon to or between roots in an arbuscular mycorrhizal symbiosis. *New Phytologist*, **163**, 617–27.

Read, D.J. and Perez-Moreno, J. (2003) Mycorrhizas and nutrient cycling in ecosystems: a journey towards relevance? *New Phytologist*, **157**, 475–92.

Reinhart, K.O. and Callaway, R.M. (2006) Soil biota and invasive plants. *New Phytologist*, **170**, 445–57.

Rice, S.K., Westerman, B. and Federici, R. (2004) Impacts of the exotic, nitrogen fixing black locust (*Robinia pseudoacacia*) on nitrogen cycling in a pine-oak ecosystem. *Plant Ecology*, **174**, 97–107.

Richardson, D.M., Allsopp, N., D'Antonio, C.M., *et al.* (2000) Plant invasions: the role of mutualisms. *Biological Reviews of the Cambridge Philosophical Society*, **75**, 65–93.

Riera, N., Traveset, A. and Garcia, O. (2002) Breakage of mutualisms by exotic species: the case of *Cneorum tricoccon* L. in the Balearic islands (Western Mediterranean sea). *Journal of Biogeography*, **29**, 713–19.

Rudgers, J.A. and Clay, K. (2005) Fungal endophytes in terrestrial communities and ecosystems. In *The Fungal Community* (eds E.J. Dighton, P. Oudesmans and J.F.J. White), pp. 423–42. M. Dekker, New York, NY.

Rudgers, J.A., Mattingly, W.B., Koslow, J.M. (2005) Mutualistic fungus promotes plant invasion into diverse communities. *Oecologia*, **144**, 463–71.

Sachs, J.L., Mueller, U.G., Wilcox, T.P. and Bull, J.J. (2004) The evolution of cooperation. *The Quarterly Review of Biology*, **79**, 135–60.

Sahli, H.F. and Conner, J.K. (2006) Characterizing ecological generalization in plant-pollination systems. *Oecologia*, **148**, 365–72.

Saikkonen, K., Faeth, S.H., Helander, M. and Sullivan, T.J. (1998) Fungal endophytes: a continuum of interactions with host plants. *Annual Review of Ecology and Systematics*, **29**, 319–43.

Sala, O.E., Chapin, F.S., Armesto, J.J., *et al.* (2000) Biodiversity: global biodiversity scenarios for the year 2100. *Science*, **287**, 1770–4.

Scherer-Lorenzen, M., Olde Venterink, H. and Buschmann, H. (2007) Nitrogen enrichment and plant invasions: the importance of nitrogen-fixing plants and anthropogenic eutrophication. In *Biological Invasions; Series: Ecological Studies*, vol. 193 (ed. W. Nentwig), pp. 163–80. Springer Verlag, Berlin.

Schmitt, R.J. and Holbrook, S.J. (2003) Mutualism can mediate competition and promote coexistence. *Ecology Letters*, **6**, 898–902.

Schwartz, M.W. and Hoeksema, J.D. (1998) Specialization and resource trade: biological markets as a model of mutualisms. *Ecology*, **79**,1029–38.

Schwinning, S. and Parsons, A.J. (1996) A spatially explicit population model of stoloniferous N-fixing legumes in mixed pasture with grass. *Journal of Ecology*, **84**, 815–26.

Selosse, M., Richard, F., He, X. and Simard, S.W. (2006) Mycorrhizal networks: des liaisons dangereuses? *Trends in Ecology & Evolution*, **21**, 621–8.

Simard, S.W., Perry, D.A., Jones, M.D., *et al.* (1997) Net transfer of carbon between ectomycorrhizal tree species in the field. *Nature*, **388**, 579–82.

Simberloff, D. and Von Holle, B. (1999) Positive interactions of nonindigenous species: invasional meltdown? *Biological Invasions*, **1**, 21–32.

Simms, E.L., Taylor, D.L., Povich, J., *et al.* (2006) An empirical test of partner choice mechanisms in a wild legume-rhizobium interaction. *Proceedings of the Royal Society B*, **273**, 77–81.

Simon, L., Bousquet, J., Levesqé, R.C. and Lalonde, M. (1993) Origin and diversification of endomycorrhizal fungi and coincidence with vascular land plants. *Nature*, **363**, 67–9.

Smith, M.J. and Szathmáry, E. (1995) *The Major Transitions in Evolution*. Oxford University Press, Oxford.

Smith, S.E. and Read, D.J. (1997) *Mycorrhizal Symbiosis*, 2nd edn. Academic Press, London.

Stanton, M.L., Palmer, T.M., Young, T.P., *et al.* (1999) Sterilization and canopy modification of a swollen thorn acacia tree by a plant-ant. *Nature*, **401**, 578–81.

Stebbins, G.L. (1970) Adaptive radiation of reproductive characteristics in angiosperms. I. Pollination mechanisms. *Annual Review of Ecology and Systematics*, **1**, 307–26.

Stinson, K.A., Campbell, S.A., Powell, J.R., *et al.* (2006) Invasive plant suppresses the growth of native tree seedlings by disrupting belowground mutualisms. *PLoS Biology*, **4(5)**, e140.

Sutton, W.D. and Paterson, A.D. (1980) Effects of the host plant on the detergent sensitivity and viability of Rhizobium bacteroids. *Planta*, **148**, 287–92.

Tandon, R., Shivanna, K.R. and Mohanram, H.Y. (2003) Reproductive biology of *Butea monosperma* (Fabaceae). *Annals of Botany*, **92**, 1–9.

Tedershoo, L., Pellet, P., Koljag, U. and Selosse, M.E. (2007) Parallel evolutionary paths to mycoheterotrophy in understorey Ericaceae and Orchidaceae: ecological evidence for mixotrophy in Pyroleae. *Oecologia*, **151**, 206–17.

Thrall, P.H., Hochberg, M.E., Burdon, J.J. and Bever, J.D. (2007) Coevolution of symbiotic mutualists and parasites in a community context. *Trends in Ecology & Evolution*, **22**, 120–6.

Tillberg, C.V. (2004) Friend or foe? A behavioral and stable isotopic investigation of an ant–plant symbiosis. *Oecologia*, **140**, 506–15.

Tilman, D. (1988) *Plant Strategies and the Dynamics and Structure of Plant Communities*. Princeton University Press, Princeton, NJ.

Traveset, A. and Richardson, D.M. (2006) Biological invasions as disruptors of plant reproductive mutualisms. *Trends in Ecology & Evolution*, **21**, 208–16.

Urcelay, C. and Diaz, S. (2003) The mycorrhizal dependence of subordinates determines the effect of arbuscular mycorrhizal fungi on plant diversity. *Ecology Letters*, **6**, 388–91.

van der Heijden, M.G.A. (2002) Arbuscular mycorrhizal fungi as determinant of plant diversity: in search of underlying mechanisms and general principles. In *Mycorrhizal Ecology. Ecological Studies*, vol. 157 (eds M.G.A. van der Heijden and I.R. Sanders), pp. 243–65. Springer-Verlag, Berlin.

van der Heijden, M.G.A. (2004) Arbuscular mycorrhizal fungi as support systems for seedling establishment in grassland. *Ecology Letters*, **7**, 293–303.

van der Heijden, M.G.A., Klironomos, J.N., Ursic, M., *et al.* (1998a) Mycorrhizal fungal diversity determines plant biodiversity, ecosystem variability and productivity. *Nature*, **396**, 69–72.

van der Heijden, M.G.A., Boller, T., *et al.* (1998b) Different arbuscular mycorrhizal fungal species are potential determinants of plant community structure. *Ecology*, **79**, 2082–91.

van der Heijden, M.G.A., Bakker, R., Verwaal, J., *et al.* (2006a) Symbiotic bacteria as a determinant of plant community structure and plant productivity in dune grassland. *FEMS Microbiology Ecology*, **56**, 178–87.

van der Heijden, M.G.A., Streitwolf-Engel, R., Riedl, R., *et al.* (2006b) The mycorrhizal contribution to plant productivity, plant nutrition and soil structure in experimental grassland. *New Phytologist*, **172**, 739–52.

van der Heijden, M.G.A., Bardgett, R. and van Straalen, N. M. (2008) The unseen majority: soil microbes as drivers of plant diversity and productivity in terrestrial ecosystems. *Ecology Letters*, **11**, 296–310.

Vandermeer, J.H. and Boucher, D.H. (1978) Varieties of mutualistic interactions in population models. *Journal of Theoretical Biology*, **74**, 549–58.

van der Putten, W.H., Klironomos, J.N. and Wardle, D.A. (2007) Microbial ecology of biological invasions. *The ISME Journal*, **1**, 28–37.

Vitousek, P.M. and Walker, L.R. (1989) Biological invasion by *Myrica faya*: plant demography, nitrogen fixation, ecosystem effects. *Ecological Monographs*, **59**, 247–65.

Vitousek, P.M., Walker, L.R., Whiteaker, L.D., *et al.* (1987) Biological invasion by *Myrica faya* alters ecosystem development in Hawaii. *Science*, **238**, 802–4.

Vogelsang, K.M., Reynolds, H.L. and Bever, J.D. (2006) Mycorrhizal fungal identity and richness determine the diversity and productivity of a tallgrass prairie system. *New Phytologist*, **172**, 554–62.

West, H.M. (1996) Influence of arbuscular mycorrhizal infection on competition between *Holcus lanatus* and *Dactylis glomerata*. *Journal of Ecology*, **84**, 429–38.

West, S.A., Murray, M.G., Machado, C.A., *et al.* (2001) Testing Hamilton's rule with competition between relatives. *Nature*, **409**, 510–13.

West, S.A., Kiers, E.T., Pen, I. and Denison, R.F. (2002a) Sanctions and mutualism stability: when should less beneficial mutualists be tolerated? *Journal of Evolutionary Biology*, **15**, 830–7.

West, S.A., Kiers, E.T., Simms, E.L. and Denison, R.F. (2002b) Sanctions and mutualism stability: why do rhizobia fix nitrogen? *Proceedings of the Royal Society of London B*, **269**, 685–94.

West, S.A., Griffin, A.S. and Gardner, A. (2007a) Evolutionary explanations for cooperation. *Current Biology*, **17**, R661–R672.

West, S.A., Griffin, A.S. and Gardner, A. (2007b) Social semantics: altruism, cooperation, mutualism, strong reciprocity and group selection. *Journal of Evolutionary Biology*, **20**, 415–32.

Whitfield, J. (2007) Underground networking. *Nature*, **449**, 136–8.

Wilkinson, D.M. and Sherratt, T.N. (2001) Horizontally acquired mutualisms, an unsolved problem in ecology? *Oikos*, **92**, 377–84.

Yamamura, N. (1993) Vertical transmission and evolution of mutualism from parasitism. *Theoretical Population Biology*, **44**, 95–109.

Zhou, J.C., Tchan, Y.T. and Vincent, J.M. (1985) Reproductive capacity of bacteroids in nodules of *Trifolium repens*, L. & *Glycine max* (L)Merr. *Planta*, **163**, 473–82.

Zimmer, K., Hynson, N.A., Gebauer, G., *et al.* (2007) Wide geographical and ecological distribution of nitrogen and carbon gains from fungi in pyroloids and monotropoids (Ericaceae) and in orchids. *New Phytologist*, **175**, 166–75.

Chapter 14

Beckerman, A.P., Petchey, O.L. and Warren, P.H. (2006) Foraging biology predicts food web complexity. *Proceedings of the National Academy of Sciences of the United States of America*, **103**, 13745–9.

Brown, J.H., Gillooly, J.F., Allen, A.P., *et al.* (2004) Toward a metabolic theory of ecology. *Ecology*, **85**, 1771–89.

Brown, R.L. and Peet, R.K. (2003) Diversity and invasibility of southern Appalachian plant communities. *Ecology*, **84**, 32–9.

Chase, J.M. and Leibold, M.A. (2002) Spatial scale dictates the productivity-biodiversity relationship. *Nature*, **416**, 427–30.

Chao, L., Levin, B.R. and Stewart, F.M. (1977) A complex community in a simple habitat: an experimental study with bacteria and phage. *Ecology*, **58**, 369–78.

Colwell, R.K., and Hurtt, G.C. (1994) Nonbiological gradients in species richness and a spurious Rapoport effect. *The American Naturalist*, **144**, 570–95.

Colwell, R.K., and Lees, D.C. (2000) The mid-domain effect: geometric constraints on the geography of species richness. *Trends in Ecology & Evolution*, **15**, 70–6.

de Roos, A.M., Leonardsson, K., Persson, L. and Mittelbach, G.G. (2002) Ontogenetic niche shifts and flexible behavior in size-structured populations. *Ecological Monographs*, **72**, 271–92.

de Ruiter, P.C., Neutel, A.M. and Moore, J.C. (1995) Energetics, patterns of interaction strengths, and stability in real ecosystems. *Science*, **269**, 1257–60.

Drossel, B., Higgs, P.G. and McKane, A.J. (2001) The influence of predator-prey population dynamics on the long-term evolution of food web structure. *Journal of Theoretical Biology*, **208**, 91–107.

Elton, C.S. (1958) *The Ecology of Invasions by Animals and Plants*. Chapman & Hall, London.

Fargione, J., Brown, C.S. and Tilman, D. (2003) Community assembly and invasion: an experimental test of neutral versus niche processes. *Proceedings of the National Academy of Sciences of the United States of America*, **100**, 8916–20.

Fridley, J.D., Stachowicz, J.J., Naeem, S., *et al.* (2007) The invasion paradox: reconciling pattern and process in species invasions. *Ecology*, **88**, 3–17.

Fukami, T. and Morin, P.J. (2003) Productivity-biodiversity relationships depend on the history of community assembly. *Nature*, **424**, 423–6.

Holyoak, M. and Lawler, S.P. (1996) Persistence of an extinction-prone predator-prey interaction through metapopulation dynamics. *Ecology*, **77**, 1867–79.

Huffaker, C.B. (1958) Experimental studies on predation: dispersion factors and predator-prey oscillations. *Hilgardia*, **27**, 343–83.

Ings, T.C.J.M.M., Bascompte, J., Blüthgen, N., *et al.* (2009) Ecological networks: beyond food webs. *Journal of Animal Ecology*, **78**, 253–69.

Jiang, L. and Morin, P.J. (2004) Productivity gradients cause positive diversity-invasibility relations in microbial communities. *Ecology Letters*, **7**, 1047–57.

Kennedy, T.A., Naeem, S., Howe, M.K., *et al.* (2002) Biodiversity as a barrier to ecological invasion. *Nature*, **417**, 636–8.

Kerr, B., Neuhauser, C., Bohannan, B.J.M. and Dean, A.M. (2006) Local migration promotes competitive restraint in a host-pathogen 'tragedy of the commons'. *Nature*, **442**, 75–8.

Kerr, J.T., Perring, M. and Currie, D.J. (2006) The missing Madagascan mid-domain effect. *Ecology Letters*, **9**, 149–59.

Knops, J.M.H., Tilman, D., Haddad, N.M., *et al.* (1999) Effects of plant species richness on invasion dynamics, disease outbreaks, insect abundances, and diversity. *Ecology Letters*, **2**, 286–93.

Leibold, M.A., Holyoak, M., Mouquet, N., *et al.* (2004) The metacommunity concept: a framework for multi-scale community ecology. *Ecology Letters*, **7**, 601–13.

Levine, J.M. (2000) Species diversity and biological invasions: relating local process to community pattern. *Science*, **288**, 852–4.

Lockwood, J.L., Cassey, P. and Blackburn, T. (2005) The role of propagule pressure in explaining species invasions. *Trends in Ecology & Evolution*, **20**, 223–8.

Lockwood, J.L., Hoopes, M.F. and Marchetti, M.P. (2007) *Invasion Ecology*. Blackwell, Malden.

Loeuille, N. and Loreau, M. (2005) Evolutionary emergence of size-structured food webs. *Proceedings of the National Academy of Sciences of the United States of America*, **102**, 5761–6.

Lonsdale, W.M. (1999) Global patterns of plant invasions and the concept of invasibility. *Ecology*, **80**, 1522–36.

Mack, R.N., Simberloff, D., Lonsdale, W.M., *et al.* (2000) Biotic invasions: causes, epidemiology, global consequences, and control. *Ecological Applications*, **10**, 689–710.

May, R.M. (1973) Stability and complexity in model ecosystems. Princeton University Press, Princeton, NJ.

May, R.M. (1976) Models for single populations. In *Theoretical Ecology: Principles and Applications* (ed. R.M. May), pp. 4–25. Saunders, Philadelphia, PA.

McGrady-Steed, J., Harris, P.M. and Morin, P.J. (1997) Biodiversity regulates ecosystem predictability. *Nature*, **390**, 162–5.

Montoya, J.M., Pimm, S.L. and Sole, R.V. (2006) Ecological networks and their fragility. *Nature*, **442**, 259–64.

Moore, J.L., McCann, K., Setälä, H. and de Ruiter, P.C. (2003) Top-down is bottom-up: does predation in the rhizosphere regulate aboveground dynamics? *Ecology*, **84**, 846–57.

Naeem, S., Knops, J.M.H., Tilman, D., *et al.* (2000) Plant diversity increases resistance to invasion in the absence of covarying extrinsic factors. *Oikos*, **91**, 97–108.

Neutel, A.M., Heesterbeek, J.A.P. and de Ruiter, P.C. (2002) Stability in real food webs: weak links in long loops. *Science*, **296**, 1120–3.

Neutel, A.M., Heesterbeek, J.A.P., van de Koppel, J., *et al.* (2007) Reconciling complexity with stability in naturally assembling food webs. *Nature*, **449**, 599–602.

Paine, R.T. (1966) Food web complexity and species diversity. *The American Naturalist*, **100**, 65–75.

Petchey, O.L., Beckerman, A.P., Riede, J.O. and Warren, P.H. (2008) Size, foraging, and food web structure. *Proceedings of the National Academy of Sciences of the United States of America*, **105**, 4191–6.

Pianka, E.R. (1988) *Evolutionary Ecology*, 4th edn. Harper & Row, New York, NY.

Pimm, S.L., and Lawton, J.H. (1977) Number of trophic levels in ecological communities. *Nature*, **268**, 329–31.

Planty-Tabacchi, A.-M., Tabacchi, E., Naiman, R.J., *et al.* (1996) Invasibility of species-rich communities in riparian zones. *Conservation Biology*, **10**, 598–607.

Ricklefs, R.E. (2004) A comprehensive framework for global patterns in biodiversity. *Ecology Letters*, **7**, 1–15.

Ricklefs, R.E. (2008) Disintegration of the ecological community. *The American Naturalist*, **172**, 741–50.

Robinson, G.R., Quinn, J.F. and Stanton, M.L. (1995) Invasibility of experimental habitat islands in a California winter annual grassland. *Ecology*, **76**, 786–94.

Scheffer, M., Hosper, S.H., Meijer, M.-L., *et al.* (1993) Alternative equilibria in shallow lakes. *Trends in Ecology & Evolution*, **8**, 275–9.

Shea, K. and Chesson, P. (2002) Community ecology theory as a framework for biological invasions. *Trends in Ecology & Evolution*, **17**, 170–6.

Solé, R.V. and Montoya, J.M. (2001) Complexity and fragility in ecological networks. *Proceedings of the Royal Society of London Series B–Biological Sciences*, **268**, 1–7.

Stachowicz, J.J., Whitlatch, R.B. and Osman, R.W. (1999) Species diversity and invasion resistance in a marine ecosystem. *Science*, **286**, 1577–9.

Stohlgren, T.J., Binkley, D., Chong, G.W., *et al.* (1999) Exotic plant species invade hotspots of native plant diversity. *Ecological Monographs*, **69**, 25–46.

Stohlgren, T.J., Barnett, D.T. and Kartesz, J.T. (2003) The rich get richer: patterns of plant invasions in the United States. *Frontiers of Ecology and the Environment*, **1**, 11–14.

Storch, D., Davies, R.G., Zajicek, S., *et al.* (2006) Energy, range dynamics and global species richness patterns: reconciling mid-domain effects and environmental determinants of avian diversity. *Ecology Letters*, **9**, 1308–20.

Symstad, A. (2000) A test of the effects of functional group richness and composition on grassland invasibility. *Ecology*, **81**, 99–109.

van der Weijden, W., Leewis, R. and Bol, P. (2007) *Biological Globalisation. Bio-invasions and their Impacts on Nature, the Economy and Public Health*. KNNV Publishing, Utrecht.

Waide, R.B., Willig, M.R., Steiner, C.F., *et al.* (1999) The relationship between primary productivity and species richness. *Annual Review of Ecology and Systematics*, **30**, 257–301.

Wiser, S.K., Allen, R.B., Clinton, P.W. and Platt, K.H. (1998) Community structure and forest invasion by an exotic herb over 23 years. *Ecology*, **79**, 2071–81.

Wootton, J.T. (1992) Indirect effects, prey susceptibility, and habitat selection: impacts of birds on limpets and algae. *Ecology*, **73**, 981–91.

Yoshida, T., Jones, L.E., Ellner, S.P., *et al.* (2003) Rapid evolution drives ecological dynamics in a predator-prey system. *Nature*, **424**, 303–6.

Index

Note: Page numbers in italics refer to Figures, Tables and Boxes.